Soju

Hyunhee Park offers the first global historical study of soju, the distinctive distilled drink of Korea. Searching for soju's origins, Park leads us into the vast, complex world of premodern Eurasia. She demonstrates how the Mongol conquests of the thirteenth and fourteenth centuries wove together hemispheric flows of trade, empire, and scientific and technological transfer and created the conditions for the development of a singularly Korean drink. Soju's rise in Korea marked the evolution of a new material culture through ongoing interactions between the global and local and between tradition and innovation in the adaptation and localization of new technologies. Park's vivid new history shows how these cross-cultural encounters laid the foundations for the creation of a globally connected world.

Hyunhee Park is Associate Professor of History at the City University of New York, John Jay College of Criminal Justice, and the CUNY Graduate Center.

ASIAN CONNECTIONS

Asian Connections is a major series of ambitious works that look beyond the traditional templates of area, regional or national studies to consider the trans-regional phenomena which have connected and influenced various parts of Asia through time. The series will focus on empirically grounded work exploring *circulations, connections, convergences and comparisons* within and beyond Asia. Themes of particular interest include transport and communication, mercantile networks and trade, migration, religious connections, urban history, environmental history, oceanic history, the spread of language and ideas, and political alliances. The series aims to build new ways of understanding fundamental concepts, such as modernity, pluralism or capitalism, from the experience of Asian societies. It is hoped that this conceptual framework will facilitate connections across fields of knowledge and bridge historical perspectives with contemporary concerns.

A list of books in the series can be found at the end of the volume.

Soju

A Global History

Hyunhee Park

City University of New York

CAMBRIDGE
UNIVERSITY PRESS

University Printing House, Cambridge CB2 8BS, United Kingdom

One Liberty Plaza, 20th Floor, New York, NY 10006, USA

477 Williamstown Road, Port Melbourne, VIC 3207, Australia

314-321, 3rd Floor, Plot 3, Splendor Forum, Jasola District Centre, New Delhi - 110025, India

103 Penang Road, #05-06/07, Visioncrest Commercial, Singapore 238467

Cambridge University Press is part of the University of Cambridge.

It furthers the University's mission by disseminating knowledge in the pursuit of education, learning and research at the highest international levels of excellence.

www.cambridge.org
Information on this title: www.cambridge.org/9781108816113
DOI: 10.1017/9781108895774

© Hyunhee Park 2021

First published 2021
First paperback edition 2022

A catalogue record for this publication is available from the British Library

Library of Congress Cataloging in Publication data
Names: Park, Hyunhee, 1972– author.
Title: Soju : a global history / Hyunhee Park, City University of New York.
Description: Cambridge ; New York : Cambridge University Press, 2021. |
Series: Asian connections | Includes bibliographical references and index.
Identifiers: LCCN 2020042864 (print) | LCCN 2020042865 (ebook) | ISBN
9781108842013 (hardback) | ISBN 9781108895774 (ebook)
Subjects: LCSH: Soju.
Classification: LCC TP607.S65 P37 (print) | LCC TP607.S65 (ebook) | DDC
663/.5–dc23
LC record available at https://lccn.loc.gov/2020042864
LC ebook record available at https://lccn.loc.gov/2020042865

ISBN 978-1-108-84201-3 Hardback
ISBN 978-1-108-81611-3 Paperback

For my husband, Fumihiko

Contents

Maps

Illustrations

Acknowledgments

Some have asked me how much I must love alcoholic beverages, given that I have suddenly turned to research on them. In fact, when my colleague Paul Buell proposed that I join his project, A Comparative Investigation of Distillation Technologies, Wine Production, and Fermented Products, and investigate the case of Korean soju, I never remotely imagined that I would write a monograph on the topic. Once I had delved into it further, however, I was struck by the potential richness of the subject. I became particularly interested in the significance alcoholic drinks have had in human lives and cultures since ancient times, as key components of both food and medicine. I discovered that in terms of the history of alcoholic drinks, the invention of distillation technology became a key innovation since it enabled longer preservation of fermented alcohols and, by extension, of other kinds of food and medicine. The study of distillation also helps us understand exchanges of technology among different societies in premodern times, when there was no refrigeration or other forms of reliable preservation, other than traditional ones such as drying. Indeed, my confrontation with the history of distillation produced a kind of "eureka" moment for me. For me, in any case, studying the history of cross-cultural contacts is always a humbling experience, and studying this new topic affected me even more so. I was also surprised to discover that the history of the main Korean distilled liquor, soju, worked well too as a means for positioning Korea in a global history extending throughout Afro-Eurasia in its premodern connections. This topic also suggests how much present-day societies owe to the long-term developments, and the exchanges, of the past.

In such a context, without my colleagues who invited me to explore this new topic and who have given me their continuous support, I would not have been able to write this book. Of course, my deepest gratitude goes first to Paul Buell, who first introduced me to the research topic in 2013. Since I first joined his project, Paul could have not been more generous with his endless support, and has always responded promptly to my queries and shared many important thoughts and insights about the topic. I also appreciate the help of my initial co-research project members, including Angela Schottenhammer, Ana Valenzuela, Maria de la Paz Solano-Pérez, Batdorj Batjargal, Dashdondog Bayarsaikhan, and Moldir Oskenbay, who provided various forms of assistance and inspiration ever since we met for the first conference at Salzburg University funded by a Eurasia Pacific Uninet grant in 2015.

Because it was my first entry into serious research on both food history and Korean history, the substantial support of experts in relevant fields was crucial. As a renowned expert on the history of Korean and East Asian food histories, Joo Young-ha helped me to critically approach relevant topics, including recent developments of Korean soju that often have been exaggerated by nationalist rhetoric. As the topic involves the history of science and technology, it was also crucial to find support from experts in that field. I was so fortunate to receive the warm support and encouragement from Shin Dongwon and Lim Jongtae after I presented my first foray into the topic at the 14th International Conference on the History of Science in East Asia (ICHSEA) in Paris in 2015, and the 15th ICHSEA in Jeonju in 2019; their sharp insights and valuable comments proved critical to developing my initial manuscript. I would also like to thank many other scholars who read the entire text or parts of my book manuscript and gave me important advice and feedback, including Morris Rossabi, Eugene Anderson, Alexandr Gorokhovskiy, Linda Feng, Seol Paehwan, Ana Valenzuela, Maria de la Paz Solano-Pérez, Limor Yungman, and Gideon Shelach-Lavi. I also thank the two anonymous readers for the press, whose suggestions, corrections, and bibliographical advice saved me from many embarrassing mistakes. The responsibility for any remaining gaffs and blunders is entirely my own.

I would also like to thank many other colleagues who shared their wisdom with me at conferences where I shared my developing work and otherwise gave generously of their ideas, insights, resources, and helpful criticism. These include Kim Hodong, Yi Eunjeung, Kaveh Hemmat, Michael Hope, Oh Young-Ju, Kim Choong Hyeon, Yoon Sungje, Joseph Lee, Kang In Uk, Michal Biran, Yuri Pines, Lee Jung, Seo Jung-Min, Anne Gerritsen, Shim Jae-hoon, Kim Young-jae, Choi Duk-Kyung, Jou Kyung-Chul, Chung Chungkee, Jeong Myung Hyun, Kim Janggoo, Lee Kang Hahn, Sung Baik-yong, Nam Jong Kuk, Park Yong-jin, Cho Wonhee, Kang Changhwa, colleagues at the Institute for Textual and Oral Histories of Food (ITOHF), the Korean Research Institute of Science, Technology and Civilization (KRISTC) at Chonbuk National University, and the Atlantic World Workshop in New York University. I also wish to express many thanks to my dear colleagues at John Jay College, in particular Allison Kavey and David Munns, for their unstinting help, advice, and encouragement. Financial and institutional support that expedited this research includes a Korea Foundation Field Research Fellowship, the fellowship from the Central Eurasian Studies Institute of the Institute of Humanities at Seoul National University, the PSC-CUNY awards from the City University of New York, a CUNY scholar incentive award, and a book publication grant from the Office for the Advancement of Research at CUNY John Jay College. I cannot overlook thanking Tansen Sen and Lucy Rymer for recommending that my initial research topic be listed among the titles of the Asian Connections series at Cambridge University Press, Danielle McClellan for providing intelligent and knowledgeable English academic editing, and Matilde Grimaldi for providing professional illustrations of many figures and maps that would help readers follow the book's arguments easily.

Finally, I wish to express my profound gratitude to all the members of my family. My study would not have been made possible without the continuous encouragement given to me by my parents, Park Dongho and Choi Bonghwa, and my little brother Park Jun Hee and sister-in-law Oh Yeon Joung. Jun Hee's encouragement to travel to Andong in person to visit the Andong Soju museum and collect relevant sources during my stay in Korea in the summer of 2014 helped me decide to launch this project; while Jun Hee was not able to accompany me for this trip as he had hoped, my parents were willing to accompany me instead and encouraged me to work on this project. My in-laws in Japan were also a source of true encouragement and support; my brother-in-law Takahiko and his wife Rie brought me to a local sake factory and museum during my trip to visit their family in Iwaki city. Finally, I would like to give my utmost thanks to my husband, Fumihiko Kobayashi, who has been with me all this time and has supported me in so many ways, through his valuable ideas, insights, and happiness.

Note on Transliteration and Other Conventions

In citing English-language materials, I copy titles and the names of authors as published. When transliterating Asian-language names and titles, I use the McCune–Reischauer system of romanization for Korean, the Pinyin system for Chinese, the modified Hepburn system employed by the Library of Congress for Japanese, and the Library of Congress system for Arabic. As for Korean authors' names, I give their preferred English spellings in the body text; however, for the notes and Works Cited, I use the McCune–Reischauer romanization for primary spellings and provide their preferred English spellings in parentheses. I have also treated terms of Persian and Turkish origin as if they were Arabic. Common words and place names, such as *arak* and "Ryukyu," appear in the generally accepted English form without diacritics. Unlike Korean and Arabic, Chinese writing consists of morphosyllabic characters, therefore the book includes characters in the cases of important words not easily found in the book's bibliography. Names and terms of Mongolian origin have been transliterated according to Antoine Mostaert's scheme as modified by F. W. Cleaves, except for these deviations: č is rendered as ch; š as sh; γ is gh, q is kh, and ĵ is j.

Citations from some Korean official sources such as the *Chosŏn wangjo sillok* include a source's date (presented as reign year, lunar month, and day) and relevant document numbers in the online database (see the Works Cited for a web address).

Introduction

Soju, the "national" distilled alcoholic drink of Korea,[1] has now become one of the world's most popular drinks, most recently thanks to the recent pop-cultural phenomenon of Korean Wave (*Hallyu*), represented by the growing popularity of K-pop, Korean dramas, and Korean foods in today's globalizing world.[2] Clear and colorless, with a taste similar to vodka, soju is a kind of spirit, or distilled liquor, which obtains a high percentage of alcohol content by means of a distillation technology that separates alcohol from the water and other compounds of fermented material. Though not as famous as such spirits as whiskey and vodka, soju currently enjoys increasingly widespread popularity worldwide. Like many other modern spirits today, the companies behind many modern soju brands produce on a mass scale in factories using modern technologies. As a consequence, few people today are aware of the drink's ancient origins or how it became popular in Korea.[3] Nor are they aware of its rich history. Indeed, the search for soju's origins leads us into the larger complex world of premodern Afro-Eurasia, where hemispheric movements of trade, empires, transfers of scientific and

---

[1] While one could question who legitimately can determine what a "national" liquor is, many people, including those in Korean and international media, commonly refer to soju as the Korean "national" liquor, in a way that is compared to similar references in Russia to vodka. For example, see Sam Dangremond, "Here's Everything You Need to Know about Soju, the National Drink of South Korea," *Town & Country*, February 8, 2018, www.townandcountrymag.com/leisure/drinks/a16752958/soju-korean-liquor (accessed September 25, 2019).

[2] For example, see Tom Dreisbach, "Move over Vodka; Korean *Soju*'s Taking a Shot at America," *National Public Radio* (*NPR*) online, September 22, 2013, www.npr.org/sections/thesalt/2013/09/22/224522548/move-over-vodka-korean-soju's-taking-a-shot-at-america (accessed May 18, 2016). A 2014 guidebook to the world's most famous spirits introduces soju as "actually the most widely consumed spirit in the world." Joel Harrison and Neil Ridley, *Distilled: From Absinthe & Brandy to Vodka & Whisky, the World's Finest Artisan Spirits Unearthed, Explained & Enjoyed* (London: Mitchell Beazley, 2014), 206.

[3] An American journalist asked owners of some Korean restaurants that sell soju in the United States what was the origin of the drink, and they replied that they didn't know. See Joshua Schenkkan, "What Is 'Traditional' Soju? A Spirited Debate," *Serious Eats*, October 3, 2017, www.seriouseats.com/2017/10/what-is-traditional-soju-korea-tokki-brandon-hill.html (accessed October 2, 2018).

technological know-how, and other forms of cross-cultural exchange created the conditions for the development of what we now regard as a singularly Korean drink. Through soju, we see signs of Korea's deep engagement with the world at large, especially Afro-Eurasia, and more importantly for this story the consequent impact of this interaction on Korean society and culture – here, in the form of a drink now identified with today's Korean nation.

This book about Korean soju is part of a larger study of the history of distillation worldwide, as it relates to the production of medicinals, distillates, and tinctures of various sorts. Liquors such as soju were used in a variety of medical, spiritual, and other social practices. The aspiration in many societies to create stronger alcoholic drinks with a higher level of preservability led to a refinement of the distillation technology that produces them. This development began early; however, the popularity of distilled liquors appears to have grown widespread during the age of the Mongol Empire, which ruled most of Eurasia during the thirteenth and fourteenth centuries. Korea is particularly important to this research because several documentary sources strongly suggest that the Mongols who ruled neighboring China introduced soju and its distillation technology to Korea, and similar forms of evidence may provide further clues to other forms of distillation technology transfers from Asia to other areas of the world. Such a study elucidates not only the history of an increasingly well-known form of liquor but also the role of Korea in world history. Despite the historical significance of soju, no Western-language study of it has been undertaken, nor has any detailed study been done from a comparative perspective in any language. This book corrects that oversight.

To undertake this history of soju, we have to start with the origin of its name, which interestingly hints at the complex dynamics that shaped its early development. The name, pronounced *soju* in Korean but written in the Chinese script as 燒酒, uses the same Chinese characters that apply to the name for a similar Chinese distilled liquor called *shaojiu* 燒酒, nearly identical to the characters used in *shōchū* 燒酎 (known as *shochu* in English without diacritics), its Japanese counterpart.[4] The existence in East Asia of three variations of a traditional liquor of similar make, all bearing the same written name, suggests a shared origin in cross-cultural exchange;

---

[4] Ishige Naomichi, "Higashi yūrashia no jōryūshu: jōryūki wo motomete" (Distilled Alcohol in East Eurasia: Seeking the Distiller), in *Shōchū higashi mawari nishi mawari* (Shochu around the World), ed. Tamamura Toyo'o (Tokyo: TaKaRa Alcohol Beverage and Life Research Institute, 1999), 122.

indeed, the evidence for this transfer of a shared recipe and the necessary technology can be found sometime during the premodern period.

This is not the only name relevant to soju's identity and history. Although many have not yet recognized this fact, people in Korea, China, and Japan have to a lesser extent also referred to this form of distilled liquor as *arak*.[5] As attested in literary sources, the name *arak* was adopted upon its arrival in East Asia as an exotic foreign term; local inhabitants then used Chinese characters to mimic the pronunciation of such foreign terms for the purpose of transliteration. This word probably originated with the Arabic word *'araq* – aka "arak" – which literally means "perspiration," and which serves as a name for many similar forms of distilled alcohol found in the Middle East as well as South and Southeast Asia. *Arkhi*, a form of distilled liquor popular in Mongolia and Central Asia, is also connected to arak in both its origin and the technology used to make it. This name, disseminated through historical trade routes, reveals another possible historical root name shared among the various distilled alcohols developed and consumed in diverse Asian societies, suggesting too that they may also share to some degree a common history of origins. Indeed, this book shows that the consumption of arak began to spread during the Mongol period, beginning with a Mongol-period cookbook written in Chinese that introduced the distilled liquor for the first time. Thus arak plays a kind of ancestral role in soju's evolution.

By closely examining all the documentary and archaeological sources available to scholars so far, we can begin to map out the trajectory of soju's introduction to Korea. For example, a highlight of my study explores the transfer of distillation technology from China to Korea during the Mongol period (1206–1368), an age that scholars now regard as the first stage of globalization's history.[6] In medieval times, the fact that the distillation of arak and shaojiu (Korean *soju*; Japanese *shōchū*) developed in China and spread to Korea can be verified by both circumstantial evidence and a variety of sources. Due to its alcoholic nature, arak was prohibited on religious grounds in Islamic societies, and therefore its development was not well documented during premodern times.[7] In

---

[5] Some modern brand names for Korean soju are called *arak*, as in the case of this news article: www.thinkfood.co.kr/news/articleView.html?idxno=43694 (accessed April 20, 2017).

[6] Among the first scholarly syntheses of the Mongol Empire from global historical perspectives are Timothy May, *The Mongol Conquests in World History* (London: Reaktion Books, 2012); Morris Rossabi, *The Mongols: A Very Short Introduction* (Oxford: Oxford University Press, 2012).

[7] Other possible factors might be local/home production (undocumented), and/or production by minorities, which could be one of the reasons why arak was mainly popular in the Middle East in regions where significant communities historically Christian lived (Lebanon, Syria,

contrast, the propagation of soju was relatively well documented in both China and Korea. In other words, exploring soju's origin can provide a concrete case for the general possibility that these distilled liquors and their newly developed technological methods spread to far-flung regions in Afro-Eurasia through cross-cultural exchanges, mostly at a gradual and continuous pace yet at times at larger scales through intensive contacts, as was the case in Mongol times.

In short, the conditions of soju's transmission to Korea make compelling history, given the extent of its remarkable transmission. In order to search for this shared origin of distilled liquors, it is important to first place the topic of soju in the context of premodern Afro-Eurasian history. This hemispheric context helps to make sense of the rise of soju and arak in Korea. Conversely, Korean history provides a useful vantage from which to detect the vectors through which a particular form of material culture spread throughout premodern Eurasia and the processes by which soju or arak production technologies, techniques, and terminology localized. Again, a hemispheric perspective helps; the documentation of not only transfers of tools, know-how, and names but also their localization in the process of adaptation in the regions that lay between the Middle East and East Asia elucidates comparable phenomena in contemporaneous Korean society as well. The development of soju in Korea suggests an interactive relationship between technology and culture. On the one hand, technology drives cultural change – consider, for example, the impact of new forms of alcohol on social practices such as medicine, gift giving, and even rituals to a certain degree. On the other hand, culture drives technological development in a local context as the boom in consumption of the new type of liquor facilitated export, import, and further developments of related technologies through exchange with other culture such as Japan and Europe.[8]

This is not a book about soju in the usual sense. Instead, we use the history of a spirit to understand how certain groups of people in specific historical contexts created, consumed, and benefited from a cultural

Egypt, and Iraq). But clearly the main influence was religion in the Islamic world. Based on personal conversations with Limor Yungman.
[8] This helps us think through how technology and culture relate through a particular example of material culture. Since the initial call to promote the scholarly study of the history of technology, focusing on the relations between technology and other elements of culture using interdisciplinary approaches developed by Melvin Kranzberg, many studies have been published on the relations between technology and culture in history. See Melvin Kranzberg, "At the Start," *Technology and Culture* 1, no. 1 (Winter 1959): 1–10. For a discussion of the influence of technology on culture from various theoretical perspectives, see Andrew Murphie and John Potts, *Culture and Technology* (New York: Palgrave Macmillan, 2003).

object, and how, through cross-cultural exchanges, they transferred it and its related technologies – not to mention cultural ideas – to other regions of Afro-Eurasia, leading to the development of premodern connections that ultimately led to our global society. As such, the book focuses on two issues: transfers and localizations. To search for signs of transfer, we trace the origin and development of distilled liquors to premodern Eurasia and follow this history until we reach the Mongol period, when the possibilities for cross-cultural exchange and the propagation of new cultural practices and technologies reached new heights. In particular, soju – whose spread to the Korean peninsula coincided with both the Mongol-run Yuan dynasty in China and Mongolia (1271–1368) and Korea's Koryŏ dynasty (918–1392) – provides a representative case study of a cultural item that transferred into Korea from Eurasia at large during premodern times thanks to cross-cultural exchange. After soju settled into Korean society, the materials and methods used to make the spirit localized; this made possible the development of its unique characteristics, which contributed to the conditions that encouraged its counter-transmission into the wider world in the tumultuous twentieth and twenty-first centuries. In short, soju, an icon of Korean food culture, first developed in a thriving milieu characterized by trans-Eurasian exchanges of culture and technology. Indeed, such characteristics make soju an ideal cultural object to observe from a global historical perspective. The example of soju ultimately suggests that the process of globalization developed in Asia earlier than it did in the West. It serves as a useful case study representing a larger historical trend, in which premodern cross-cultural encounters profoundly influenced and reshaped cultural patterns in different societies, as Jerry Bentley argued, a pioneer in the field of world history.[9] Here, let us focus on specific issues that clarify the limitations and problems inherent in existing research and then elucidate the book's key components.

A Global History of Alcohol in Food, Medicine, Science/Technology, and Culture

This book deals primarily with alcoholic drink as a kind of food, and therefore its narrative first frames its subject within the context of food history. In particular, it borrows methodological frames of "the cultural history of food" proposed by Deborah Valenze, who took food beyond the realm of anthropology by situating food and cultural change "into

---

[9] Jerry H. Bentley, *Old World Encounters: Cross-Cultural Contacts and Exchanges in Premodern Times* (New York: Oxford University Press, 1993).

frameworks related to historical change," bringing it squarely into the center of historical analysis.[10] (Valenze also stresses the importance of migrations of food and people in world history as avenues for future research, pointing to the precedent set by Alfred W. Crosby's works on the Columbian exchange.)[11] This opened the way for the relatively young field of food history to develop many of the interdisciplinary methods and studies that address issues of globalization in world history.[12]

Soju has also attracted popular attention as a useful means to understand how traditional culture has survived and even become emblematic of national identity in the present global age. Korean kimchi, Japanese sushi, Italian pasta, and Indian curry exemplify foods that successfully developed into national, and then ultimately global, dishes.[13] Societies in modern nation-states, as a part of their efforts to secure national identities, have increased their interest in the history of foods deemed emblematic of their nation.[14] Of course, as with other historical subjects, closer scrutiny reveals many cases in which larger sets of dynamic influences were mediated through cross-cultural exchanges to shape development.[15] Soju provides just such a specimen. Likewise, evidence demonstrates a number of cases in which cross-cultural transmissions and influences moved the other way, leading society on the Korean peninsula to contribute to changes in patterns of food culture in premodern East Asia. In recent times, Koreans have tried to reverse the trajectory of influence in order to globalize Korean food, which they have done successfully so far. Some government-sponsored books advertise Korean foods globally. The academic world has seen the publication of Korean-language monographs on Korean food in addition to studies by scholars

---

[10] Deborah Valenze, "The Cultural History of Food," in *Routledge International Handbook of Food Studies*, ed. Ken Albala (London: Routledge, 2013), 101–102.

[11] Ibid., 108–109.

[12] Raymond Grew, "Food and Global History," in *Food in Global History*, ed. Raymond Grew (Boulder, CO: Westview Press, 2000), 1–32.

[13] Emiko Ohnuki-Tierney, "We Eat Each Other's Food to Nourish Our Body: The Global and the Local as Mutually Constituent Forces," in *Food in Global History*, ed. Raymond Grew (Boulder, CO: Westview Press, 2000), 271–272. For the cases of pasta and curry, see Silvano Serventi and Françoise Sabban, *Pasta: The Story of a Universal Food*, trans. Antony Shugaar (New York: Columbia University Press, 2002); Colleen Taylor Sen, *Curry: A Global History* (London: Reaktion Books, 2009).

[14] Chu Yŏngha (Joo Young-ha), *Ŭmsik inmunhak: ŭmsik ŭro pon han'guk ŭi yŏksa wa munhwa* (The Cultural Anthropology of Food: Korean History and Culture Viewed from the Perspective of Food) (Seoul: Hyumŏnisŭt'ŭ, 2011), 457.

[15] For example, there was a one-day symposium entitled Global Food History at Leiden University in 2016. Several papers presented there, including my paper about soju, discuss various issues crucial to the theme. These include migration of people, circulation of technology, and global and cross-cultural contact.

like Michael Pettid.[16] However, the origin and history of food culture in Korean history at a level of detail that includes soju is a topic yet to be undertaken.

Korea's case reflects the general situation for food studies in the world. Despite its fundamental importance to human life, food history as a distinct field, particularly global and comparative approaches to the subject, began to receive large-scale scholarly attention only recently. Most recently, scholars began to pay more attention to exchanges in material cultures, including cultures of food and medicines.[17] While interest in the transfers of foodways and culinary technology in global history grows, an interdisciplinary field called *food studies* has developed simultaneously thanks to the efforts of scholars in different fields working on related topics.[18] The tendency to discuss food culture from humanistic perspectives has been expanding worldwide, according to food anthropologist Joo Young-ha. International conferences on food science over the past few decades, such as the 1987 meeting sponsored by the Association for the Study of Food and Society (ASFS), a US group, and more recent meetings such as one held by the Institut européen d'histoire et des cultures de l'alimentation (IEHCA) in 2001, have helped to motivate this expansion.[19] Meanwhile, scholars in recent years have published significant articles and books on various foods and foodways, many of which pay attention to the overall impact of foods and food culture on human lives. Going beyond simple examinations of what people ate at certain places and times, these scholars have examined cases of cultural, material, and technology transfers in their relevant

---

[16] Michael J. Pettid, *Korean Cuisine: An Illustrated History* (London: Reaktion Books, 2008).

[17] For example, see Paul D. Buell and Eugene N. Anderson, *A Soup for the Qan: Chinese Dietary Medicine of the Mongol Era as Seen in Hu Sihui's Yinshan Zhengyao*, introduction, translation, text, notes, appendix by Charles Perry, 2nd revised and expanded edition, Sir Henry Wellcome Asian Series 9 (Leiden and Boston: E. J. Brill, 2010); John Kieschnick, *Impact of Buddhism on Chinese Material Culture* (Princeton: Princeton University Press, 2003).

[18] For a recent collective work of scholarship that provides key definitions, methodologies, and theoretical approaches in the field, see Ken Albala, ed., *Routledge International Handbook of Food Studies* (London: Routledge, 2013).

[19] For example, the Association for the Study of Food and Society (ASFS), established in 1985, aims at interdisciplinary research on food and society (see www.food-culture.org). They organize annual conferences and publish the academic journal *Food, Culture & Society* in order to promote their relatively new research field. Its European equivalent, the European Institute for the History and Culture of Food (IEHCA), has been active since its foundation in 2001 and is a key actor in the international and interdisciplinary research community surrounding food studies, with annual conferences and their multilingual academic journal, *Food & History*, which they have been publishing since 2003 (see http://iehca.eu/en). Joo Young-ha argues that the background to this academic trend has been the growing ethnic food culture in the US consumer economy since the 1980s. Chu Yŏngha, *Ŭmsik inmunhak*, 31.

historical and social contexts, including the introduction through cross-cultural exchanges of new food ingredients and cooking ways. In Korea, Joo Young-ha has helped to advance the cause of food studies by developing the Institute for Textual and Oral Histories of Food (ITOHF), a research institute that focuses on food and humanities research, and expanding his research into the field of cultural anthropology. He now has published several major works in Korean on the history of Korean foods, some of which have been translated into Chinese, and continues to advocate for its development through his persistent efforts to facilitate interdisciplinary scholarly exchanges.[20]

Despite its many fruits, the food studies project also faces limits. Most importantly, nearly all books and articles about foodways written by Western scholars in Western languages have focused on the Western food cultures that are familiar to them and introduce Asian or other non-Western equivalents only to make certain comparisons.[21] In contrast, studies of Asian food culture written in Asian languages tend to focus specifically on Asian topics, such as the histories of foods and alcohol distillation in places like China or Korea. Moreover, within the global scholarly community, the literature of food history in Asian countries suffers limited readership, which certainly hinders the development of a balanced global perspective. Therefore more academic studies in Western languages from broad and comparative perspectives are needed in order to create wider global access to this topic.[22] This book about soju strives to do just that by placing liquor in a larger context of space and time, namely a long history of Eurasia that stretches from the deep past to immediate present.

Alcoholic beverages are an important part of the history of food. Alcohol has always functioned as an important cultural item in human societies, since it has always comprised a key element of food culture. For example, in his book about the history of alcoholic drinks in China, He Manzi argues that alcohol has accompanied foods at feasts and rituals since ancient times and inspired numerous works of literature

---

[20] See many pioneering works by Joo Young-ha, including the following, which was translated into Chinese: *Han'gugin ŭn wae irŏk'e mŏgŭlkka: siksa pangsik ŭro pon han'guk ŭmsik munhwasa* (Why Do Koreans Eat Like This? A Cultural History of Food in Korea Investigated through the Dining Custom) (Seoul: Hyumŏnisŭt'ŭ, 2018).

[21] For example, see Raymond Grew, ed., *Food in Global History* (Boulder, CO: Westview Press, 2000); Adam Rogers, *Proof: the Science of Booze* (Boston: Houghton Mifflin Harcourt, 2014).

[22] A few pioneering works include Eugene N. Anderson, *The Food of China* (New Haven: Yale University Press, 1988); and Paul D. Buell, Eugene N. Anderson, Montserrat De Pablo Moya, and Moldir Oskenbay, *Crossroads of Cuisine: The Eurasian Heartland, the Silk Roads and Food* (Leiden: Brill, 2020).

and art.[23] Alcoholic brews also found utility as a medicine of major importance during the premodern era. Indeed, modern life would be incomparably different without such alcohol-based treatments, for they have played an important role in the realms of both folk and conventional medicine. This is true particularly for Chinese medicine (while not only for China), where alcohols are used both externally and internally, on the premise that alcohol possesses both preventive and therapeutic effects.[24]

In recent times, academic interest in globalization has led to several studies of alcohol in world history. While focusing primarily on Western cases, recent works on the history of alcoholic beverages, such as *Uncorking the Past* by Patrick McGovern, are noteworthy for their attempt to bring a comparative form of understanding to topics with a global scope by introducing non-Western examples like China's Yellow River basin, the Silk Roads, the New World and Africa.[25] In her attempt to analyze the development of alcohol from a holistic, global perspective, *Alcohol in World History*, Gina Hames argues that alcoholic beverages developed continuously and in accord with patterns common to different world societies.[26] Hames's succinct developmental history is too general in scope to investigate alcohol's cross-cultural origins in any depth or detail; however, it has opened further stages for deeper discussion on these topics.

Deeper examinations of the history of alcohol from comparative and connective perspectives would indeed help us contemplate further related issues. One such issue is the implications of long-distance connections among different societies for our understanding of world history. It is true that, since ancient times, many societies developed similar alcoholic drinks, particularly fermented ones like wines and beers that were universal to most societies. Yet we see that, over time, people exchanged these items through cross-cultural contacts, and that people in different societies began to enjoy alcoholic beverages with different (foreign) characteristics, some of which would be popular there, too, and would be assimilated into local cultures (such as French-style wines produced in Japan). We see this phenomenon more often in modern times, but we can find similar cases in the premodern period as well. We also similarly notice that people developed certain technology, like distillation, to

---

[23] He Manzi, *Zuixiang riyue: Zhongguo jiu wenhua* (Drunken Sun and Moon: Chinese Wine Culture) (Shanghai: Shanghai guji chubanshe, 1991), 35–41, 50–61.

[24] Ibid., 62–69.

[25] For example, see Patrick E. McGovern, *Uncorking the Past: The Quest for Wine, Beer, and Other Alcoholic Beverages* (Berkeley: University of California Press, 2009).

[26] Gina Hames, *Alcohol in World History* (London: Routledge, 2012), 1–4, 129–134.

develop the existing fermented beverages further, and the new forms of alcoholic beverages began to spread further to wider areas. From certain periods of time, people in different regions began to share similar distilled beverages made of similar yet different forms of distillation technology adapted to local environments.

Here we see another issue: the interaction between culture on the one hand and science and technology on the other. The history of distilled liquors provides clear examples of interactions between alcoholic beverages as particular items of material culture and science and technology. In most cases, developments of science and technology in different societies are uneven, and so is the case of distillation technology, which enabled the creation of distilled liquors. This topic requires comparative examination on a global historical scale. As a specialized topic in the field of food studies, the history of distilled liquors requires researchers to possess both a good command of relevant primary sources and also specialized technical knowledge of the relevant details that could elucidate continuities and differences across space and time – such as in, say, distillation techniques that enable the separation of materials using heat, variable boiling points, and condensation-utilizing stills. This creates an important challenge for this current book.

Toward a Global History of Distillation and Distilled Liquors

Scholars have assumed that human beings began to learn about alcoholic drinks by accident; that is, beginning with their interactions with naturally fermented grapes or grains.[27] Even in China, whose many written records include myths and legends about the origins of alcoholic drinks, people began to develop a realistic understanding in ancient times. This is evident in works like *Shuowen* 說文, an early second-century Chinese-character dictionary, which says, "if the color of an alcoholic drink becomes white, we call that *sou* 醙 [white wine],[28] because this is one

---

[27] Rogers, *Proof*, 20.

[28] Nowadays, words for alcoholic drinks are tightly defined due to taxation and regulation. For example, historical words for wine (the latter of which probably derives from the Georgian word *ghvino* via Latin *vinum*; cf. Hebrew *wainos*) refer to fermented alcoholic drinks made from any fruit. Therefore Eugene Anderson argues that, while alcoholic drinks of China, called *jiu*, are usually lumped under the term "rice wine," this is not correct. Anderson, *The Food of China*, 120. This book, however, sometimes uses the term "wine" to refer to any kind of fermented alcoholic beverage, because some earlier studies – particularly on those in China and Korea – used it as a general term in this way, such as the Korean clear strained wine called *ch'ŏngju* 清酒 (fermented drink based on rice).

that rice spoiled and transformed."[29] People in early societies all over the world produced a variety of alcoholic drinks by using a diversity of ingredients. Most agricultural societies based their process of making alcohol on grains; this is the case with beers, which have the longest history of all alcoholic beverages. Many early peoples also produced wines based on fruits or plants such as grapes, coconuts, and agave, depending on what predominated in their environments.[30] Steppe nomadic societies also made alcoholic beverages by fermenting animal milk.[31] Beginning in early modern times as global networks began to grow in earnest – about 1500 – distilled spirits won favor among people all over the world. Whether a brandy based on grape wines or a shaojiu/soju based on grains, spirits in all their varieties have long shared a common trait: all are created by means of a scientific process called *distillation*, which separates alcoholic from non-alcoholic ingredients to produce a distilled drink that contains a high concentration of alcohol.

Many people enjoy spirits, yet few people are familiar with how they have been made. One distillation technique, based on a technology commonly known in English as a *still*, enables the separation of the high-alcohol component of a liquid by exploiting the different boiling points of its constituent parts. People nowadays commonly apply this principle to distillation techniques that, for example, isolate the distilled alcohol component from fermented wine.[32] Today's industrial producers of distilled alcohols carefully ferment raw material, typically with fermentation starters like yeast, in order to prepare it for the next stage of distillation. Scholars also began to pay attention to another independent and very old tradition of distillation, namely freeze distillation, a process in which a distilled spirit is drawn from a fermented liquid using cold rather than heat; the raw material is first cooled into a cold slush, from which distillers

---

[29] *Shuowen jiezi* 說文解字 (lit. "Explaining Graphs and Analyzing Characters") was written sometime during the Han dynasty (206 BCE–220 CE); see He Manzi, *Zuixiang riyue*, 11. For more details about the stories and theories about the invention and development of alcoholic beverages in Chinese documents, see ibid., 8–34.

[30] Ibid., 121–124. Recent studies demonstrate that grape wines developed in many places in the world, even in China, independently from the ancient period, refuting a previous theory about the transfer of grape wine from the West. Björn Kjellgren, "Drunken Modernity: Wine in China," *Anthropology of Food* (online) 3 (December 2004): 12, http://journals.openedition.org/aof/249 (accessed December 3, 2018); Hsing-tsung Huang, *Science and Civilization in China*, vol. 6: *Biology and Biological Technology*, part 5: *Fermentation and Food Science* (Cambridge: Cambridge University Press, 2000), 153 and 243.

[31] He Manzi, *Zuixiang riyue*, 122–123.

[32] By "wine" we mean generally the non-distilled alcoholic beverage. We must be careful about using the term "wine," because that can specifically mean grape wine. Still, we will occasionally use the term to refer to the grape-fermented alcoholic beverage whenever the context is clear.

can remove the portion that has neither frozen nor congealed. This approach has long been used in cold steppe regions like Northern or Central Asia.[33] Although the boiling and freezing processes differ, both always require an initial round of fermentation, especially in the case of fruits or berries, which generally ferment naturally and do not need yeast or starters.[34] Unlike non-distilled alcoholic beverages, whose alcohol content measures around 10 percent, distilled spirits report alcohol levels higher than 20; indeed, many traditional ones range between 40 and 70. Such high levels of alcohol content have appealed to many people, of course. There were other, more practical reasons to produce spirits as well. Distillation helped to improve the efficiency of food consumption, since fermented and distilled alcohols can be made from a wide variety of foods, including grains, fruits, and milks; this creates an opportunity to utilize surpluses and prevent waste.

Despite their special features and general importance to food history, distilled alcohols have yet to receive adequate academic attention, and as a consequence big gaps appear in the historiography of alcoholic beverages published to date. Previous studies of distillation's history focused mainly on Europe; one authority even regarded distillation as a thoroughly European invention that disseminated to other cultures after its invention.[35] However, studies like Joseph Needham's celebrated volumes *Science and Civilisation in China*, which demonstrate Asia's early knowledge of distillation technology and the use of devices entirely different from European mainstream pot stills with worm tubs, have rendered such a view inappropriate and laid the foundations for current global and world-historical approaches.[36]

In order to advance toward a global history of distilled spirits, we must overcome the following hurdles. First, we must use comparative study to

---

[33] Joseph Needham, Ho Ping-Yü, and Lu Gwei-djen, *Science and Civilisation in China*, vol. 5: *Chemistry and Chemical Technology*, part 4: *Spagyrical Discovery and Invention: Apparatus, Theories and Gifts* (Cambridge: Cambridge University Press, 1980), 133. See also the discussion in Buell and Anderson, *A Soup for the Qan*, 120–121.

[34] Paul D. Buell, "Mongol Empire and Distillation: Technology and Popularization," an unpublished paper presented at the Science and Technology Transfer workshop held at the Hebrew University of Jerusalem, June 10–11, 2015.

[35] Robert James Forbes, *Short History of Distillation: From the Beginnings up to the Death of Cellier Blumenthal* (Leiden: E. J. Brill, 1948).

[36] Joseph Needham and Lu Gwei-djen. *Science and Civilisation in China*, vol. 5: *Chemistry and Chemical Technology*, part 2: *Spagyrical Discovery and Invention: Magisteries of Gold and Immortality*. Cambridge: Cambridge University Press, 1974; Joseph Needham, Ho Ping-yü, and Lu Gwei-djen. *Science and Civilisation in China*, vol. 5: *Chemistry and Chemical Technology*, part 3: *Spagyrical Discovery and Invention: Historical Survey, from Cinnabar Elixirs to Synthetic Insulin*. Cambridge: Cambridge University Press, 1976; Joseph Needham, *Science and Civilisation in China*, vol. 5: *Chemistry and Chemical Technology*, part 5: *Spagyrical Discovery and Invention: Physiological Alchemy*. Cambridge: Cambridge University Press, 1983; Huang, *Science and Civilisation in China*, vol. 6, part 5.

adjudicate the many conflicting theories about the origins of distillation technology that have proliferated and continue to confuse people. Second, we must transcend the issue of origins by redefining the term to refer to when a variety became popular in the society in question. For example, while distillation traces back to ancient times, it does not appear in historiography until the fourteenth and fifteenth centuries. This is significant, however, for the two centuries overlap with the era of the Mongol Empire, which spanned the thirteenth and fourteenth centuries. Most recently, Paul Buell proposed a co-operative project focusing on the early history of Eurasian distillation formulated in terms of technological exchanges during the Mongol era (approximately 1200 to 1400), which were unprecedented in terms of range, scope, and cross-cultural influence.

This current book was inspired in the course of my participation in Buell's initial project entitled A Comparative Investigation of Distillation Technologies, Wine Production, and Fermented Products, a collected study of the topic undertaken with several coworkers from different academic institutions, including in Germany, Mongolia, and Mexico, over the course of three years (2014–2017).[37] Scholars have yet to examine the histories of many globally popular distilled spirits, like cognac or vodka; it may be that all distilled liquors share a common origin, not in antiquity, but in the age of the Mongols.

The Mongol Role in Distillation's Trans-Eurasian Spread to Korea

Studying the history of distillation matters. Distilled beverages and other distillates have significantly influenced the course of history in nearly all world societies. Because of this, distillation history can serve as a vehicle for expanding our knowledge and understanding of the large- and small-scale patterns of cultural interaction that occur in world societies. For example, distillation has a distinct and dynamic history of rise, prosperity, and decline, which can point to undetected general trends in society as a whole. The most prominent for the advancement of distillation was the development of alchemy in the medieval Islamic world, the basis of the type of distillation used by the Arabs in making perfumes and other products employing alcohol tinctures. A recent study by Ahmad Y. al-Hassan also strongly suggests that the early Arabs used distillation to

---

[37] Scholars collaborating on this project, including Paul Buell and myself, have already presented and published a number of papers at international conferences, such as the 14th International Conference on the History of Science in East Asia (Paris, July 6–10, 2015), and in journals, such as *Crossroads*, volumes 13 and 14.

make the distilled wines that came to be called *'araq* (along with other names).[38] The Mongols who rose from the steppes in the thirteenth century to change the world probably did not invent distillation, but they certainly did popularize a great variety of spirits best known by the Arabic word *arak* (*'araq*) across the empire and the successor states they created during the thirteenth and fourteenth centuries.

In such a situation, a few studies raise questions about the possible Mongol role in the development and spread of distilled alcohols called arak in Asia. In his pioneering project, Paul Buell examines the key role that the Mongols played in improving a pre-existing form of Chinese distillation technology by developing an easily transportable apparatus, disseminating that improved device throughout Eurasia, and creating an environment that encouraged the development of local varieties using locally available ingredients. Distilled liquors called arak begin to appear in documents written in a variety of places beyond the Islamic world, in fact soon all over the Afro-Eurasian world, beginning with a fourteenth-century Chinese document, plausibly the result of interconnections made through contacts and trade routes under the Mongol aegis.[39] It is precisely at this time that shaojiu, which had existed in China as a strong form of liquor for centuries, first appears in documentary sources as a distilled liquor in conjunction with the foreign distillate arak. Soon after, it appears in Korean sources simultaneously as *soju* and arak. This suggests a clear connection between shaojiu/soju's origins and the arak tradition spreading into China and across Eurasia at that time.

Looking at these connections, we should consider the Mongols' customs that facilitated the spread of distillation technology during the Mongol era and played an important part in distilled-alcohol propagation. Among these were both drink and drinking rituals. As Buell postulates, this first incorporated the consumption of *kumiss* and *shubat* (the fermented milk of mares and camels respectively), but soon moved on to imported liquors such as grape wine before moving on again, for various reasons, to distilled beverages.[40] A successful ruler or chief had to supply

---

[38] See Ahmad Y. al-Hassan, "Alcohol and the Distillation of Wine in Arabic Sources from the 8th Century," in *Studies in Al-Kimya': Critical Issues in Latin and Arabic Alchemy and Chemistry* (Hildesheim: Georg Olms, 2009), 283–298.

[39] The following chapters discuss descriptions of *alaji* 阿剌吉, in Mongolian *arajhi*, a palatalized Turkic intermediation of *arak* in an official dietary manual written for the Mongol court of China (*Yinshan zhengyao* 飲膳正要, 1330; see Chapter 1), *alaji* 阿剌吉 (read *aralgil* in modern Korean transliteration) by the poet Yi Saek 李穡 (1328–1396; see Chapter 2), and *araki shu* 荒木酒 (pronounced *huanmu jiu* in modern Mandarin Chinese) or *araki shu* 阿剌吉酒 (*alaji* in Mandarin) in Japan in the eighteenth-century (see Chapter 6).

[40] One of these included the need to preserve kumiss after it was prepared for reasons of social prestige. See Chapter 1.

his followers with both booty and also the right kinds of food (such as banquet soups) and drink, and this effort involved considerably numerous social rituals. For example, Marco Polo's (1254–1324) vivid descriptions of court drinking in China under Mongol rule help us to delve into Mongol social ritual.[41] It is worth noting here that Koreans participated in the imperial bodyguard of the Mongol-run Yuan dynasty in China and Mongolia to a considerable extent, and among their activities they would have engaged in ceremonial drinking as part of the Mongols' integrative rites. Many Persian manuscript paintings depict Mongols feasting and drinking in Iran under their rule; such artifacts help us to make cross-cultural comparisons between Mongol China and other places within the Mongol domain where assimilation was underway.

In part, given their primarily nomadic lifestyle, for the practical reason of portability, we can assume that wherever the Mongols arrived in the course of their political expansion, they probably showed an active interest in producing spirits locally. According to anthropological researchers, Mongol stills have always been easily portable, although the wooden ones of the past have now given way to metal ones.[42] To make them quickly and to maximize their portability during the decades of military expansion, the Mongols appropriated stills used by Chinese to produce a simplified, portable still that they could take with them. As a consequence, Mongol stills easily moved throughout Eurasia, including Korea. Joseph Needham, adopting Hommel's models of the Mongolian and Chinese stills for his notable volume on distilling technology, categorizes such a still as a Mongolian type – for example, the inside-pot, catch-alcohol-style still, which differs from the more complicated Chinese type that uses a tube to remove the alcohol from the pot.[43] We do not know the history of these various kinds of stills or how one kind related to others. For the most part, one can only guess that Mongol-type stills improved thanks to the influence of other kinds of stills. Certainly, they endured, remaining popular above other varieties thanks to the needs they served. Politically, Mongol rulers had to provide kumiss to their followers or lose influence. Culturally, alcohol was popular among this nomadic people, who found it convenient and practical, as it served the need for liquid. Moreover, they could provide distilled kumiss when the real stuff was out of season, a need that grew as

---

[41] Marco Polo (1254–1324), *The Description of the World*, trans. Sharon Kinoshita (Indianapolis: Hackett Publishing Company, 2016), 77.

[42] In Central Mongolia, Paul Buell and Montserrat de Pablo saw an old wooden still, not long out of use, and a large yak during their field research there in 2015. Based on personal conversation with Buell.

[43] See Needham's reconstructions after Hommel in Needham, Ho, and Lu, *Science and Civilisation in China*, vol. 5, part 4, 62–63; Huang, *Science and Civilisation in China*, vol. 6, part 5, p. 216.

the drink's popularity grew more widespread. Before we discuss the still varieties that developed throughout Eurasia in the following chapters, let us note here that Needham failed to investigate other important cases involving this kind of technology transfer, like those that moved from China under Mongol rule to Korea. This omission has led to a persistent oversight in subsequent studies in Western languages. This left Korean and Japanese scholars to provide the only detailed scholarship on the history of soju.[44]

The case of Korea is worth particular attention for comparative purposes because, contrary to other cases (including the Asian-type Mexican stills of a later time), extant written sources describing transfers of distillation technology based on Asian prototypes into Korea through direct contacts with China and the larger Mongol world after the Mongol invasions (1231–1258) are relatively numerous and well preserved. This helps us to draw connections between this popularization in Korea and the unprecedented level of cross-cultural interaction that existed throughout thirteenth- and fourteenth-century Asia, where Mongol rulers actively promoted the transfer of distillation technology along with a host of other technologies, ideas, and goods. After its sudden rise, soju differentiated itself from other liquors and developed into its Korean form. Still, traces of its non-Korean origins remain. Notably, soju differs from other alcoholic beverages developed on the Korean peninsula because it is made using a special technique that was propagated from China, and Eurasia beyond, during the Mongol period. This makes soju's history more complex, dynamic, and global. Since then, it has undergone a constant process of localization, often in response to bigger changes in world history.

Korean History in Global Context

While the book mainly focuses on the transfer of distillation technology as a concrete example of both food history and technology transfer and innovation in world history, it also views this as a historical phenomenon in the context of global history – that is, in the broader context of long-distance, cross-cultural contacts and exchanges, material and non-material, between Korea, China, and the world at large during Mongol and post-Mongol times as part of the grand history of intra-Eurasian interactions that led ultimately to the creation of a global society. Just as historians have shed new light on the histories of the Mongol Empire and

---

[44] No works about alcohol and distillation history in Western languages discuss Korea, including those by Needham and Hames. The Korean and Japanese works that deserve translation include those by Yi Seong-wu (Yi Sŏngu), Joo Young-ha (Chu Yŏngha), and Ishige Naomichi. See the following chapters for more details.

the Yuan dynasty through which the Mongols governed China and Mongolia,[45] which have resulted in radically revised depictions of their histories and their respective interpretations of significance, the concurrent epoch in Korean history – namely the Koryŏ dynasty – deserves an equivalently significant re-evaluation.[46] This book seeks to help begin that important progress by looking at the history of Korea from a global perspective that privileges the factor of cross-cultural exchange.[47] Existing studies that have made similar efforts have mostly focused on Korea's political and economic relations, so the sociocultural emphasis of this study will enhance their achievements by demonstrating the equal significance of an important feature of Korean cultural history.

In addition to food history and global history, it is important to understand the development of soju in the context of Korean history, especially the latter Koryŏ dynasty, whose dates roughly coincide with the Mongol age. After all, the earliest evidence of distilled spirits like soju and arak in Korean documents appears in late Koryŏ-era documents. While the field of Koryŏ studies has thrived in Korea, most books and studies focusing on Korean history and culture published in Western countries deal with the Chosŏn period (1392–1897), for which more primary sources are extant. This leads to a divergence between two parts of the world in the depiction of Korean history. However, considering the growing trend to rethink regional history from a global-history perspective, we should pay attention to the Koryŏ period, particularly its final century, when the people of the Korean peninsula were most closely connected to the premodern networks of Eurasia. In other words, we cannot place Korea into the broader context of the Mongol Empire without conversely trying to understand the place of the Mongol Empire in Korean history.

[45] For examples, see chapters that provide several concrete examples of Mongol influences in China in Morris Rossabi, ed., *Eurasian Influences on Yuan China* (Singapore: Institute of Southeast Asia Studies, 2013); and Qiu Yihao, *Menggu diguo shiyexia de Yuanshi yu dongxi wenhua jiaoliu* (Studies on the History of Yuan Dynasty and Trans-Eurasian Culture Exchanges from the Perspective of Mongol World Empire) (Shanghai: Shanghai guji, 2019).

[46] For a pioneering monograph, see David M. Robinson, *Empire's Twilight: Northeast Asia under the Mongols* (Cambridge, MA: Harvard University Press, 2009).

[47] Several monographs and textbooks about Korean history have been published in Western languages during the past several decades. The most authoritative and comprehensive study of Korean history translated into English is Lee Ki-baik, *A New History of Korea*, trans. Edward W. Wagner and Edward J. Shultz (Cambridge, MA: Harvard University Press, 1984). More recent books have begun to approach Korean history from global perspectives, in keeping with the pace of globalization. For a concise survey of Korea's history from ancient times to the present that provide foreign readers with macroscopic insights into the nation's long historical development as well as recent scholarly findings, see Kim Jinwung, *A History of Korea: From "Land of the Morning Calm" to States in Conflict* (Bloomington: Indiana University Press, 2012).

Following the tenth-century collapse of Great Silla (also known as Unified Silla or Later Silla, 668–935 CE), renowned for unifying the Korean peninsula after the end of Korea's Three Kingdoms period (first century BCE–676 CE), the Koryŏ dynasty expanded its territory northward, securing most of the Korean peninsula as its domain. Lasting several centuries until its fall in 1392 to Yi Sŏnggye, the founder of the Chosŏn dynasty, the Koryŏ dynasty witnessed many important cultural developments, including the rise of Buddhism and the development of literature, that help us to understand the ancient Korean history of both Koryŏ and pre-Koryŏ times. It also experienced several dynamic political changes, particularly during the era of Koryŏ's military regime (Musin Jŏnggwŏn, 1170–1270), not to mention its greatest challenge, which occurred at the time of the Mongol invasions and subsequent Yuan domination (1216–1356) of the peninsula.[48] During this period, Koryŏ became a close vassal state of the Yuan through its royal princes' intermarriage with Mongol imperial princesses, and ultimately became an important part of the broader Mongol world. Because of its special status, various political, economic, and cultural exchanges took place at this time and gave rise to many traditions and innovations in a short period.

While the Koryŏ dynasty has remained less familiar to historians of Korea, in recent years it has attracted growing attention that has led to a number of important studies by historians like David Robinson who are well versed in the requisite Korean, Chinese, and other languages needed to study it. These new endeavors have produced investigations of Koryŏ's relations with the wider world from political, social, and economic vantage points. Such accomplishments undergird our effort in this book to pursue Koryŏ's cultural aspects.[49] Some pioneering studies have examined the general state of trade relations that existed between Korea and China at this time, while others have provided case studies of some important science and technology transfers, including astronomy.[50] More case studies are needed to fully understand the transfer of material goods and technologies that took place during this time. This book will make a small early step toward fulfilling that need, which will help us to

---

[48] Michael J. Seth, *A History of Korea: From Antiquity to the Present* (Lanham: Rowman & Littlefield Publishers, Inc., 2011), 103–125.

[49] For example, see Robinson, *Empire's Twilight*. Many primary sources written in China and Korea show active political and cultural relations between the courts of the Koryŏ dynasty of Korea and the Yuan dynasty of China. For Chinese sources about Koryŏ, see Chang Tong'ik (Jang Dong-Ik), *Wŏndae Yŏsa charyo chimnok* (Collection of Historical Sources from the Yuan on Koryŏ History) (Seoul: Sŏul taehakkyo ch'ulp'anbu, 1997).

[50] For the trade relations between Koryŏ and the Yuan empire, see Yi Kanghan (Lee Kang Hahn), *Koryŏwa Wŏnjegugŭi kyoyŏgŭi yŏksa* (History of Trade Relations between Koryŏ and the Yuan Empire) (Seoul: Ch'angbi, 2013).

understand the general environment that facilitated the transfer of technologies like distillation from China to Korea in the final century of Koryŏ rule.

After the fall of Koryŏ in 1392, soju continued to develop, and did so throughout the Chosŏn period into modern times, to eventually become Korea's national drink.[51] Recently, most of the large companies that manufacture soju brands on a mass scale actually produce an industrial form of the drink by using the column stills based on advanced technology instead of the traditional Korean stills called *soju kori*. Still, this modern method of making industrially produced soju evolved from the traditional and classic method, so interest in traditional Korean alcoholic drinks and their methods of production have grown in tandem with the spirit's popularity. Thus, along with co-researchers on the larger project entitled Recovery of Traditional Technologies: A Comparative Study of Past and Present Fermentation and Associated Distillation Technologies in Eurasia, this book contributes to "bridging an internal 'Great Divide' between the histories of the premodern and modern sciences," a task proposed by Professor Lim Jongtae for promoting scholarship on East Asian science and technology within the combined fields of the history of science and global history.[52]

Methodologies and Structure of the Book

Using interdisciplinary methods including food studies, history of science and technology, and Korean and global history, the study of soju's history ultimately provides a rich case study for a newly rising field, the role of foods in global history – in particular, the growing discussion about how transfers of foods and medicinal technologies in premodern Eurasia could have significantly impacted people's lives in a broader social context. This book examines the history of soju using a two-track approach: first, by identifying soju in the history of global distillation, and second, by

---

[51] After that, we will look at soju's continuous development in Korea from the early modern period to the present.

[52] Lim Jongtae, "Historiographical Dependency and a Prospect beyond It: EAHSTM's Position in Regard to Ever Changing Trends of HPS," unpublished paper presented at the 14th International Conference on the History of Science in East Asia (ICHSEA), Ecole des hautes études en sciences sociales (EHESS), Paris, France, July 6–10, 2015. The direction of the current book was first inspired greatly by the panel entitled Recovery of Traditional Technologies: A Comparative Study of Past and Present Fermentation and Associated Distillation Technologies in Eurasia that Paul Buell and I organized for this conference. This is also part of a larger project entitled A Comparative Investigation of Distillation Technologies, Wine Production, and Fermented Products, which is a collected study of the topic produced by several coworkers, including myself, from different academic institutions in Germany and Mexico from 2014 to 2017.

pinpointing soju in Korea's history of cross-cultural exchanges with the larger Eurasian world. It achieves necessary academic rigor but also aims to provide concrete examples and broad insights that aid both the scholar and the general reader interested in world history. Given the grand scale of the book's topic, the book focuses on sociocultural exchanges during the Mongol period, presented above, that directly relate to soju's importance to the transfers of distilled liquors in global history. It does so in six chapters based on periods and topics.

Chapter 1 opens the book with a brief global history of distilled liquors, focusing on current debates about their origins and early development, and the possible transfers of knowledge that linked major Afro-Eurasian societies in ancient times, including Greece, the Middle East, South and Southeast Asia, China, and Mongolia. This establishes the context against which soju's beginnings can be analyzed. The evidence so far shows that distillation never developed in one place. Instead, it appears, the technology developed in multiple locations, though it is difficult to know yet for certain because reliable documentary sources are few. However, rather than developing at the same time and pace in many places, there is a high probability that distillation cultures flourished in some areas and spread their technology and know-how to other areas. Naturally, this circulation pattern would have been more likely to appear in Eurasia, where premodern occurrences of cross-cultural, inter-civilizational exchange are well documented. So far, such evidence has been identified in ancient West Asia, South Asia, and China. A close examination of distillation processes in the Middle East, South Asia, China, and Central Asia reveals that they bear different characteristics with regard to both their ingredients and distillation methods. However, one cannot overlook the fact that all the distilled liquors in these countries were originally called arak (*'araq* meaning "sweat" or "perspiration" in Arabic), which suggests a common agent of transference – namely the Mongols. While the Arabs probably developed distilled liquors including *'araq*, the Mongols contributed to a mass-produced arak with portable stills and then popularized the word. Given this possibility, the chapter ends with an overview of the Mongols' role in the widespread dissemination of arak-type spirits to different parts of Eurasia, including Korea.

Chapter 2 examines the history of alcoholic drinks in Korea from ancient times and explores Korean sources to ascertain how distilled liquors like soju and arak suddenly appeared when they did in the late Koryŏ period in the fourteenth century. Of course, their appearance by means of cross-cultural transfer means that, in contrast to cases of direct imports of alcoholic drinks like mare's milk liquor from China to Korea through diplomatic channels, no documentation explicitly mentions transfer routes. Nevertheless, a few documents about soju dated to the end of the Koryŏ period strongly hint that such a transfer occurred during the time of Mongol rule in China. In order to gain better focus on this possible phenomenon, we examine debates on soju's origins and its process of development on the Korean peninsula from comparative perspectives that also adopt the long view. From this expanded horizon, a critical eye on both early scholarly discourse and pioneering studies of the late twentieth century on the history of distilled liquors in Korea reinforces the foundation for this study. Despite their irrefutable contributions, the early studies face two major limits. First, they focus mostly on documentary sources, which later studies have simply recycled. Second, they have not examined sufficiently the broader historical context of the Mongol period that facilitated the transfer and rapid rise of soju in Korea. Hence the new evidence from more recent pioneering efforts in historical, archaeological, and anthropological studies has shed considerable light on discussions of Eurasian distillation transfers generally, which suggest that the time has come to revise the standard approaches to the history of distillation technology in Korea in cross-disciplinary ways by taking advantage of these new sources.

Chapter 3 examines cross-cultural contacts between the Koryŏ dynasty (918–1392) in Korea and the Yuan dynasty (1271–1368) in China and Mongolia (and broader Mongol Empire), in order to discern the historical context that made the transfer of distilled liquors to Korea possible. At times, this perspective expands to include the greater Mongol world, of which Korea was just one part. As its suzerain state for nearly 150 years, the Mongols were able to exert considerable influence on Korea – leaving vestiges like soju, not to mention many other foods and cultural artifacts (even Korea's national dress). The conditions that enabled distillation's transference to the Korean peninsula during Koryŏ times are quite clear. The expanded context this creates sharpens the analysis of cultural exchanges that relate particularly to soju. Such examples also serve as clues to similar flows of ideas and tools *from* Korea to other parts of the Mongol Empire. An increasing number of studies have shown that cross-cultural interactions between the Yuan and Koryŏ realms provided the Koreans with access to genuinely cosmopolitan societies in Eurasia. For historians of the Mongol Empire, a better understanding of Korea's material culture in Koryŏ times produces a more complete picture of Korea's place in Eurasia under the Pax Mongolica.[53]

---

[53] In this book, we define the Pax Mongolica as the situation in which a transcontinental peace under the aegis of the Mongol Empire during the thirteenth and fourteenth centuries facilitated cross-cultural interaction across the empire and with countries beyond. This included the circulation of commodities and information adopted from the cultures of societies the Mongols subjugated, thanks to the empire's loose political

Chapter 4 observes the development of soju in Korea during the Chosŏn period, which is characterized by its localization with regard to methods and culture. This period is important to the history of soju, because the spirit spread rapidly throughout Korea and settled into its role as an important Korean alcoholic drink along with other kinds of alcoholic beverages that had been consumed since antiquity. Soju evolved, leading to its documentation in a variety of sources, including cookbooks that provided households with recipes using soju, medical books containing guides for medical applications of soju, and official documents testifying to the governmental use of soju as domestic and diplomatic gift giving, an important political activity in premodern government. State sources also pointed out whenever the consumption of soju grew large enough to cause problems for the Korean economy, which was possible given the use of rice, Korea's most basic staple, as its main ingredient. Indeed, because of this, the government issued prohibitions during droughts against alcoholic drinks, particularly soju. Official documents and medical books also report side effects caused by the abusive consumption of the spirit.[54] From Korea, it also traveled to other countries like Japan, as either a diplomatic gift or a trade commodity. An expanded set of relevant sources supports the notion that soju developed during this period, so we will focus on the most representative cases.

Chapter 5 explores the big transformations that soju underwent during the contemporary period, which spans the twentieth and twenty-first centuries, and the consequences of that change for the societies that have consumed the spirit. During the period of Japanese colonial rule (1910–1945), new distillation methods introduced to Korea from Japan fundamentally changed the methods of distillation from the traditional still to continuous distillation using large machines. Even after the end of Japanese colonialism and the Korean War (1950–1953), the Korean government supported factory-manufactured soju made from potatoes because the country lacked grain. Only in the 1980s did the government begin to promote traditionally distilled soju as a *minsokchu* (national folk liquor), part of its policy of promoting national culture. At the same time, the modern, industrially produced soju continued to develop in variety,

barriers and a boom, despite conflicts among the successor khanates, in traffic across the overland and maritime routes passing through its domain. See Thomas T. Allsen, *Commodity and Exchange in the Mongol Empire: A Cultural History of Islamic Textiles* (Cambridge: Cambridge University Press, 1997), 6; May, *The Mongol Conquests in World History*, 109–129, 232–256; and Hyunhee Park, *Mapping the Chinese and Islamic Worlds: Cross-cultural Exchange in Pre-modern Asia* (New York: Cambridge University Press, 2012), 19 and 193.

[54] These include *Chosŏn wangjo sillok* (Veritable Records of the Chosŏn Dynasty) and *Sŭngjŏngwŏn ilgi* (Journal of the Royal Secretariat).

contributing to its popularization at cheaper prices, at different levels of alcohol content, and with a variety of tastes. With these developments underway, producers began to export soju to other countries, in the long run making it a global brand.[55] With these dramatic changes in its production, distribution, and consumption, people began to debate what constituted traditional soju, which concludes this chapter.

Chapter 6 establishes the base from which to extend our examination of soju into the future, using the comparative cases of Japan and what is now Mexico. Japan, which has been interacting with its neighbors on the Korean peninsula and in China for centuries, long ago developed a drinking culture, one that resembles those of Korea and China. While sake, a fermented, strained wine, predominates in Japan, the spirit known as *shōchū* (written with the Chinese characters *shaozhou* 燒酎, "roasted liquor"), Japan's counterpart to soju, also developed as a unique form of distilled liquor. Theories abound about how such spirits developed in Japan. Here, we examine the possibility that transfers of distilled liquors and distillation methods occurred between China and Korea and Japan.[56] A similar form of distilled liquor interestingly also appeared in the Americas, in what later became Mexico, which is quite distant from Eurasia. This chapter thus also reviews existing theories about the development of distilled liquors in Mexico, including possibilities that the East Asian stills such as those specific types used to make soju, ones that began to develop during the Mongol period in Eurasia, were brought to Mexico through the great Columbian exchange that developed out of the rise of global sea networks after 1500. Anthropological field research on moonshine production in Mexico using stills resembling the Mongol forms from Eurasia strongly suggests the possibility that Afro-Eurasian distillation methods influenced the development of local spirits after (or even possibly before) the arrival of Europeans.[57] Even if we cannot completely deny the possibility that distillation developed independently to a certain

---

[55] For example, see Dreisbach, "Move over Vodka."

[56] On diplomatic and commercial relations between China, Korea, and Japan, see David C. Kang, *East Asia before the West: Five Centuries of Trade and Tribute* (New York: Columbia University Press, 2012). For recently published works that view the history of China within the context of a regional framework, focusing on its relations with Korea and other neighboring countries, see Robinson, *Empire's Twilight*; and Evelyn S. Rawski, *Early Modern China and Northeast Asia: Cross-border Perspectives* (Cambridge: Cambridge University Press, 2015).

[57] See Ana G. Valenzuela-Sapata, Ana G., Paul D. Buell, María de la Paz Solano-Pérez, and Hyunhee Park, "'Huichol' Stills: A Century of Anthropology – Technology Transfer and Innovation," *Crossroads* 8 (2013): 157–191; see also several pioneering works by Henry J. Bruman, including his doctoral dissertation in 1940 that was published more than a half century later under a new title *Alcohol in Ancient Mexico* (Salt Lake City: University of Utah Press, 2000).

degree in Mesoamerica, a comparative examination of the rise of distilled liquors in different parts of the world remains a worthy and rewarding endeavor. As early Mesoamerica offers no evidence other than the archaeological, the well-documented case of soju can provide us with far better clues from a comparative point of view. In sum, research focused on a well-documented Korea can help us identify similar patterns and other forms of distillation technology transfer from Asia to other areas of the world, including Mexico, but even to Europe, which has its own Mongol-style stills and bootleggers making a type of hard liquor from the time of the Mongol invasions on. Through such cultural comparisons, we can enrich our understanding of the broad dimensions of premodern cross-cultural contacts that laid down the basic networks out of which our globally connected world has evolved.

From this final chapter and its grand comparative approaches, we return to the topic discussed in Chapter 1: the global history of distilled liquors at a more expanded level. We can then conclude by calling for further study of similar historical transfers or the relationship between distilled liquors and world history, such as European brandy and Russian vodka, both of which date back close to Mongol times. The case of distilled liquors shows that the transfer of cultural items and technologies between different regions led to the consequent establishment of new traditions and innovations in enduring ways. We see that the transfers and exchanges of foodways like alcoholic drinks remain ongoing. This encourages further discussion of important issues of modernization, globalization, and changes in people's tastes affected by them. People's tastes often change; traditions change accordingly.

1   Soju and Arak
The Eurasian Roots of Distilled Liquors

Soju's origins run deep and extend wide. The manufacture of today's soju indeed began in Korea and developed over time in response to local needs and tastes. However, the context of soju's invention reaches far beyond the frontiers of Korean society, back to the multicultural milieu and hemispheric economy of Eurasia at a time of unprecedented political unity created by the Mongol Empire. There is little doubt that the liquors ancestral to today's soju first came to the Korean peninsula from China, in the form of *shaojiu* (literally "burned liquor," or "distilled liquor"), better known today as *baijiu*, as analyses of sources in the following chapters confirm.

Records also show that, since Mongol times, soju has been linked to other distillates and even identified with them. For example, it is often referred to in historical sources using a nickname of Arabic origin, arak. This suggests that a dynamic and complex process is behind soju's development. Since their first appearance in Chinese and Korean sources, shaojiu and soju have often been identified as arak (also *araki* in various transliterations). Arak, a distilled spirit of high alcohol content similar to shaojiu and soju, is popular today throughout the Middle East, Central Asia, and the Mediterranean region. Is there a direct relationship between them? On the face of it, it is hard to say. For one thing, the specific relationship between soju and modern arak brands has not yet been studied fully.[1] In addition, their similarity is not immediately

---

[1] While as prime resources there have only been the many commercial websites for different regional brands of arak, a few recent academic studies have begun to reverse this situation, and to examine the spirit's history from a less popular perspective. For example, see Joseph El-Asmar, *The Milk of Lions: A History of Alcohol in the Middle East* (London: Gilgamesh Publishing, 2020). For a groundbreaking study of distilled liquors in the medieval Islamic world, see also Al-Hassan, *Studies in Al-Kimya'*. See also "The Return of Arak," *New York Times*, December 15, 2018, www.nytimes.com/2005/01/25/travel/the-return-of-arak.html (accessed December 9, 2018). Stefanie Brinkmann argues that few significant studies of alcoholic beverages have been attempted in Islamic studies because of various factors, including the discipline's focus on "scholarly discourses and ruling elites based on text material." Stefanie Brinkmann, "Wine in Hadith: From Intoxication to Sobriety," in *Wine Culture in Iran and Beyond*, ed. Bert G. Fragner, Ralph Kauz, and

apparent, because the ingredients of these distilled liquors are not the same. Arak in western Eurasia is made with fruit wines, while traditional Chinese shaojiu and Korean soju use fermented grains like sorghum and rice. A distilled liquor brand with a similar name, arkhi, developed in modern Mongolia and Central Asia (particularly *shimiin arkhi*, a light milk liquor), is even made with fermented cow's or mare's milk.[2] These similarities in name and other features suggest evidence of cross-cultural exchanges or a common origin that makes comparisons worthwhile. In order to explore the possible historical relations between different varieties of a distilled liquor, we should investigate their origins in relevant medieval sources. This is the goal of this chapter.

While there are abundant documents that describe alcoholic beverages in Eurasia, particularly in China, we have to conduct careful and comparative analyses in order to elucidate the relationships that would point to the origins of these distilled alcohols. What may be one of the earliest extant records of arak as such is found in the *Book of Drinks* by Ibn Qutayba (828–889) in Iraq,[3] yet the first explicit description of the arak as a distilled liquor (distilled liquors under other names are older) is found in Chinese documents written in the fourteenth century. Chinese references to *shaojiu*, an older name, and to arak found their way into later records compiled in Korea and Japan. Some of the Chinese sources explicitly introduce another nickname for shaojiu, *hanjiu* 汗酒, meaning "sweat liquor."[4] This hints at a strong connection between Chinese shaojiu and West Asian arak ("perspiration" in Arabic). A fourteenth-century Chinese source supports this, stating that arak liquor did not originate in China but rather came to the Middle Kingdom from foreign countries through maritime contacts. Certain sources make it clear that distilled spirits called *shaojiu* and arak began to rise as popular beverages under the Mongol Yuan dynasty (1276–1368). If indeed distilled liquors were imported to China from foreign countries, did distillation methods come to China from outside, too?

Florian Schwarz (Vienna: Verlag der Österreichischen Akademie der Wissenschaften, 2014), 79–80.

[2] Nowadays arkhi is mass-produced in factories as Mongolian vodka. It is made with fermented grains, thanks to the influence of Russian vodka. For example, see Chinggis, the most representative brand of vodka from Mongolia, produced since 2009, at www.chinggisvodka.ch (accessed December 8, 2018).

[3] Ibn Qutayba (828–889), quoting a Bedouin, says, "We drank what was left from the bottle of ʿaraq." Ibn Qutayba, *Kitāb al-Ashriba* (Book of Drinks), ed. Yāsīn Muḥammad al-Sawwās (Damascus: al-Maṭbaʿa al-ʿilmiyya, 1420/1999), 125.

[4] See a poem by Yang Weizhen 楊維楨 (1296–1370) of the Yuan dynasty in his *Wuti xiao shang yin ti* 無題效商隱體, https://m.gushici.com/t_1150981 (accessed May 30, 2019); also, a Qing dynasty named author Di Hao 翟灝 (?–1788) says that people of the Yuan era referred to shaojiu as *hanjiu*; see Di Hao, *Tongsu bian* (Popular Culture) (Taipei: Guotai wenhua shiye youxian gongxi, 1980), juan 27, *yinshi* 飲食, "shaojiu."

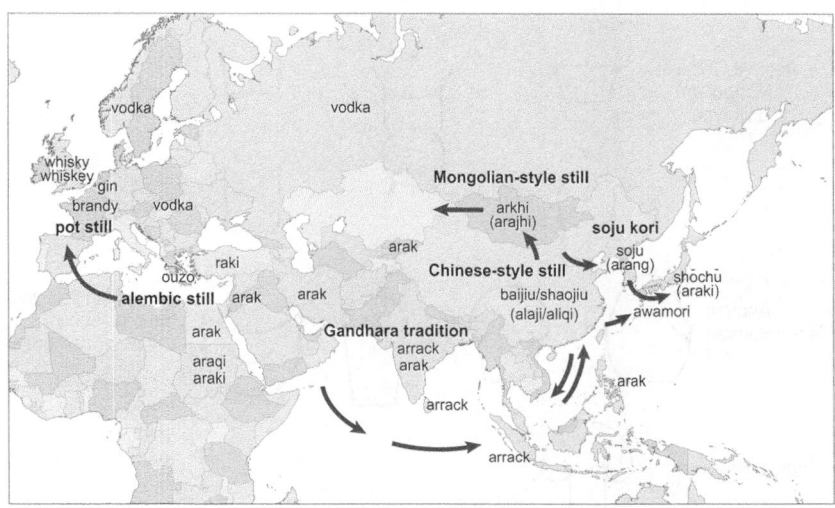

Map 1.1 Various spirits that have developed in Afro-Eurasia and possible transfer routes of major distillation technologies with distinguishing characteristics.

Some scholars, like Paul Buell and Angela Schottenhammer, have begun to argue that, while whoever first invented the idea of distillation remains uncertain, the likelihood of multiple, independent inventions in at least the ancient Near East, China, and India is becoming increasingly likely (see Map 1.1).[5] Earlier studies have also demonstrated that there have been some differences in distillation technology that developed in these different regions.[6] Yet scholars have also continuously proposed different theories of distillation technology transfer in Eurasia. While many scholars have proposed transfers of distillation from the west to the east,[7] Needham and his colleagues showed that distillation developed earlier in ancient China. In their most thorough and comparative categorization of different still types

[5] For example, see Angela Schottenhammer, "Distillation and Distilleries in Mongol Yuan China," *Crossroads: Studies on the History of Exchange Relations in the East Asian World* 14 (2016): 146–147.

[6] See the different types of still that developed from the ancient period in various regions from comparative perspectives reconstructed by Joseph Needham; see Needham, Ho, and Lu, *Science and Civilisation in China*, vol. 5, part 4, 80–81.

[7] Rogers, *Proof*, 84–90; Kevin R. Kosar, Whiskey: A Global History (London: Reaktion Books, 2010), 53–55; R. J. Forbes, *A Short History of the Art of Distillation: From the Beginnings up to the Death of Cellier Blumenthal* (Leiden: E. J. Brill, 1948), 9–11.

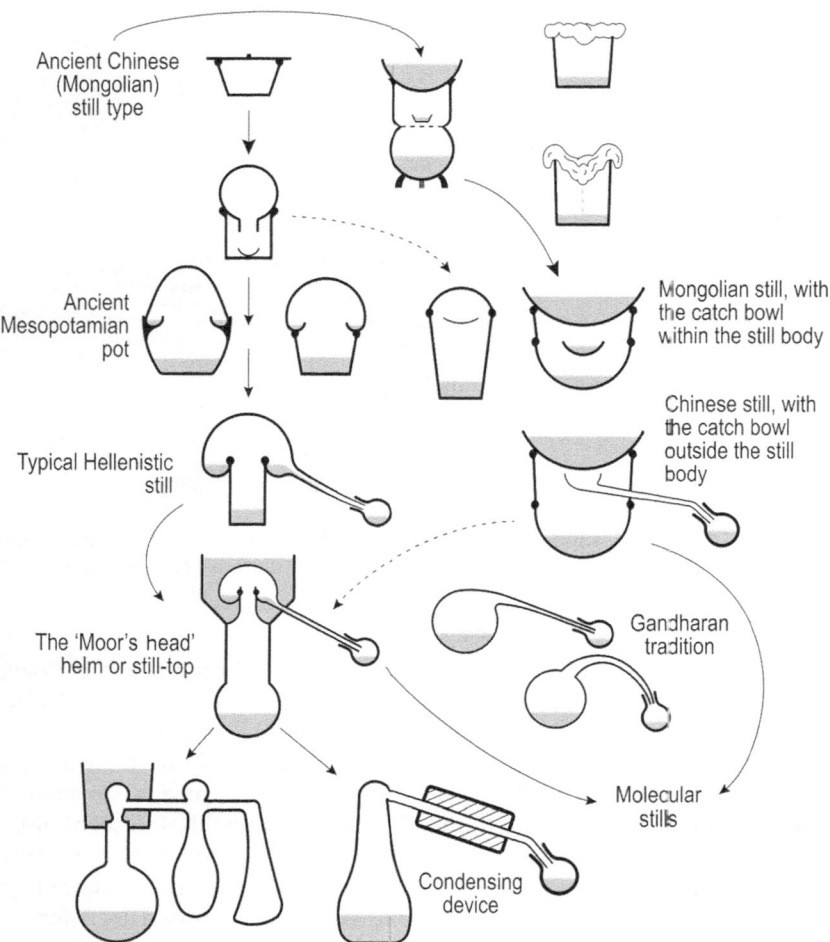

Figure 1.1 Development of different types of still according to distinctive regional features categorized by Joseph Needham. Drawing by Matilde Grimaldi, after Joseph Needham, Ho Ping-Yü, and Lu Gwei-djen, *Science and Civilisation in China*, vol. 5: *Chemistry and Chemical Technology*, part 4: *Spagyrical Discovery and Invention: Apparatus, Theories and Gifts*, Cambridge: Cambridge University Press, 1980, Figure 1454, which provides more details on this still type.

organized according to distinctive features that developed in major regions (see Figure 1.1), Needham and his colleagues claimed a possible transfer from China to the West, though they admitted that they were not able to find evidence of this alleged westward transfer.[8]

A fresh and comparative examination of sources and archaeological excavations available to us now suggests that the actual situation was more dynamic than a simple linear transfer. That is, traditions were interacting with each other in the course of transfers of ideas and goods through cross-cultural contacts and trade relations over many centuries. All of the regions that were involved in the development of distillation interconnected through premodern trade routes. They developed similar, yet unique, distilled-liquor traditions such as arak in the Mediterranean region; arak (or *arrack*) in South and Southeast Asia; shaojiu/soju/shochu (also called arak) in China, Korea, and Japan; and arkhi in Mongolia and Central Asia.

We explore these traditions one by one, investigating the origins of distillation technology and other material characteristics, and possible connections among these distilled-liquor traditions by the time of the Mongol period in the thirteenth and fourteenth centuries. The most important problem to solve is the nature of the development of arak liquor in premodern Eurasia, from the perspectives of both foodways and technology transfers, because arak became the nearly universal name for the distilled liquor spread from the Mongol period.[9]

Thus, before addressing the history of soju, its place in Korean history, or its recent globalization, we must first observe a fundamental context: the global history of distillation and distilled liquors that contributed to soju's creation. This global context will help us to discern the gap that exists between ancient distillation development and the rise of distilled alcohol in places like Korea in the thirteenth century. Such an improved understanding will establish the proper context within which to detect soju's rise in Korea.

Probable Origins of Distillation and Distilled Alcohols: Linear Transfer versus Multiple Invention and Dissemination

Distillation is a technique used to separate the constituents of a liquid by means of boiling and condensing (or controlled evaporation and freezing as an alternative method). It is well known popularly as a tool for making alcoholic spirits, but it also has many industrial or even alchemical uses,

---

[8] For example, see Thomas Ottfried Höllmann, *The Land of the Five Flavors: A Cultural History of Chinese Cuisine*, trans. Karen Margolis (New York: Columbia University Press, 2013), 96.

[9] We will see a number of alternative names for arak throughout the book, such as *araki* and arrack. The names themselves suggest in their variety a kind of geography of cross-cultural interaction.

including the production of medical alcohol solutions. The boiling method takes advantage of the variant boiling points of constituents to create a vapor free of other constituents, or impurities, and then collects the purified vapor after it returns to liquid form. Freeze distillation applies the same logic, setting the temperature of a mixed concentrate just high enough to liquefy a target component in order to separate it from its mixture. In short, distillation commonly involves fermentation, but the technique can be applied to a variety of materials. Indeed, some of the earliest stills discovered in archaeological excavations were used to distill perfumes and mercury, rather than fermented liquids like liquor.

It seems that, until the late medieval and early modern period, when the technique of distillation to create alcoholic beverages spread worldwide, people consumed fermented alcoholic beverages with low alcohol content like beer and wine far more frequently than they did distilled liquors. Most societies have valued distilled alcoholic beverages, and yet documentation of their production before the modern era is rare. Some alcoholic drinks cited in certain historical passages could be identified as distilled liquors; however, these descriptions often are quite vague and thus subject to fierce scholarly debate. While scholars have endeavored to identify and analyze documentation that serves as evidence for distillation and its by-products in ancient times, few have paid serious attention to the conditions that created distilled-alcohol production and its widespread consumption in certain regions and eras.

Because distillation is a chemical technology based on a scientific theory that shapes both the process and the tools, it would not have been easy for people lacking sufficient scientific and technological knowledge to develop it anywhere during the premodern period. Unlike fermented-alcohol production, whose causes could first have been observed in nature and then imitated, distilled-liquor production typically required not only the calculation of boiling points for a variety of ingredients but also the development of a distilling apparatus, colloquially called a *still*. (People in cold areas, particularly northern Eurasia, made spirits using freeze distillation, which requires no special equipment, but this part focuses mainly on the boiling method.) In his best-selling book *Proof: The Science of Booze*, Adam Rogers highlights the rise of distillation in civilization, which was achieved by the accumulation of science and technology addressing human needs over 2,000 years.[10]

Stills usually comprise several parts. At the base lies a heat source, above which a container holds a fermented mash of organic material (grains, etc.), to which the heat is applied. In the case of a European pot

[10] Rogers, *Proof*, 5–7, 84.

still, above the fermented mash is a boiling chamber that offers a space for vapor to form, which is associated with worm tubs of cool water (see Figure 1.2A). In the case of a Chinese-style still, above the fermented mash sits a condenser, which also accommodates the condensing vapors and is associated with a container of cool water directly above the device (see a Chinese-style still in Figure 1.2B). In both cases, the water acts to cool the rising vapor so that it will condense inside the condenser, where the resulting liquid can then be led into a separate receiving pot. The process, therefore, demands that the cooling water must be continuously replenished.[11]

Compared to other technologies, the idea of distillation does not require a considerably high level of technology to invent, as some scholars like Joo Young-ha have pointed out. Once ancient peoples with basic scientific knowledge determined the fundamental principle of distillation, they argue, it might not have been very difficult for them to develop a theory and devise a process of distillation using simple devices.[12] Such a perspective supports recent arguments that place heavier weight on the possibility of multiple points of origin rather than a linear transfer of technology from one society to another.

If so, a fundamental question arises: why did ancient peoples develop distilled alcohols? While there could have been many reasons for developing distilled alcohol in addition to "being able to get drunk," people in most of the societies generally welcomed distilled alcohol as a type of medicine and only later identified its recreational uses. Early medieval texts from the Islamic world and Europe reveal an interest in distillation for such things as distilling tinctures and solutions to obtain essential medical ingredients, and later to obtain scented oils (e.g., rose oil) for perfumes.[13] Medieval alchemists have even thought of distilled alcohol as a kind of cure-all elixir.[14] Furthermore, technical and practical aspects of such distillate products, such as preservability and transportability, could

---

[11] Forbes, *A Short History of the Art of Distillation*, 291, Figure 143A; and Luo Feng, "Liquor Still and Milk-Wine Distilling Technology in the Mongol–Yuan Period," in *Chinese Scholars on Inner Asia*, ed. Xin Luo and Roger Covey (Bloomington: Indiana University, Sinor Research Institute for Inner Asian Studies, 2012), 491, Figure 4B. According to distillation experts, it is extremely important that the distillation process produce a pure end product; that is, a distillate that contains no dangerous impurities such a methyl alcohol. We still do not entirely understand all of distillation's facets; hence, there is the need to learn more about the current state of the field in the science of distillation. For more, see Forbes, *A Short History of the Art of Distillation*.

[12] Based on personal conversations with several scholars including Joo Young-ha (Chu Yŏngha).

[13] Forbes, *A Short History of the Art of Distillation*, 101–111.

[14] Zahary A. Matus, *Franciscans and the Elixir of Life: Religion and Science in the Later Middle Ages* (Philadelphia: University of Pennsylvania Press, 2017), 40–69.

A

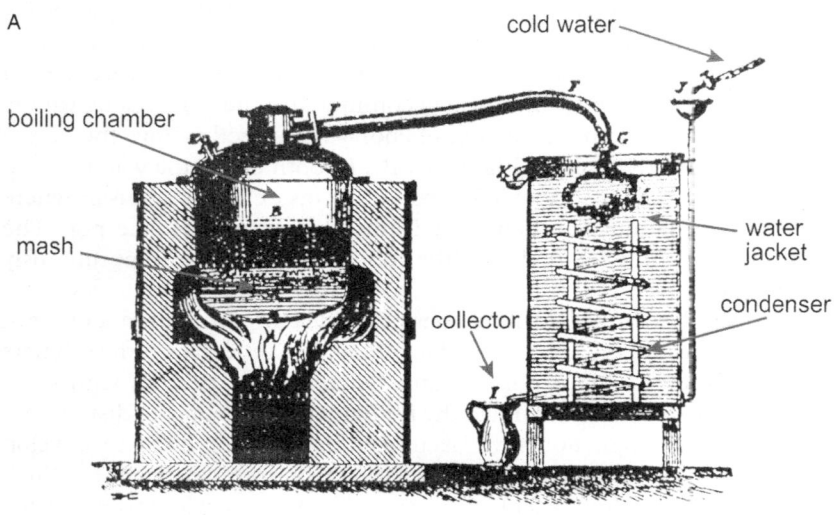

boiling chamber

cold water

mash

water jacket

collector

condenser

B

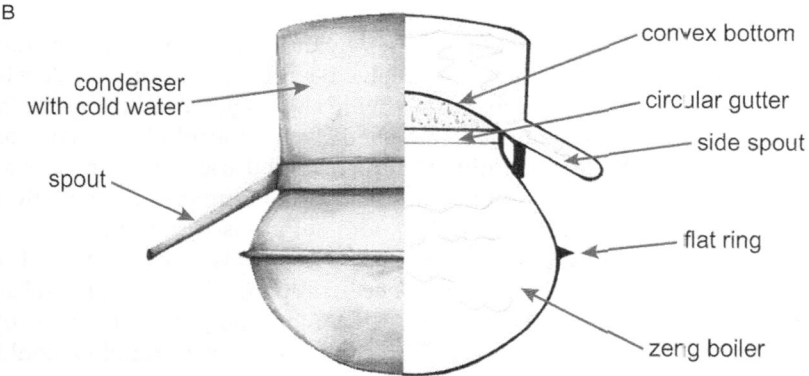

condenser with cold water

convex bottom

circular gutter

side spout

spout

flat ring

zeng boiler

Figure 1.2 A: a classic pot still with worm tubs used to distill whiskey or cognac. R. J. Forbes, *A Short History of the Art of Distillation*, 2nd edition, Leiden: E. J. Brill, 1970. B: a Chinese prototype excavated in Qinglong, Hebei, *c.* 13th century. Drawing by María de la Paz Solano-Pérez, after Luo Feng, "Liquor Still and Milk-Wine Distilling Technology in the Mongol–Yuan Period," in *Chinese Scholars on Inner Asia*, edited by Xin Luo and Roger Covey, Bloomington: Indiana University, Sinor Research Institute for Inner Asian Studies, 2012, 487–518, Figure 4.

have also increased their popularity over simple fermented wines or milks. People could have learned from accumulated collective experience that they could preserve alcoholic beverages for a longer time if they increase the level of alcohol content. Before the invention of refrigerators, it was sometimes crucial to people's welfare to improve the preservability of alcoholic beverages, even more so in hot and humid regions around the Indian Ocean and the Middle East, and among merchant societies living along the overland trade routes where long-distance travel with intensive interactions was a fact of life.

While we do not know how many societies independently invented distillation technology, we should note a more important point: the invention is less meaningful if a few people found it at certain points in time and then lost it before it could be spread further. It is also important to consider whether people developed more systematic methods of distilling alcohol in certain regions, and whether new distilling technologies caused the liquors to become widespread. For a long time, experts claimed that the Greeks invented distillation, after which it disseminated to places like China. Thanks to the accomplishments of Joseph Needham and his colleagues, we now know that distillation technology, at first used to purify mercury, not alcohol, is older in China than it is in the West, and given inherent differences in the two general methods, mostly likely developed independently. The Chinese never used worm tubs, a standard component of a classic pot still popular in the West, through which distillates were condensed and collected (compare diagrams A and B in Figure 1.2). Nor did the stills of East and Central Asia commonly use anything like that (with the exception of the Kalmucks, more a part of the West because they lived on the Volga, as discussed later in this chapter). Indian distillation also differs technologically from the other two and may be as old. More recently, a scholar has advanced the plausible claim that people in the New World also invented distillation independently using ceramic equipment. However independently they may have first developed, various distillation techniques probably interacted with each other over time through the influential movements of people, ideas, and goods through premodern trade networks. New techniques were perennially applied to traditional ones, as the number and significance of distillation's functions and benefits grew – as medicine, as a mechanism of social rituals, and, of course, as a means of intoxication – with exchanges of ideas. It would not be surprising, therefore, to find that people continually created new techniques and models by mixing elements from independent sources, including techniques and ingredients, at certain times and in societies whose reach was expanding over time, like that of Korea. We are now in a better position to study such influences and interactions than ever before, thanks to so many unprecedented studies recently made available to us, revealing dynamic and large-scale cross-cultural contacts in premodern Eurasia. Against this backdrop, let us survey a brief history of alcoholic beverages as informed by ancient records and archaeological artifacts, in order to explore the global history of distillation.

Distillation and Distilled Liquors in the Middle East and South and Southeast Asia: The Possible Origins of the Name "Arak" and Its Distillation Technology

The earliest possible evidence of distillation, argue some scholars, like Martin Levey, is pottery that was excavated in a northern part of ancient Mesopotamia. At first glance, the artifact in question resembles other pottery, but a close look shows that the upper part displays a unique structure with an upper edge and an indented trail (see Figure 1.3).[15]

Levey and later scholars like Edagawa Kōichi have argued that this could be a primitive still that people used to obtain the alcohol drops condensed at the top edge of the vessel after boiling had occurred. They dated this ancient Mesopotamian vessel and possible distilling structure to around 3500 BCE and consider it evidence of distillation's origin there. Other scholars, like Eugene Anderson, are skeptical about this view, arguing that this Mesopotamian piece represents a fermentation lock and provides no evidence of any distilling function.[16] Even scholars like Levey and Edagawa who have proposed it as a type of still assumed that the object was used to create distilled rose oil or perfumes instead of alcoholic drink, as there is no evidence for distilled beverages at that time.[17] For example, in his monograph on chemical technology in ancient Mesopotamia, Levey argues that we have textual evidence in a group of Akkadian tablets for Sumero-Babylonian distillation as perfumery operations, dated c. 1200 BCE,

---

[15] Martin Levey, *Chemistry and Chemical Technology in Ancient Mesopotamia* (Amsterdam: Elsevier, 1959), 31–36; Needham, Ho, and Lu, *Science and Civilisation in China*, vol. 5, part 4, p. 82.

[16] In a private conversation, Eugene Anderson told me that users of this alleged Mesopotamian still could fill the vessel to the brim with water and put a jar upside down on it, so that the water seals the fermenting material inside while allowing gases to escape. In short, he argues, the piece represents a brilliantly appropriate technology in history yet provides no evidence for distilling.

[17] Levey, *Chemistry and Chemical Technology*, 36–41, 132–146; Edagawa Kōichi, "Jōryūshu, higashi mawari nishi mawari" (Distilled Alcohol, Bound for East and West), in *Shōchū higashi mawari nishi mawari* (Shochu around the World), ed. Tamamura Toyo'o (Tokyo: TaKaRa Alcohol Beverage and Life Research Institute, 1999), 23–24.

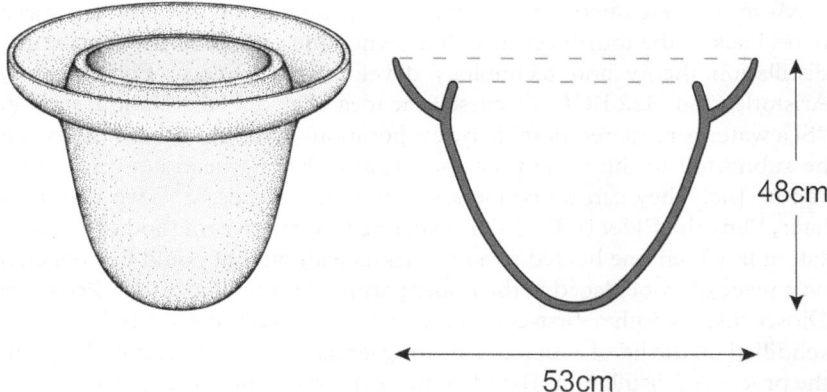

48cm

53cm

Figure 1.3 Pottery excavated in a northern part of ancient Mesopotamia
that provides the earliest evidence of distillation. Drawing by Matilde
Grimaldi, after Ishige Naomichi, "Higashi yūrashia no jōryūshu: jōryūki
wo motomete" (Distilled Alcohol in East Eurasia: Seeking the Distiller),
in *Shōchū higashi mawari nishi mawari* (Shochu around the World),
edited by Tamamura Toyo'o, Tokyo: TaKaRa Alcohol Beverage and
Life Research Institute, 1999, 75–130, Figure 7.

yet he does not detect any distilled liquor in his chapter on wine and
beer dated at that time.[18]

This place is also close to a region where some scholars have traced the
oldest fermented beverages. Since 1978, an archaeology team from the
University of Pennsylvania has excavated well-preserved jars of wine and
beer at an archaeological site named Godin Tepe, dated 3000–4000 BCE
and located in western Iran. The site was integrated into an extensive
trading network that connected to both westward and eastward travel
routes.[19] In the course of reconstructing the history of beer, they even
succeeded in restoring some ancient Sumerian barley beer.[20] Strictly
speaking, these are fermented alcoholic beverages, not distilled liquors.
There is no associated evidence of any distilled alcohol or distilling device.
In short, we have too few ancient relics to offer a generalization, and we
should wait until we find concrete evidence that demonstrates that people
in this region developed more complex technological stills from which use
for distilling alcoholic drink became widespread.[21]

---

[18] Levey, *Chemistry and Chemical Technology*, 36, 55–58.
[19] McGovern, *Uncorking the Past*, 60–71.   [20] Ibid., 69.
[21] Edagawa, "Jōryūshu, higashi mawari nishi mawari," 23–24.

More concrete documentary evidence for distillation in western Eurasia dates back to the fourth century BCE, when the earliest forms of scientific distillation theory and technology developed in Greece. For example, Aristotle (384–322 BCE) discussed the idea of alcohol distillation, stating, "Sea water is rendered potable by evaporation; wine and other liquids can be submitted to the same process, for, after having been converted into vapors [*sic*], they can be condensed back into liquids."[22] Two centuries later, Pliny the Elder (CE 23–79) explained a primitive method of condensation in which one heated rosin to produce an oil that could be collected on a piece of wool placed in the upper part of the still. The doctor Pedanius Dioscorides, another first-century Greek, reportedly discovered droplets solidified on the lid of a mercury-heating jar, and from this he could explain the process of distillation. Based on these theories, ancient Greek inventors created a device they called an *ambix*, meaning a "cup" or "beaker" in ancient Greek, because of its similarity in shape. During the golden age of Islamic science, this influenced Muslim scholars such as Jābir ibn Hayyān (also known by his Latinized name, Geber) (721–815), who borrowed the name *ambix* and its technology to develop his *al-inbīq*, a specialized device for distilling rosewater (see Figure 1.4).[23]

After the Arab Muslim conquest of the Middle East, many ancient Greco-Roman scientific theories and technologies were preserved by Arab and Persian scholars, who used them to further advance scientific fields such as chemistry and alchemy. Analyzing the name *al-inbīq*, it makes sense that the Arabs affixed *al*, the definite article in the Arabic language, to the Greek word *ambix*. From there, the transition from *al-inbīq* to its Latin name *alembic* after the name transferred to Europe through Spain in the twelfth century appears obvious.[24] We have ample documentary evidence for the continuous and systematic development of distillation technology in the medieval Islamic Middle East led by Muslim scientists. For example, many archival manuscripts document a variety of alembic devices that developed there.[25] Most were used to distill herbal alcohol and perfume. However, while Forbes has claimed that the Arabs

---

[22] In *Meteorology* (*c.* 350 BCE), Book 2, 2. The translation appears in Joseph William Mellor (1873–1938), *A Comprehensive Treatise on Inorganic and Theoretical Chemistry* (1922), vol. 1, 37.

[23] E. Wiedemann and M. Plessner, "al-Anbīḳ," *Encyclopaedia of Islam*, 2nd edition, ed. P. J. Bearman et al., 12 vols. (Leiden: Brill, 1954–2005), vol. 1, 486.

[24] The Arabic short-*a* vowel becomes *e* in Arabic dialects and also European languages, and *x* turns into *c*, probably because western Europeans believed there was a plural -*s*. Thus the English word *alembic* was originally fashioned based on knowledge and technology transferred from the Islamic world.

[25] Forbes, *A Short History of the Art of Distillation*, 31–54. Al-Hassan, *Studies in Al-Kimya'*, 283–298.

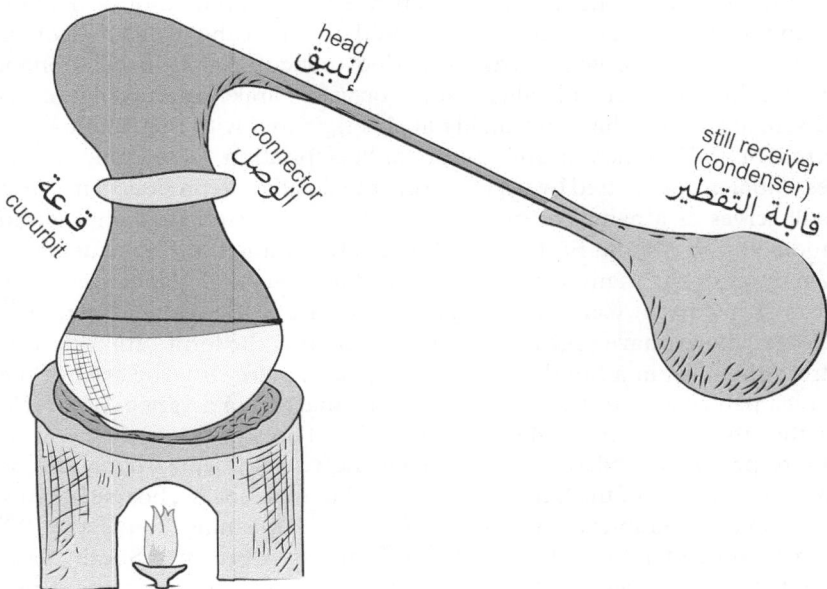

Figure 1.4 An Arabic manuscript by Jābir ibn Hayyān that depicts an alembic device and the distillation process. Drawing by Matilde Grimaldi, after Salim T. S. Al-Hassani, ed., *1001 Inventions: The Enduring Legacy of Muslim Civilization*, Washington, DC: National Geographic, 2012, p. 93.

knew distillation techniques, but did not use distillation for wine in medieval times, Ahmad Y. al-Hassan argues that various descriptions of alcoholic wine bottles in Arabic books of secrets and in military treatises suggest that Muslim chemists knew the distillation of wine and the properties of alcohol from the eighth century.[26]

Did distilled alcohols develop and become popular in the medieval Islamic world? One can consider this a possibility. One can even speculate that, after this development, distillation technology and the idea of distilling alcohols subsequently traveled further east to change the drinking culture there, just as the alembic influenced the development of distilled liquors like brandy in Europe beginning in the fourteenth century.[27]

---

[26] See Forbes, *A Short History of the Art of Distillation*, 29–54, and Al-Hassan, *Studies in Al-Kimya'*, 283–298.

[27] Forbes, *A Short History of the Art of Distillation*, 101–102. In his masterpiece *Civilization and Capitalism, 15th–18th century*, Fernand Braudel showed that distilled liquors,

While we consider the possible independent development of distillation technology in eastern Eurasia, we should note that the distilled alcohols that spread in the later medieval period (concurrent with the Mongol period in Asia) were all called "arak" or used "arak" as a nickname. As discussed above, this word highly likely originated with the Arabic word *araq* (or arak), which meant "sweat" or "perspiration," a reference to the essential drops created by vapor during the distillation process. This term still serves as a name for many similar forms of distilled alcohol found today in the Middle East, Central Asia, and South and Southeast Asia. Analyses of the many forms of the word *araq* in Altaic languages show a long history in Central Asia, particularly from the Mongol period.[28] Some scholars have suggested the possibility that the term "arak" derives from the term in a Southeast Asian language, *areca-nut* (the seed of the areca palm), a material long used for manufacturing a variety of arrack, rather than from the Arabic for "sweat."[29] The word *arcq*, however, is more precisely spelled *'araq*, in which the reversed apostrophe acts as a transliteration of the Semitic letter *'ain* that represents a hoarse breathy consonant in Arabic; the root of the word, *'-r-q*, meaning "to perspire."[30] This word, therefore, did not derive from any word in a South Asian language. That said, since some Chinese sources claim that the single nickname used to refer to distilled liquors was "sweat liquor" (*hanjiu*), it is highly likely that the Arabic word *'araq*, "perspiration," transferred somehow to China.[31] Ahmad Y. al-Hassan argues that, while we need to investigate further whether *'araq* was the common name for distilled wine among the public in the medieval Islamic world, some sources between the eleventh and sixteenth centuries do refer to distilled wine as *'araq* or *'araqi*.[32] Considering this context, let us examine the drinking

including brandy and rum, began to be consumed popularly and produced industrially in various parts of Europe after 1500. After posing the question whether or not such European developments influenced the distilled liquors found outside Europe at that time, Braudel briefly surveyed the possible development of distillation in non-European societies, but he was unable to answer the question, or to go very deeply into the debate. At the time, Braudel probably still did not have access to major studies on distillation in Asia, including the Needham volumes. Fernand Braudel, *Civilization and Capitalism, 15th–18th Century*, vol. 1, *The Structures of Everyday Life: The Limits of the Possible*, trans. Siân Reynolds (New York: Harper & Row, 1981), 241–249.

[28] See the discussion by the linguist G. J. Ramstedt below.

[29] Samuel Morewood, *A Philosophical and Statistical History of the Inventions and Customs of Ancient and Modern Nations in the Manufacture and Use of Inebriating Liquors* (Dublin: W. Curry, jun. and Company, and W. Carson, 1838), 140–141, cited in *Encyclopaedia Britannica: A Dictionary of Arts, Sciences, Literature and General Information*, 11th edition, vol. 2 (New York: The Encyclopædia Britannica Company, 1910), 642.

[30] I appreciate Eugene Anderson's insights on this.   [31] See footnote 4 in this chapter.

[32] Al-Hassan, *Studies in Al-Kimya'*, 285–286.

cultures of West Asia and the Middle East, which had an original development of arak liquor.

First of all, one may wonder whether Muslims used their distillation technology frequently to produce alcoholic drinks, given that Islam prohibits their consumption. As several studies have pointed out, however, many Muslims, particularly Iranians, actually enjoyed drinking alcoholic beverages to a certain degree.[33] Moreover, many non-Muslim ethnic groups in the Middle East faced no such prohibition, and they lived in broad areas of the region. Therefore, we can accept the *possibility* that people in predominantly Muslim societies tried to distill fermented wine to create a stronger form of alcoholic liquor, which they named "arak." It might be too far-fetched to suppose that it was difficult to speak about it openly because the rulers drank alcohol openly. However, we can assume that it could have spread gradually, at least among groups of people that did not leave much documentary evidence. As people discovered the important advantage of distilled alcohols in preservation, the distillation of alcoholic drinks might have become particularly useful to medieval Middle Eastern soldiers and merchants who traveled long distances while undertaking military expeditions or conducting long-distance trade.

Many scholars have assumed that some of the distilled alcoholic drinks produced in the Middle East could have transferred eastward through the overland Silk Routes and the Indian Ocean trade networks. As one Chinese source of the fourteenth century explicitly states, arak liquor was transferred to China via the "Southern Sea"; that is, the Indian Ocean route that connected Western and Eastern Asia.[34] This seems plausible. Considering this, we can assume that, although arak as a distilled liquor was not documented in the Islamic Middle East, it could have transferred to South and Southeast Asia, where it localized to become arrack (or arak), a variety popular in South and Southeast Asia even today, before reaching East Asia.[35] This transfer scenario seems quite likely, given the profusion of Islamic merchants traveling from West Asia across the Indian Ocean soon after the rise of Islam, catalyzing both trade and conversion. Suffice it to say here that, given the close interregional connections between the Middle East and South and Southeast Asia through a booming medieval maritime

---

[33] For example, see Rudi Matthee, "Alcohol in the Islamic Middle East: Ambivalence and Ambiguity," *Past and Present* 222, supplement 9 (2014): 100–125; Willem Floor, "The Culture of Wine Drinking in Pre-Mongol Iran," in *Wine Culture in Iran and Beyond*, ed. Bert G. Fragner, Ralph Kauz, and Florian Schwarz (Vienna: Verlag der Österreichischen Akademie der Wissenschaften, 2014), 165–209.

[34] See the discussion of *Nanfan shaojiu* 南番燒酒 called *aliqi* 阿里乞 in *Jujia biyong* below.

[35] The two araks in the two regions nowadays bear differences. While this may be the result of localization in the course of the spread of the same beverage, it is also possible that the two spirits developed independently.

trade, it is possible that technological or linguistic information regarding arak traveled with those who were forging those ties.

The above-mentioned speculation about the direct connections between arak in the Middle East and the same word in South and Southeast Asia, and the eastward migration of arak and its distillation technology, poses a problem. Distillation might have developed in both regions during the ancient period, long before arak's transference from the West to South and Southeast Asia in the medieval period.[36] Several studies, like Frank Allchin's analyses of ancient Indian stills, direct us to consider the possibility that both distillation technology and a separate arak tradition originally developed independently in these regions in ancient times and have since continued into the present.[37] In fact, some of the famous arak or arrack brands that grew popular in modern South and Southeast Asia have no active connection to the tradition of arak in West Asia, which is made of grapes and aniseed. While there is little documentary evidence, anthropological research of the past two centuries indicates that distilled liquors based on palms developed in India and Southeast Asia, not West Asia. Additionally, scholars have also discovered a tenth-century inscription made in Java that contains a list of liquors, including arrack. This suggests that people of the Chao Phraya river region of Thailand consumed spirits, although it is not yet clear when they existed in the inner areas of mainland Southeast Asia and China.[38]

An important piece of concrete evidence for the possible independent development of distillation in South and Southeast Asia is some portions of artifacts that archaeologists have excavated in the region of ancient Gandhara (which spanned modern-day Afghanistan and Pakistan): one in Taxila, Punjab, dating as early as the first century BCE–first century CE, and the other in Shaikhan Dheri, dating between the second century BCE to the fourth century CE. After reconstructing entire apparatus based in large part on their imagination (see Figure 1.5), they concluded that it typified what developed there, and, moreover, it differed from West and East Asian still types.[39] This was included in Joseph Needham's chart for the evolution of the still as a particular Gandharan style, which later

---

[36] See the discussions of the development of distillation in ancient India and China below.

[37] Frank R. Allchin, "India: The Ancient Home of Distillation?", *Man*, new series 14, no. 1 (1979): 55–63.

[38] Takayama Takumi, "Tōnan ajia tairikubu no jōryūshu" (Distilled Beverages on the Southeast Asian Continent), *Daigokai kokusai sake bunka gakujutsu kentōkai ronbunshū* (Collected Volume of the Fifth Conference of International Alcohol Beverage Research) (Tokyo: Nihon jōzōgakkai, 2006), 282, cited in Yi Sanghun, "Urinara kangwa parhyoju ŭi chŏn'gae wa t'ŭkching" (The Development and Characteristics of Korean Fortified Fermented Wine), MA thesis, Seoul Venture University, 2014, 33.

[39] See Figures 3 and 5 in Allchin, "India: The Ancient Home of Distillation?", 58, 60.

A

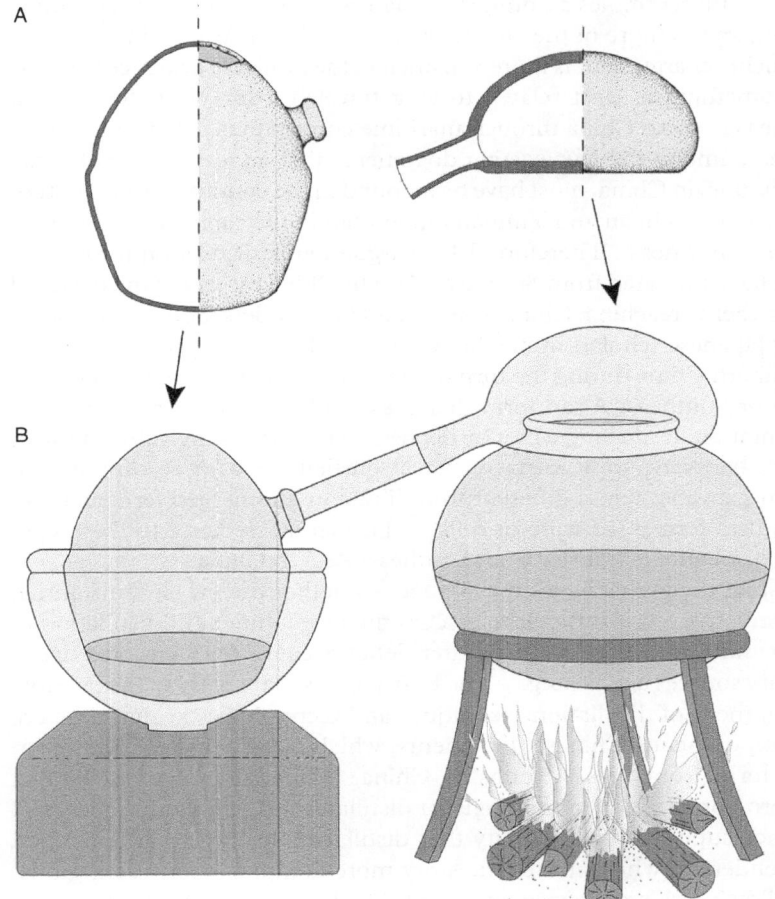

B

Figure 1.5 A: the upper portion of a still-like artifact excavated in
Taxila, Punjab. B: a reconstruction of the entire apparatus based in
large part on imagination. Drawing by Matilde Grimaldi, after Ishige,
"Higashi yūrashia no jōryūshu," Figure 9.

scholars have often followed in the course of their studies (see Figure
1.1).[40] Observing the still and its distillation technology as described in
a fourteenth-century Chinese cookbook, Ishige Naomichi argues that the

[40] Needham, Ho, and Lu, *Science and Civilisation in China*, vol. 5, part 4, pp. 86–87; Ishige,
"Higashi yūrashia no jōryūshu," 100–107.

Chinese still resembles a Southeast Asian still, while other scholars argue that it displays more of the characteristics of a West Asian still.[41]

Whichever argument is more convincing, the available evidence suggests that something at least related to arak traveled from West, South, and Southeast Asia to China through maritime connections. Ishige also points out that, among the few existing documents that appear to describe distilled liquors in China, most have been found in the country's southwestern provinces of Sichuan and Yunnan rather than in the capital area far to the country's northeast. Therefore, Ishige argues, arak distillation technology might have migrated from Southeast Asia to China by way of the overland routes there, reaching China where Yunnan borders Southeast Asia.[42] Other Japanese scholars argue that Chinese stills, which began to develop in southern China during the turn of the second millennium, resemble the South or Southeast Asian form that uses solid-state and semi-solid-state fermentation for making wines by depositing a solid culture substrate such as rice, barnyard grass, corn, or wheat on flatbeds after seeding it with microorganisms; this is different from liquid or submerged fermentation, which uses fermented wine or milk.[43] This lends credence to the theory about distillation's transfer from Southeast Asia to China.

Despite the archaeological evidence – whether direct, circumstantial, or speculative – it is difficult to discuss the development of distillation in South and Southeast Asia in greater detail because relevant local documentary sources are lacking.[44] The best sources that testify to the possible development of distillation techniques and technologies at that time are contemporaneous Chinese documents, which clearly suggest transfer in some form from those regions to China. While some sources hint at Western connections to the origin of distillates in China, other pieces of evidence support the possibility that distillation in East Asia developed independently, which makes the story more complex. Considering this, the following section reviews cases of distillation in China that provide clues to the possible origins of distillation in East Asia and the further exchanges of relevant elements, such as a name for the form of distilled liquor that developed and popularized there, namely arak.

[41] Ishige, "Higashi yūrashia no jōryūshu," 106; Huang, *Science and Civilisation in China*, vol. 6, part 5, 227–228.

[42] Ishige, "Higashi yūrashia no jōryūshu," 125–127.

[43] Kosaki Michio, "Tōnan ajia no sake" (Alcoholic Beverages in Southeast Asia), in *Daigokai kokusai sake bunka gakujutsu kentōkai ronbunshū* (Collected Volume of the Fifth Conference on International Alcohol Beverage Research) (Tokyo: Nihon jōzōgakkai, 2005), 61; Takayama, "Tōnan ajia tairikubu no jyōryūshu," 271, cited in Yi Sanghun, "Urinara kangwa parhyoju ŭi chŏn'gae wa t'ŭkching," 33–34, 44–45.

[44] Some scholars are skeptical about using the archaeological and textual evidence for Indian/Pakistani distillation.

Distillation and Distilled Liquor in China, and the Possible Origins of Shaojiu and Arak

As He Manzi discusses in his book about the culture of alcoholic drinks in the oldest continuous society in world history, China has a history of making and drinking alcoholic beverages that is at least 4,000 years old and yields the richest historical documentation of the practice.[45] Based on new archaeological findings, Gideon Shelach-Lavi argues that Late Neolithic network-oriented elites used alcoholic beverages in ceremonial feasting events such as funerals in order to boost their prestige and gain power.[46] Putting aside the many myths about who invented the alcoholic beverage, many historical documents of ancient provenance testify to the different kinds of beverage that people made and drank before the advent of the Common Era, including medicinal wines like "chrysanthemum wine" and "jujube wine," which were made by soaking flowers or herbs in a grain-fermented alcohol.[47] Analyses of residues in Neolithic pot shards revealed evidence of brewing fermented alcoholic beverages composed of rice, honey, and various fruits.[48] However, the remains of these detected alcoholic drinks apparently lacked a high alcohol content, no more than ten percent, including the stronger varieties.[49]

An archaeological excavation in 1956 sparked scholarly debates about the origin of distilled alcohols in China.[50] The dig recovered an ancient still, which some scholars have claimed was made during the Han dynasty (206 BCE–220 CE). Many use this hypothesis to argue that China's distillation technology has ancient origins.[51] However, debates have arisen about the still's origin because its discoverers found it in a rubbish heap at a construction site in 1950, rather than in the

[45] In his book, He Manzi states that he provides only a list of specialized Chinese documents relating to alcoholic beverages and their consumption because a list of documents for all the beverages mentioned in the book would become as large as half the size of the Chinese classical canon. He Manzi, *Zuixiang riyue*, 200–201.
[46] Gideon Shelach-Lavi, *The Archaeology of Early China: From Prehistory to the Han Dynasty* (New York: Cambridge University Press, 2015), 157–158.
[47] He Manzi, *Zuixiang riyue*, 13.
[48] Shelach-Lavi, *Archaeology of Early China*, 158; Patrick E. McGovern, Anne P. Underhill, Hui Fang, Fengshi Luan, Gretchen R. Hall, Haiguang Yu, Chen-shan Wang, Fengshu Cai, Zhijun Zhao, and Gary M. Feinman, "Chemical Identification and Cultural Implications of a Mixed Fermented Beverage from Late Prehistoric China," *Asian Perspectives* 44, no. 2 (2005): 249–275; Li Liu, Jiajing Wang, and Huifang Liu, "The Brewing Function of the First Amphorae in the Neolithic Yangshao Culture, North China," *Archaeological and Anthropological Sciences* 12, no. 118 (2020): 1–15.
[49] He Manzi, *Zuixiang riyue*, 29.
[50] Huang, *Science and Civilisation in China*, vol. 6, part 5, pp. 209–213.
[51] Scholars argued about whether this could function as a still but, after conducting tests, found that it could be used to distill alcohol. Huang, *Science and Civilisation in China*, vol. 6, part 5, pp. 213–214.

ground.[52] Even if we accept its Han dynasty origin, the size of the still is too small to distill an amount of alcoholic drink larger than a small cup, which therefore makes it an unlikely apparatus for making alcoholic drink. We do not know whether the technology enjoyed wider use, either. Thereafter, researchers continued to exhume stills that resemble in shape the bronze stills found in other parts of China, like Anhui province, although the exact dating lacks supporting documentary evidence.

Indeed, we should note the world of difference between distillation as a concept and distillation in practice, which is exemplified by the case of the Islamic Middle East. Needham and his colleagues argue that ancient Daoists (also spelled Taoists) developed distillation methods in their pursuit of alchemy. Most likely, they distilled mercury in seeking to invent an elixir of life, although it is possible that they prepared limited quantities of distilled wine in secret.[53] Among the many relevant Chinese documents, it is difficult to find any record that suggests the common consumption of distilled alcohols in antiquity. For example, well-preserved agricultural and culinary books such as *Qimin yaoshu* 齊民要術 (Essential Techniques for the Welfare of the People) (*c.* 533–544), written by Jia Sixie 賈思勰, an official at the court of China's Northern Wei dynasty (386–534), and *Beishan jiujing* 北山酒經 (Classic of Alcoholic Drinks in Northern Mountain), compiled by Zhu Gong 朱肱 (1068–1165) of the Northern Song dynasty, delineate fermentation recipes; however, they do not provide any descriptions of distilled alcohols.[54] Many of the specialized records of alcoholic drinks demonstrate the long-standing development of winemaking technology in China, while other sources confirm the existence of liquor markets all over the country. In these markets, sources state, one could find a variety of alcoholic drinks made from grains of excellent quality, using good leaven and water, and improved winemaking technology worthy of praise by both writers and drinkers.[55] However, it is difficult to find references to widespread consumption of distilled liquors dated before the turn of the second millennium.

Huang Hsing-tsung's chapters on the evolution of alcoholic drinks in Needham's volume on biology and biological technology in *Science and Civilisation* argue that, once scholars began to suggest the possibility of early origins, researchers began to discover records that hint at the possible existence of distilled alcohols as early as the Tang dynasty (618–907).[56] For example, several poems written during that era mention

[52] Ishige, "Higashi yūrashia no jōryūshu," 110.
[53] Huang, *Science and Civilisation in China*, vol. 6, part 5, p. 207.   [54] Ibid., 169, 191.
[55] He Manzi, *Zuixiang riyue*, 32.
[56] Needham, Ho, and Lu, *Science and Civilisation in China*, vol. 5, part 4, 144–146; Huang, *Science and Civilisation in China*, vol. 6, part 5, pp. 221–224.

both shaojiu and *shaochun* 燒春, a contemporaneous nickname for shao-jiu. In fact, shaochun gained notoriety as a popular brand in Sichuan province.[57] Some have their doubts, however: if they were so popular, why then did Bai Juyi, the legendary Tang poet who loved both making and drinking alcoholic beverages, not celebrate them in his poems?[58] Surely, if the product had existed, he would have known and written about it, they say. Other scholars point to another great Tang poet, Li Bo, whose poems mention shaochun, arguing that the Chinese probably began to make and consume shaojiu during the late Tang period.[59] Whatever its actual trajectory, the development of distilled beverages in China and their popularization at that time remains in doubt.

Whatever the situation for the Tang era, evidence abounds for distilled-liquor production during the Song period. A few scholars still have their doubts about it. In his book about the history of Chinese science, Liu Guangding argues that liquors with a high alcohol content documented in pre-Yuan sources are not necessarily distilled; so, too, does Wang Saishi in his article about the evolution of the names for Chinese shaojiu in premodern texts.[60] Huang Hsing-tsung reminds us that the ancient Chinese used means other than distillation to make a powerful liquor that they called *shaojiu*.[61] Indeed, looking at the evidence of various documentary sources, it is clear that there existed a stronger, pasteurized wine that differed from the clear strained wine that was more commonly consumed during the Tang period.

Indeed, additional literary works strengthen the case for the existence of distilled-alcohol production during the Song era. For example, the famous Song dynasty official and literatus Su Shi (1037–1101) once described a wine that, when it catches fire, can be smothered with a piece of blue cloth, which sounds like a distilled liquor with an alcohol concentration high enough to catch fire.[62] Other sources testify to the fact that many people suddenly died after drinking shaojiu – only distilled, not fermented, liquor could have done that. Nonetheless, it

---

[57] For Jian Nan Chun 剑南春 liquor produced in Sichuan, see Wang Yanfang, "Top 10 Chinese Wines," *China.org.cn*, June 20, 2011, http://china.org.cn/top10/2011–06/20/co ntent_22822126_5.htm (accessed December 4, 2018).

[58] Yi Sanghun, "Urinara kangwa parhyoju ŭi chŏn'gae wa t'ŭkching," 17–18.

[59] Shang Binghe (1870–1950), *Chūgoku shakai fūzokushi* (A History of Folk Culture in Chinese Society], trans. Akita Shigeaki (Tokyo: Heibonsha, 1969), 221, cited in Yi Sanghun, "Urinara kangwa parhyoju ŭi chŏn'gae wa t'ŭkching," 21.

[60] See Liu Guangding, *Zhongguo kexue shi lunji* (Collected Essays on the History of Chinese Science) (Taipei: Guoli Taiwan daxue chuban zhongxin, 2002), 318–333; Wang Saishi, "Zhongguo shaojiu mingshi kaobian" (A Study of the Names of Chinese Shaojiu), *Lishi yanjiu* 6, no. 3 (1994): 73–85.

[61] Needham, Ho, and Lu, *Science and Civilisation in China*, vol. 5, part 4, p. 133.

[62] Huang, *Science and Civilisation in China*, vol. 6, part 5, p. 206.

is undeniable that liquors called *shaojiu* became popular during the Song era. Lastly, a Song literary work entitled *Mengliang lu* 夢粱錄 (Dreaming over a Bowl of Millet), written during the Southern Song dynasty (1127–1279), explicitly attests to the popularity of a mild form of shaojiu in southern China.

Next, let us connect shaojiu as described in literary works and depicted in paintings to the distillation forms reflected in recent archaeological findings. Since the discovery of the Han dynasty still, fieldworkers have excavated other stills and distilleries in China, including in the north, that date back to the twelfth and thirteenth centuries. This is significant, because China was divided into north and south at that time: that is, the north, ruled by the Khitan Liao (907–1125) and Jurchen Jin (1115–1234) dynasties, and the south, ruled by the Southern Song dynasty, which was actually a continuation of the Song dynasty that began in 960.[63] These stills and distilleries include the bronze steamer still discovered in 1975 in a Jin tomb near Chengde, Qinglong county, Hebei province; the 2002 discovery of a 700-year-old spirit distillery in an ancient township called Lidu, south of Nanchang city in southern China's Jiangxi province; and the 2006 discovery of the Da'an distillery site in northern China's Jilin province.[64]

These excavations and the subsequent studies of them have provided us with valuable insights into distilleries and distillation technologies, not to mention the widespread distribution of distillates throughout China already in evidence in the early Song. As explained below, the late Southern Song relics at the Lidu site are mainly residential, and thus offer little to confirm commercial production in the early years of the Southern Song. However, the excavation of the large-scale distillery in Da'an, given its unsuitability for either home-kitchen or military use, suggests that commercial production indeed developed when the Liao and Jin dynasties ruled north China; the commercialization occurred earlier than it did in south China. Feng Enxue has argued that, because shaojiu was alleged to repel cold and to tone (increase the flow of energy

---

[63] On the period, see Morris Rossabi, ed., *China among Equals: The Middle Kingdom and Its Neighbors, 10th–14th Centuries* (Berkeley: University of California Press, 1983); Valerie Hansen, *Open Empire: A History of China to 1800*, 2nd edition (New York: Norton, 2015), 237–307.

[64] Nanchang 南昌 city, Lidu 李渡 county, Jiangxi province. For the Qinglong bronze still in Hehei, see Huang, *Science and Civilisation in China*, vol. 6, part 5, pp. 208–210. For the Lidu distillery site, see Schottenhammer, "Distillation and Distilleries," 153–155. For the Da'an distillery site, see Feng Enxue, "Zhongguo shaojiu qiyuan xintan" (Preliminary Analysis of the Origin of Distilled Spirit in China), *Jilin daxue shehui kexue xuebao* (Jilin University Journal, Social Sciences Edition) 55, no.1 (January 2015): 163–170.

to) blood circulation, its appearances in the northeast became multi-plied, and it spread quickly by commercialization.[65]

Specialists lack adequately detailed evidence that explains how trad-itional Chinese distillation technologies – shown by both the ancient still that dates back to the Han dynasty and those developed by Daoists – were applied to the production of distilled beverages over time. Given the evidence of diversely shaped stoves and distilling pots used in various forms in Da'an, we can assume that the Chinese developed a variety of distillation techniques.

At that period of time, a new distillation method was brought to China from other countries through maritime contacts. The Song dynasty text *Qu bencao* 麴本草 (Ferment Cultures and Basic Herbals) mentions a distilled liquor called *Xianluo jiu* 暹羅酒, which means "Thailand liquor," a name that indicates the arrival of distilled liquors from Southeast Asia.[66] The author explains the characteristics of spirits in detail, including its production, which depends upon distillation, or *shao*, which literally means roasting or otherwise cooking by directly applying high heat as opposed to boiling – in liquid – or frying – in oil. This is done twice, adding fragrance, before the resulting liquid is then buried in the soil for two to three years to remove the odor, after which it is dug up and is ready to consume. The alcohol is strong, taking only two to three cups to intoxicate the drinker. The price was high, dozens of times higher than ordinary liquor. Interestingly, people were allegedly quite fond of consuming large quantities of it while sailing. Takayama Takumi argues that the merchants engaged in maritime trade with Thailand were drinking this a lot during their voyages and in the elev-enth century shipped it to the international seaport of Guangzhou.[67] Such a transfer of distilled liquor to China through overseas markets makes sense, because many merchants from West and South and Southeast Asia frequented the major port cities in China during the Tang and Song periods, when China's foreign trade flourished. Muslim merchants from West Asia actively plied the long Indian Ocean sealanes and populated its seaports, looking to sell spices to the Chinese while buying their silks and porcelains; we can assume that arak was part of their cargo. Oddly enough, the name *arak* does not appear in any maritime handbook of the Song period, most notably Zhao Rukuo's *Zhufan zhi* (Description of Foreign Lands, 1225), or in earlier Chinese

[65] Ibid., 170.
[66] Needham, Ho, and Lu, *Science and Civilisation*, vol. 5, part 4, pp. 144–145.
[67] Takayama, "Tōnan ajia tairikubu no jyōryūshu," 282, cited in Yi Sanghun, "Urinara kangwa parhyoju ŭi chŏn'gae wa t'ŭkching," 28.

sources.[68] Even names that might imply distilled alcohol are absent from any Song or pre-Song record about foreign trade.

However, a later source – *Jujia biyong shilei quanji* 居家必用事類全集 (Essential Things for Living at Home) (hereafter *Jujia biyong*), probably written in the fourteenth century during the Yuan (1276–1368) or early Ming (1368–1644) dynasties – suggests that the spirit called "Thailand liquor" (*Xianluo jiu*) from South or Southeast Asia could have been related to arak, and suggests that traders shipped it to China via sea routes of the Song dynasty. This account also discusses a Chinese liquor transliterated as *aliqi* 阿里乞, a variant of the name *arak*. This text provides promising clues not only about the source of arak's name and its trade, but also about its technology, in a section about foreign recipes entitled "Nanfan shaojiu fa" 南番燒酒法 (The Burnt-Wine Method of the Southern Barbarians). It describes a method of distillation that derived from the so-called "Southern Barbarians," which could refer to foreigners who arrived in China by way of maritime routes – meaning anyone from South, Southeast, and West Asia.[69] The entry states that the foreign name of this spirit is *aliqi jiu* 阿里乞酒 and describes a rudimentary Arab-style retort-still type. In the following recipe, this suggests a Western derivation at the very least (see Figure 1.6):

"Method of making Southern Barbarian shaojiu" (*Nanfan shaojiu fa* 南番燒酒法)
    The barbarian name is aliqi (arak). As for an ingredient, it does not matter if it is sour or sweet, insipid, and thin; any liquor lacking a proper flavor will do. Make ready a pot eighty percent full. Place another empty pot obliquely over the top. Bring the mouths of the pots together. First, make a hole in the side of the empty pot. Secure it with a bamboo tube [which functions] as a beak. Again, secure the empty pot below. Fill its mouth, mounting with the bamboo beak. Towards the area around the openings of the two pots fill in the holes with pieces of white porcelain bowl. Make it snug by covering it up. One can also use pieces of earthenware.
    Take paper fiber and pulverized lime and apply abundantly, as thick as four fingers. Place into a new large urn and reposition. Take paper ashes and fill up the ash with about two or three catties of glowing hard coals. Place along the sides of the pot. Inside the pot, set the liquor to boil. Its vapor will rise into the empty pot. Then, inside the bamboo tube of the empty pot, you must siphon off what has filled the empty pot. Its color will be very white and will look no different than clear water. Sour liquor will have a bittersweet flavor, and the

---

[68] For a full English translation, see Zhao Rukuo (1170–1228), *Chau Ju-Kua: His Work on the Chinese and Arab Trade in the Twelfth and Thirteenth Centuries Entitled Chu-fan-chi (Description of Foreign Peoples)*, trans. Friedrich F. Hirth and W. W. Rockhill (St. Petersburg: Printing Office of the Imperial Academy of Sciences, 1911). The name Zhao Rukuo has also become widely known in the literature as Zhao Rugua.

[69] Ishige, "Higashi yūrashia no jōryūshu," 123–124.

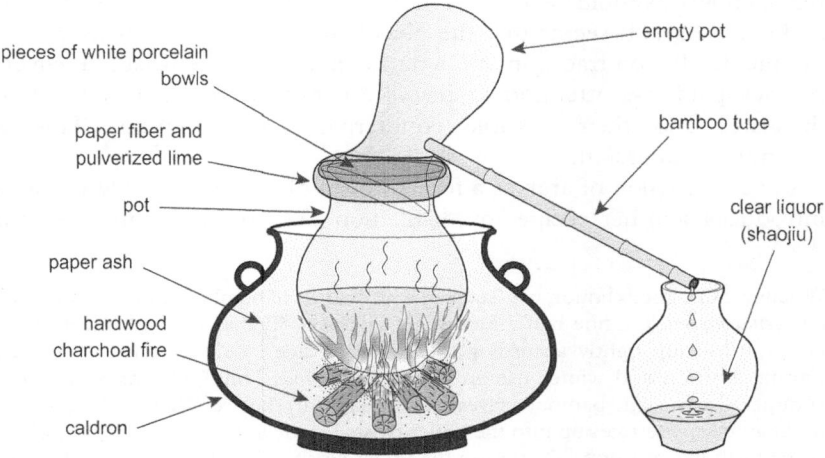

pieces of white porcelain
bowls

paper fiber and
pulverized lime

pot

paper ash

hardwood
charchoal fire

caldron

empty pot

bamboo tube

clear liquor
(shaojiu)

Figure 1.6 Still used to make a "Southern Barbarian" form of shaojiu called aliqi (arak). Drawing based on description in *Jujia biyong*, c. fourteenth century. Drawing by Matilde Grimaldi, after Takayama Takumi, "'Tōhō kenbunroku' ni okeru sake ni kansuru ichi kōsatsu" (Consideration of the Alcohol Beverages in The Travels of Marco Polo), *Nihon jōzo kyōkaishi* (Journal of the Japanese Brewery Association) 102, no. 3 (2007), 172–186, Figure 15.

insipid liquor will [now] have a sweet flavor. One can obtain one-third "good liquor." You can use this method to cook liquors decocted during the winter, all of them.[70]

In the reference above, we note two facts: (1) By the fourteenth century, the Chinese were already consuming shaojiu, a grain-based spirit similar to arak that is sometimes found in Song dynasty literature; (2) they also began to use a new technology imported from foreign countries through the maritime trade route to make a new type of distilled liquor called "burnt liquor of the Southern Barbarians," whose original foreign name is arak. The author of the text above probably wanted to distinguish its special distillation method from the Chinese one. As mentioned above, a Japanese scholar has argued that this might be an example of Gandharan-style distillation, which originally developed in South and Southeast Asia. Huang, in the Needham volume, introduces these iterations of the word *arak* in Chinese sources; however, he does not pursue

[70] *Jujia biyong shilei* (Essential Things for Living at Home) (Tokyo: Chūgoku shokukei sōsho, 1973), 12: 42b–43a, translation by Paul Buell.

further how this could be related to the arak tradition in Southeast, South, and West Asia. It seems that the Needham team tried to focus on the unique distillation tradition in China as an anti-Eurocentrist maneuver, so they paid less attention to possible connections between Chinese distillates and their possible counterparts in the West. This is a significant omission.

In its discussion of arak as a foreign form of shaojiu, *Jujia biyong* also introduces another recipe for distillation or "cooked" liquor (*zhujiu* 煮酒):

Whenever one cooks liquor, use 2 *qian* of wax, 5 slices of bamboo leaf, and "official" *Arisaema japonica*, a fine half a kernel for each *dou*. Transform and put into the liquor. Close up tightly according to method. Place inside a boiler. ([subtext] During autumn and winter use an *Arisaema japonica* "pill." During spring and summer, use wax and bamboo leaves). After that, start the fire. Wait until the aroma of the liquor penetrates up into the boiler twists [of the apparatus]. The liquor will come forth in profusion. Then raise the boiler again. Then take up the entire pot [with the liquor], open [it] up and look. If the liquor is boiling, then it is ready. Place into the fire for a long time. When you take it down, put it into the lime. One should not move continuously. One wants the white liquor to expel [in order] to obtain the clear [distilled] liquor. Afterwards, when cooking again and again, use mulberry leaves to repose. This is to prevent the aroma *qi* [vapor] from being cut off.[71]

The method explained here can be interpreted as another method of distilling forms of alcoholic liquor. The "boiler twists" described here could suggest a kind of still, but we cannot be sure of its shape. This could be a still from an older tradition, one that had been used traditionally before the introduction of the foreign shaojiu distillation method. We can assume from this that distillation grew widespread in the second half of the fourteenth century, produced many variations, and mixed different distillation technologies and names in the process. In fact, such descriptions signal only the tip of the iceberg, as far as information goes. A recently published general history of China under the Yuan dynasty verifies the existence of a thriving brewing industry, which included the manufacture of exotic liquors such as kumiss (also spelled koumiss or qumyz; known as *airag* in Mongolia), which is made based on a form of fermented mare's and cow's milk popular among the Mongols and in other nomadic societies.[72]

---

[71] *Jujia biyong*, 11: 35a, translation by Paul Buell. A *qian* 錢 today equals 3.13 grams; a *dou* 斗 equals 5161.9 milliliters

[72] Yang Yinmin, *Diguo shang yin: Yuandai jiuye yu she hui* (The Empire's Popular Drinks: Wine Industry and Society during the Yuan Dynasty) (Tianjin: Tianjin guji chubanshe, 2009); Christopher P. Atwood, *Encyclopedia of Mongolia and the Mongol Empire* (New York: Facts on File, 2004), 321–322.

In short, it is Huang Hsing-tsung, in his study of distillation in China, who rekindled the debate on the origins of distillation. In this debate, scholars paid more attention to the origins of distillation and less to the origins of its popularity. The important point that Huang did not discuss fully is the spread of distillation during the Mongol period and the historical context that facilitated that spread. We should look at the case of the Mongols because, although distillation in Eurasia probably developed independently in China, India, and the Hellenistic world, it was the Mongols who actually spread the technology and popularized its products.

The Mongols and Distillation: Arak, *Shaojiu*, and Arkhi

The first clear description of distilled liquors and evidence of a boom in their consumption throughout China is dated to sometime between the thirteenth and fourteenth centuries; that is, during the Mongol era. As mentioned earlier, the first clear evidence of alcohol distillation comes from an official dietary manual for the Mongol court in China, entitled *Yinshan zhengyao* 飲膳正要 (Proper and Essential Things for the Emperor's Food and Drink), compiled by Hu Sihui 忽思慧 (fl. 1314–1330) and presented to the court in 1330. It also provides the first extant documentation of the term *arak* in any source:

Arajhi liquor is sweetish in flavor and piquant. It is very heating and makes a great poison. It is good for dispersing chill hard accumulation. It removes cold *qi*. Good wine is distilled to procure a dew. This is the Arajhi.[73]

This passage clearly describes a brandy that is obtained by accumulating small drops of liquor with higher alcohol content during the vaporization of good-quality alcohol through a distillation process. This source refers to arak in Chinese as *alaji* 阿刺吉, in Mongolian arajhi, a palatalized Turkic intermediation of *arak*.[74] Such a palatalized form is typical of the text, which displays substantial Turkic influence. This influence is not strange because many of the foreigners who worked for the Mongol court were Uighurs and because this source probably was written with their active participation.

The term *arak* appears in several diverse transliterated forms in contemporaneous sources. For example, a later Yuan dynasty work, "Poem about Arajhi Liquor" (*Yalaiji jiu fu* 軋賴機酒賦) by Zhu Derun 朱德潤, uses a set of homophonic Chinese characters – *yalaiji* 軋賴機 – to describe

---

[73] Buell and Anderson, *A Soup for the Qan*, 499.
[74] Ibid., *juan* 3, P. 6B (original text), 499 (translation).

the liquor.[75] After explaining its good fragrance, Zhu vividly describes how to acquire distilled drops of liquor using a pot-like still with two parts, one for collecting condensed spirits and one for cooling.[76] We have *arak* introduced in another source using the Chinese transliteration *aliqi* in *Jujia biyong*, which suggests that a distilled liquor of foreign origin had already been introduced to China and received its Chinese name, "Southern Barbarian *shaojiu*," which is the same as or probably related to the one that appeared in the Song-period source as "Thailand liquor."

By the time the name *arak* began to appear regularly in Chinese writings, China had been under Mongol rule for more than a century. By then, the country's Mongol rulers had long been in the practice of recruiting to their court different groups of foreigners, many of whom were Uighurs and Persians from Central Asia. Surely, distilled liquor using a foreign name with an Arabic root must have been circulating through elite circles by then. At the same time, pieces of evidence indicate that these elites were also enamored of the traditional Chinese culture of making alcohol, including the culture that surrounded shaojiu.

Archaeological artifacts exhibit evidence of the influence that these Chinese distillation traditions exerted on northern nomadic societies one to two centuries before the Mongol period. In addition to more recent discoveries like the Da'an distillery site, two other bronze stills were discovered, one in 1983 at Shi'erduan village in Inner Mongolia and another in 1975 at Xishanzuo village in northern China's Hebei province.[77] Their basic forms resemble the older Han dynasty still, with the upper and lower parts attached together. However, these two stills are more important to our study. Unlike the smaller Han still, they are large enough to distill alcoholic beverages for consumption. Moreover, they date back to the Song period when sources began to reveal the spread of distilled liquors in China.[78] The fact that these stills were found in northern China hints at the gradual spread of distillation to northern China, and beyond to the nomadic and semi-nomadic peoples living further north, like the Mongols and Jurchens.

---

[75] Liu, *Zhongguo kexue shi lunji*, 317.

[76] Luo Feng, "Liquor Still and Milk-Wine Distilling Technology," 513–514; Huang, *Science and Civilisation in China*, vol. 6, part 5, p. 229.

[77] For the one discovered in 1983 at Shi'erduan 十二段 village, Longchang 隆昌 township, from Bairin Left Banner 巴林左旗, Inner Mongolia, see Luo Feng, "Meng-Yuan shiqi de niangjiuguo yu zhengliu naijiu jishu" (Wine-Making Cauldrons and Technology of Distilling Milk Wine in the Mongolian Khanate and Yuan Dynasty Period), *Kaogu* 450 (2008): 66. For the one discovered in 1975 at Xishanzuo 西山嘴 village, Qinglong 青龍 county, Hebei, see ibid., 69. Cited in Schottenhammer, "Distillation and Distilleries," 144–145.

[78] Schottenhammer, "Distillation and Distilleries," 150–152.

Because many pieces of evidence show that the consumption of distilled alcohols called *shaojiu* began to appear during the Song era, we can
speculate that the distillation technology that produced it gradually transferred from Song China to the Khitans, Jurchens, and Mongols. These
findings raise many questions. Is it possible that these stills, which relied
on Chinese distillation technology of the Song period, developed first in
north China and then influenced distillation practices among the northern nomadic peoples? This argument seems particularly compelling in
view of the major non-Chinese peoples who began to occupy and rule
north China after the tenth century, pushing the Song state south. They
included Khitans, Jurchens, and lastly the Mongols, although there was
a substantial Turkic presence as well. What materials did distillers in
question use with these stills? How did they relate to stills later developed
by northern nomadic peoples? If people in northern China with nomadic
backgrounds made liquor using the Chinese stills, then perhaps they
localized the ingredients – for example, distilling horse or camel milk
instead of fermented grains.

While Huang does not pay much attention to the history of distillation
among the nomadic peoples of China and its neighbors (including the
Mongols), he does introduce the freezing method of distillation that was
popular among them. However, in his categorization of the world's
traditional stills, Needham divides East Asian stills, which differ from
Middle Eastern (Hellenistic-style) stills, into Mongolian and Chinese
styles: the former more primitive with its suspended catch bowl, the latter
more advanced and complex, having a "large spoon with a hollow handle
leading into a jar." In making this categorization, Needham cited
a sixteenth-century work of technical literature from Mughal India that
describes the above Mongolian and Chinese styles, as well as the Greek
still-head form with two pipes and two receivers (the *dibikos*).[79] He
concludes, "At such a late date, of course, it is difficult to trace anything
of technical inter-change in earlier times," implying that interregional
transfers of distillation technology were not yet possible. But, in fact,
the two stills discovered near Mongolia belong to the Chinese outside-
receiving types, according to Needham's categorization. This shows more
dynamic influences at play than he allowed. Today, it is much easier to
accept the possibility that people in different societies exchanged distillation concepts and tools over time through cross-cultural exchange, a fact
that renders his fixed categorization untenable. As this book shows, this
mixing of traditional still technology that developed in China and Korea
forces us to revise this categorization from Needham's Mongolian and

[79] Needham, Ho, and Lu, *Science and Civilisation in China*, vol. 5, part 4, pp. 106–107.

Chinese types to a different categorization, inside-receiving and outside-receiving types, because the latter reflects truer and less biased forms of interconnection between different distillation traditions.[80]

Recently, scholars have analyzed additional variants of stills and distinguished them by specific functional differences among selected methods, such as solid-state fermentation of grains (China) versus semi-fermentation of milks (nomads), or receiving alcohol drops at the center versus the edge of a still pot. Therefore, while this book sometimes uses Needham's categorization, because of its utility in distinguishing types of premodern stills, we should remember the deep, complex structure of cross-cultural technology transfers underlying it.

Looking at the actual history, it appears that nomadic peoples like the Mongols developed a drinking culture quite different from that found in agricultural society. To understand it, we should look at the historical context surrounding the development of the liquor called arkhi. Chinese histories testify that, in ancient times, the nomadic peoples of northern China did not engage in cultivation, but instead stocked and consumed milk from cows and horses (and camels) and meat such as beef and mutton. As for the milk, once they fermented it, they made it into drinks that contained as little as one to three percent alcohol. They were always preparing mare's milk and similar fermented dairy products that have long been popular in both Inner and Outer Mongolia and Central Asia (and continue to be so today).[81] For example, the entry on the Xiongnu people in the biography section of Sima Qian's *Shiji* (Records of the Historian), written in about 94 BCE, reports that this people of the western border region of ancient China awarded a cup of liquor to any soldier who killed or captured the most enemy soldiers. Another episode in this passage states that a Chinese turncoat advised the Xiongnu king to reject China's grain-based alcoholic beverages sent to him as tribute by the court of the Han emperor so that they could show their milk-based alcoholic beverages were superior to Chinese grain-based spirits. Based on these documentary records, it is clear that alcoholic drinks were important to nomadic peoples and also a major part of their diet, when female horses and camels provided milk.[82]

It should have been spirits based on their milk-based alcoholic beverages, therefore, that the Mongols and other nomadic peoples preferred to make once they learned the technique of distillation. Of all their

---

[80] As for the stills that developed in Korea, see Chapter 4.

[81] See Moldir Oskenbay, "Fermented Dairy Products in Central Asia: Methods for Making Kazakh Qurt and Their Health Benefits," *Crossroads: Studies on the History of Exchange Relations in the East Asian World* 14 (2016): 205–218.

[82] He Manzi, *Zuixiang riyue*, 122–123.

fermented milks, the Mongols preferred kumiss, a fermented mare's milk, low in alcohol content, whose popular name derives from a Turkic, not Mongolian, tongue (Mongolian is *airag*, also spelled *ayrag*).[83] As biologist Batdorj Batjargal shows, a wide variety of fermented milk products make up an integral part of the Mongolian heritage, have developed over a long period of time, and have great social, religious, cultural, economic, and medicinal importance in Mongolian society.[84] As the Mongols began to gradually expand their political power across Eurasia in the early thirteenth century, they brought with them alcohol like kumiss. Bayarsaikhan Dashdondog, in her examination of the drinking habits and drinking culture of medieval Mongols, argues that overindulgence in alcohol among Mongols markedly affected their lifestyle, fertility, and longevity.[85] Kumiss does not easily cause intoxication because of its low alcohol content; however, Mongols drank large quantities of the beverage, not to mention other alcoholic beverages, according to a description of Mongol behavior written by William of Rubruck.[86] Morris Rossabi points out that alcoholic excesses and overindulgence as a cause of the Mongol Empire's decline have been overstated. He convincingly argues that many Mongol leaders were careful about drinking during military campaigns, and that we should consider several other social and economic factors for the empire's decline.[87] Indeed, while Ögedei (*c.* 1185–1241), the second Great Khan of the Mongol Empire, died because of his addiction to

---

[83] The term for fermented mare's milk in Mongolian is *airag.* Joseph A. Kurmann, Jeremija L. Rasic, and Manfred Kroger, *Encyclopedia of Fermented Fresh Milk Products: An International Inventory of Fermented Milk, Cream, Buttermilk, Whey, and Related Products* (New York: Springer, 1992), 174.

[84] Other kinds of fermented milk include *tarag* (yoghurt), which is made from cow, goat, or sheep milk. Another is a fermented camel's milk named *khoormog*, a kind of kefir. For a scientific explanation of the probiotic effects of fermented foods and the important role they have played in the Mongol diet since ancient times, see Batdorj Batjargal, "Probiotic Properties of Lactic Acid Bacteria Isolated from Mongolian Fermented Mare's Milk," *Crossroads: Studies on the History of Exchange Relations in the East Asian World* 14 (2016): 257–268.

[85] See Bayarsaikhan Dashdondog, "Drinking Traits and Culture of the Imperial Mongols in the Eyes of Observers and in a Multicultural Context," *Crossroads: Studies on the History of Exchange Relations in the East Asian World* 14 (2016): 161–172.

[86] When he analyzed William of Rubruck's famed travelogue in detail, American historian John Masson Smith Jr. calculated the amount of liquor drunk by guests at a great drinking festival that Möngke Khan hosted on June 24, 1254, and found that each guest was provided with the equivalent of nineteen shots of 80-proof whiskey. John Masson Smith Jr., "Dietary Decadence and Dynastic Decline in the Mongol Empire," *Journal of Asian History* 34, no. 1 (2000): 46, cited in Dashdondog, "Drinking Traits," 170.

[87] Morris Rossabi, "Alcohol and the Mongols: Myth and Reality," in *Wine Culture in Iran and Beyond*, ed. Bert G. Fragner, Ralph Kauz, and Florian Schwarz (Vienna: Verlag der Österreichischen Akademie der Wissenschaften, 2014), 211–223.

alcohol,[88] some Mongol rulers like Chinggis Khan enjoyed longer lives because they drank less alcohol, while Toghon Temür (1320–1370), the last emperor of the Yuan dynasty, is depicted as sober in Chinese sources.[89] Despite the occasional caution about alcohol overindulgence, Mongols considered alcoholic drinks like kumiss and airag vital to their royal rituals, feasts, and festivals like the Quriltai (the "Great Royal Assembly" that decided major issues and elected new rulers), and had been so since the formation of their empire.[90] Writers who observed the traits of the Mongols during their age of empire, including John of Plano Carpini (c. 1185–1252) and William of Rubruck (fl. 1253–1255), made clear that alcohol formed an essential part of the Mongol lifestyle, and as they expanded militarily across Eurasia, they introduced their drinking culture to the people they subjugated. They considered it a vital part of their lives and, apparently, to the lives of their imperial subjects as well. A situation like this makes the cross-cultural transference of these fermented milks likely, if not inevitable.

According to contemporaneous European travel writers, the Mongols produced two kinds of kumiss. Some dismissed one variety as white, cloudy, and sour tasting; however, writers described another, superior form of fermented mare's milk called in Turkic qara kumis ("black kumiss") as very sweet and potent. Mares' milk does not curdle, so winemakers churned the milk until everything solid sank to the bottom, leaving a liquid that on its top layer was very clear. According to Rubruck, the slaves were given the white dregs, which reportedly had a soporific effect on them.[91] The clear black kumiss that remained was then presented to the Mongol lords for their consumption. Offering high-quality kumiss to one's guests during the feasts repeatedly held by Mongols reflected the importance the Mongol host attached to his guest.[92]

The Mongols, however, encountered a big problem with transporting kumiss whenever they embarked on their military campaigns. Kumiss did not keep and was also available only during certain times of the year, at the very least during the spring and summer when milk was abundant. Even the clarified black kumiss quickly spoiled, creating a real problem for

---

[88] Ibid., 214–215; Thomas T. Allsen, "Ögedei and Alcohol," *Mongolian Studies* 29 (2007): 3–7; Atwood, *Encyclopedia of Mongolia and the Mongol Empire*, 413.

[89] Dashdondog, "Drinking Traits," 167.

[90] Rossabi, "Alcohol and the Mongols," 219; Sŏl Paehwan argues this in detail in his study of the importance of the Quriltai in the history of the Mongol Empire; see Sŏl Paehwan, "Mongwŏn Cheguk K'urilt'ai (*Quriltai*) Yŏn-gu" (A Study of the *Quriltai* in the Mongol Empire), PhD dissertation, Seoul National University, 2016, pp. 3, 22, 25, 37, 119, 131, and 201.

[91] Dashdondog, "Drinking Traits," 168.

[92] Schottenhammer, "Distillation and Distilleries," 144.

Mongols on the move. For any Mongol ruler, a lack of abundant kumiss during campaigns was politically unthinkable, as it was even during more peaceful pasturing or during events such as the Quriltai.

However, the Mongol khans were used to following a nomadic lifestyle, so they and their armies were well suited to long-distance campaigns and thus developed strategies for ensuring the supply of spirits to their armies. Thus Batu Khan, lord of the Golden Horde of Russia and Ukraine, received daily a supply of black kumiss supplied by thirty men stationed one day's ride away from his *ordu* or camp. Each rider was supplied with the kumiss of 100 mares, meaning that each station could supply the yield of 3,000 mares. This anecdote gives no indication of the number of mares that were producing ordinary kumiss, nor the volume of their output. Marco Polo claimed that Khubilai Khan kept thousands of mares just to satisfy his *ordu*'s demand for black kumiss. Polo noted that the Mongols drank kumiss as a common beverage and that it probably played an important role in religious ceremonies, such as honoring and praying to their earthly god who protected their sons, animals, and crops.[93] As a consequence, we can speculate that Mongols sought to preserve their kumiss for as long as possible. This could be achieved through distillation, since high-alcohol distillates preserve easily. The Mongols very likely distilled alcohol such as kumiss in order to provide it to followers in order not to lose their influence over them.

We have no evidence that the Mongols developed any distillation technology. However, we have to pay attention to the fact that they were good at using the skills of various peoples and societies under their rule and facilitating cultural exchange with them. It is highly likely, therefore, that they adopted a form of distillation technology that existed in China. The Mongols could have let some of their prisoners or soldiers with knowledge of distillation technology distill kumiss in order to render it both strong and easy to preserve. We already possess evidence from William of Rubruck that the Mongols relied on Europeans to make beer – so why would they not have used specialists from China to distill their kumiss? Buell argues that by adapting to improved traditional Chinese stills, the Mongols gradually concocted at least two new kinds of kumiss – which differed slightly in how they collected the distillate, one kind using a pot, the other a pipe – primarily to distill milk liquors made from the mare and camel milk available to them.

Luo Feng argues that two stills discovered in Mongolia and northern China and built during or before the Mongol dynasty were created to distill airag or kumiss because they both had the capacity to produce pure,

---

[93] Marco Polo, *The Description of the World*, 56–57.

high-quality kumiss.[94] Indeed, both stills were designed to distill fermented liquid, not fermented mash, as was typically done in southern China to make shaojiu. The coolant water in the container at the top of the apparatus – the condenser – allowed more steam to condense, and the spout permitted the heated water to drain easily. The dome-shaped top with a convex bottom and the circular gutter made a thorough collection of the distillate possible (see Figure 1.2B).[95] Luo Feng supported his claim with eyewitness accounts of the stills and distillation practices of the Volga Kalmucks and Mongols written by the late eighteenth-century German zoologist and botanist Peter Pallas (1741–1811), to which Paul Buell and Montserrat de Pablo have recently added an extensive comparative analysis.[96] These European eyewitness accounts of the stills and distillation practices of the Volga Kalmucks and the Mongols show that they used stills to distill airag or kumiss. Kalmucks at that time used a long-wrapped pipe, air-cooled, which served as a condenser–collector that both collected and released vapors and condensate, the latter of which collected in a fairly large closed pot. This type of still is a variation on a European model that uses worm tubs, through which water circulates, leading through a pipe to a collector of some sort, the water being continuously recycled in order to keep the pipe cool. This type using worm tubs distinguishes European-style stills from Asian ones. Despite the possible influence of European-style stills and European distillation methods using fermented grains on the distillation of the Kalmucks of Russia, we can assume that they used fermented milks, and not grain-based ferments, to obtain their distilled liquors. People were able to modify their distillation devices at their convenience; however, they had to use ingredients that were available to them. Little doubt, then, that such a practice of distilling fermented dairy products has continued to develop, to be consumed as arkhi, a distilled liquor made from cow's milk in distilled form in modern Mongolia and other nomadic societies inhabiting northern Eurasia – although due to the influence of Islam, most Kazakhs no longer distill mare's or cow's milk.[97] Considering

---

[94] Luo Feng, "Liquor Still and Milk-Wine Distilling Technology," 517–518.

[95] Ibid., 501–504.

[96] Ibid., 502–503; Paul D. Buell and Montserrat de Pablo, "Distilling of the Volga Kalmucks and Mongols: Two Accounts from the 18th Century by Peter Pallas with Some Modern Comparisons," *Crossroads: Studies on the History of Exchange Relations in the East Asian World* 13 (2016): 115–123.

[97] Experts argue that modern arkhi is made from cow's milk because mare's milk is not good for distilling. Luo Feng, "Liquor Still and Milk-Wine Distilling Technology," 502–503; Ishige, "Higashi yūrashia no jōryūshu," 83. According to Paul Buell, this is untrue, because he saw the mares milked, saw the milk being fermented, and tasted the final vodka (Buell and de Pablo, "Distilling of the Volga Kalmucks and Mongols," 118–120).

the archaeological findings and historical evidence for the development of distillation in Mongolia, we can conclude that the arajhi consumed by the Mongols – documented in *Yinshan zhengyao* in the fourteenth century – was based on fermented milk from mares or camels. This is an important piece of evidence that bridges the Mongol stage of distillation history and modern arkhi.[98] On the other hand, a later source written during the Ming era (1368–1644), *Bencao gangmu* 本草綱目 (Compendium of Materia Medica) by Li Shizhen 李時珍 (1518–1593), a work of considerable influence on later books of medical literature in East Asia, clearly states that his contemporaries made distilled spirit from grain-fermented liquor.[99] Intriguingly, Li Shizhen says that *shaojiu*, "burnt liquor," is in fact *huojiu* 火酒, namely "fire liquor," and *alaji jiu* 阿剌吉酒, "arajhi wine." As its primary source, the book cites Hu Sihui's *Yinshan zhengyao*, which introduces arak for the first time as a distilled liquor using a different Chinese transliteration. This suggests that the arak liquor presented in the fourteenth-century official dietary manual for the Mongol court in China, in just a few decades, would be identified as *shaojiu*, which is based on grain-fermented liquor. In fact, other liquors made of fermented grains are also mentioned in *Yinshan zhengyao*, such as *boza*, an old Turkic word, and *sürmä*, "cooked," apparently an Uighur word, although it can also be found in Tibetan as well, apparently as a loan word: "*Sürmä* Liquor: It is also called *Boza*. It is slightly sweet and piquant in flavor. It is good for augmenting *qi* 氣 and controlling thirst. If too much is drunk it makes a person fat and produces phlegm."[100] The name and description of the title make us wonder whether this liquor was a distilled one, although *boza* is now a fermented millet beverage with a lower alcohol content of just 1 percent, not distilled, popular in modern-day Central and West Asia and in Eastern Europe. Likewise, another possible distilled liquor is recorded in *Yinshan zhengyao* as "small coarse grain" liquor, described in terms similar to arajhi and *sürmä*.[101] This might be a sorghum distillate, since sorghum in later times was very

---

[98] Scattered archeological and documentary sources (including a Korean travel account of the seventeenth century, discussed in Chapter 4) suggest that some arajhi (arak) consumed in the northern part of China before its transfer to Korea was based on cow's and mare's milk (or even camel's milk).

[99] As Chang Chi-Hyun suggests, it is possible that the Mongols used the milk of mares or cows when they first popularized the distilled wine called arajhi. However, when it spread throughout China, people began using fermented grains to make distilled liquors, as documented in the Ming dynasty account. See Chang Chihyŏn (Chang Chi-Hyun), *Hanguk oeraeju yuipsa yŏngu* (A Study of the Influx Of Foreign Alcoholic Liquors to Korea) (Seoul: Suhaksa, 1989), 61–62.

[100] Buell and Anderson, *A Soup for the Qan*, 499.   [101] Ibid., 498.

popular in China for manufacturing whisky (named after its base grain, *gaoliang* 高粱).

In his testimony to the different kinds of alcoholic drink that Mongols enjoyed in the thirteenth century, William of Rubruck leaves little doubt that the Mongols drank grain-fermented alcoholic beverages, and mead as well as kumiss and *qara kumis* (black kumis).[102] Contact with the sedentary Islamic west and Chinese south allowed the Mongols to adapt to more, and more varied, drinking habits. In time, the Mongols drank rice mead, rice ale, honey mead (*bal*), fermented millet (*buza* or *boza*), and red grape wine, which Rubruck compared to the French wine La Rochelle. Indeed, it was the Mongols who facilitated the growing popularity of grape wine in China.[103] In turn, the Chinese introduced rice wine to Persians, who called it *tardsun*. The Mongols drank both. We should also remember that *Jujia biyong*, a Ming dynasty source, also introduces a kind of distilled alcohol called "foreign " that was directly influenced by a foreign distillation method, and states that any kind of liquor lacking a proper flavor (if it is sour or sweet, or insipid and thin) can be used to make the foreign shaojiu called arak. How far back this went among the Mongols is unclear. Rubruck does not mention distillation, but when arak spread to southern China, the Chinese must have begun to consume large quantities of fermented grains in order to make distilled liquors, a well-documented fact.

At the same time, Li Shizhen clearly traces the origins of distillation back to the Yuan period, stating that the making of burnt wine, *shaojiu*, was not an ancient art. The technique (*fa* 法), Li says, was first developed in Yuan times. This led to a big debate on the origins of shaojiu and distilled liquors in China, because some other sources like *Qu bencao* (Ferment Cultures and Basic Herbal) from the Song dynasty also claim that distilled wines from Southeast Asia were introduced to China before the Yuan period, most likely during the Song period.[104] Ming author Li Shizhen, however, probably remembered that distilling of alcoholic

---

[102] William of Rubruck, *The Mission of Friar William of Rubruck: His Journey to the Court of the Great Khan Möngke 1253–1255*, trans. Peter Jackson (London: Hakluyt Society, 1990), 178–179, 191.

[103] According to Dashdondog, "Drinking Traits," 169, "a colony of Muslim artisans originally from Samarqand settled in Simali just north of Beijing, cultivated grapes, and provided wine for the imperial court throughout the thirteenth century." The Onggud were also growing grapes and sent some wine to Chinggis Khan, who enjoyed it.

[104] A different passage about grape wine in the same chapter of Li Shizhen's *Materia Medica* also implies that grape-wine distillation came from the western regions during the Tang dynasty (Schottenhammer, "Distillation and Distilleries," 145–146). However, I think we can also interpret from this text that its author meant that grape wine, not distillation itself, was first introduced as early as the Tang dynasty. This fact in the passage is also incorrect, because recent research shows that the Chinese produced grape wines before

beverages rapidly spread during the Mongol period. Therefore we should look at some major historical changes that occurred in China regarding the development of alcoholic beverages during the Mongol Yuan period.

It is highly likely that, in the course of integrating different brewing and distilling cultures, the Mongols not only created portable stills to bring with them on their military campaigns but also promoted distilleries in China. This aided the development of a Mongol-influenced drinking culture, which became a part of traditional ceremonies in which many Chinese and other foreigners also participated at the Khan's court in particular.

Promoting such integration, the Mongol emperors of Yuan China propagated distillation methods of different kinds, including those from the south that produced shaojiu and arak based on a combination of earlier Chinese distillation technologies (maybe as far back as the Han dynasty, if we believe the scholars who made that claim), and new foreign distillation technology was introduced from maritime contact. This is how traditional shaojiu and arak made with foreign methods developed so rapidly in the Yuan dynasty, thanks to mutual technological exchange. We can speculate that, by the time *Yinshan zhengyao* was written, arak ("arajhi" in Turkish) was well on the road to becoming a general name for distilled liquors of every sort. Without question, the arak phenomenon was a Mongol-era development. The term simply did not exist in wide circulation before. Although the Mongols drank almost any kind of hard alcohol, among them was *sayin darasun*, "good wine," most likely one of the first brandies.[105]

Chinese and later Korean sources offer intriguing hints about a Mongol aristocrat who probably gained fame as a shaojiu brewer during the Yuan dynasty (as discussed in detail in the following chapter). According to *Yuanshi* 元史 (The History of the Yuan), Ananda 阿難答 (?–1307) – a grandson of Khubilai Khan, the Anxi Prince after his father,[106] and a competitor for the title of Khan after Temür Khan died in 1307 – received, along with other Mongol princes, an exemption from a 1303 imperial ban on alcohol brewing. Although a devoted Muslim, he very likely promoted the importation and production of shaojiu, or arak, which

the beginning of the Common Era. Because premodern authors like Li Shizhen had limited sources available to him, we should also review the contents in text critically.

[105] On a Yuan dynasty wine jar is glazed the message "*sayin darasun*" (good wine) in the 'Phags-pa script used in the Yuan dynasty. Morris Rossabi, *Khubilai Khan: His Life and Times* (Berkeley: University of California Press, 2009), 158–159; "Good, excellent," as in *sayin darasun*, "good wine": This was a term that appeared in Chinese plays of the era. It apparently was a borrowing from the Mongolian *sayin darasun*, "good wine," also meaning distilled liquor.

[106] Song Lian, *Yuanshi* (The History of the Yuan) (Beijing: Zhonghua shuju, 1976), vol. 14. Ananda was the Anxi Prince's 安西王 grandson and Manggala's 忙哥剌 (?–1280) third son, and was given the title Anxi Prince 安西王 after his father.

merchants introduced to China by sea. Ananda's case, the first documentary evidence of a large-scale brewery and distillery in China monopolized by powerful Mongol nobles, clearly suggests that the Yuan court promoted a brewery and distillery industry of unprecedented size. When the Mongols conducted expeditions to Java in the 1280s, it is also possible that they brought back distilled liquor and the distillation method from Southeast Asia directly.

Archaeological data now supports the documentary evidence. Many new archaeological excavations in the historically wealthy areas of southern China show that distilled liquors dispersed more widely than before across China between the late Song and Yuan dynasties. Moreover, they enjoyed Mongol support. In fact, Angela Schottenhammer argues that large-scale production of consumer goods – for both domestic and foreign markets – already existed in Southern Song China, with most likely privately run distilleries operating particularly in the commercially well-developed southeast.[107]

After the establishment of "alcohol and vinegar bureaus" (jiucu wu 酒醋 務) throughout north China in 1231–1232, the Yuan court extended its alcohol monopoly, at least theoretically, to encompass both production and distribution, which became an important source of revenue for the dynasty.[108] At around the same time, workshops run by wealthy households, merchants, and village communities began to invest in the production of distilled wines and spirits, a response in part to China's increasingly commercialized economy, which was becoming especially intense in the larger cities of the commercially vibrant southeastern coastal regions. This led to the creation of large distilleries – state-run and private. Thus we see the growth of large-scale production sites for spirits and wine (shaofang 燒 坊) like arak and shaojiu, as new enterprises grew to meet the demands of both the Mongol ruling elite and the ordinary people for such commodities. This developed hand in hand with the expansion of cities and market systems that is a hallmark of late Song and Yuan history.

The Lidu distillery site provided the first solid material evidence of a large-scale distilled-beverage industry in Yuan China.[109] Archaeologists

---

[107] Schottenhammer, "Distillation and Distilleries," 155–155.

[108] The terminology for the liquor monopoly traces back to Northern Song times, as Herbert Schurmann demonstrates. They were supervised by officials (fangchang guan 坊場官) charged with overseeing the production and sale of alcoholic beverages and the collection of taxes on them. See Herbert Franz Schurmann, Economic Structure of the Yuan Dynasty: Translation of Chapters 93 and 94 of the Yuan shih (Cambridge, MA: Harvard University, 1956), 204.

[109] The UNESCO World Heritage Centre has ranked Lidu as one of the top ten archaeological discoveries in the world. See whc.unesco.org/en/tentativelists/5320 (accessed January 3, 2017). See also Fu Jinquan, "Cong Lidu yizhi kan woguo baijiu shi" (Discussion of Liquor History in China through the Study of the Lidu Memorial Site), Jiuliang keji 3 (2003): 95.

excavated a site that covers about 15,000 square meters and unearthed a total of eleven cultural levels that span six time periods, from the later Southern Song to modern times. Unlike the Da'an site, the Lidu distillery has been dated as early as the Yuan era. Further excavations and study will provide us with new and valuable insights into Yuan-era distilleries and distillation technologies.[110] Certainly, a site in Sichuan province called the Shuijingjie, in the city of Chengdu, which also dates back to late Yuan or early Ming times, proves the premodern origin of a now world-famous Chinese spirit.[111]

All these pieces of evidence help us to solve the question we raised at the beginning of the chapter: how the name *arak* found its way into Chinese documents in the fourteenth century, continued to be used there, and then transferred to other countries like Korea and Japan. Based on the points raised above, we can conclude that much cultural mixing occurred during a short period of time that roughly parallels the Mongol period.

Here we consider two interlinked possibilities: one, the importation of distilled alcohols through the maritime trade; and two, the subsequent adaptation of pre-existing distillation technology by the Mongols. Moreover, the original Mongol arak, based on milk, might have continued in its production in northern China, Mongolia, and Central Asia after the Yuan dynasty collapsed in 1368. Indeed, a form of arak based on fermented cow's milk or other materials continued to develop in these regions and then transferred to Manchuria, as texts written in Altaic languages including Manchu demonstrate. Renowned linguist G. J. Ramstedt detected many Altaic forms of the word *araq* in historical sources in order to trace the etymology of the word in Central Asia. The resemblances are clear: *araq* in Persia and Arabia, *araqy* in Turkey, *araka* among the Tungus people, *araki* in Mongolia, *arki* in Manchuria, and *arang* and *araegi* in Korea.[112] Later scholars used Ramstedt's research results to propose that arak and its distillation methods and technology traveled to East Asia by way of Eurasia's overland route.[113] Yet, as we have seen in this chapter, earlier documentary evidence already confirms the transfer of arak from western Eurasia to China by sea. It is more likely that the Mongols expanded arak's domain of production and

[110] Schottenhammer, "Distillation and Distilleries," 153–154.
[111] Ibid., 155. *Shuijingjie* 水井街, workshop from Jinjiang 錦江, Chengdu city.
[112] Gustaf John Ramstedt, *Studies in Korean Etymology*, vol. 1 (Helsinki: Suomalais-ugrilainen Seura, 1949), 13. Arak becomes *raki* in modern Turkey and the eastern Mediterranean. The stuff is usually a white, tasteless brandy or vodka but is generally flavored with anise seeds or oil, or something similar.
[113] Chang Chihyŏn, *Hanguk oeraeju yuipsa yŏngu*, 75–88.

consumption further west by land after the fourteenth century. Ramstedt's progression of word forms adds further to this history. We have already noted the word *arajhi*, the prevalent Turkic form of *arak*; clearly, only a historical context in which Turkic people were active in the Mongol court in the fourteenth century could explain this.

In addition to the development of an arak economy and culture in northern Eurasia, the Chinese continued to develop distilled liquors using grains, their traditional base ingredients, albeit at a larger scale built on the earlier foundations that were laid down under Mongol auspices. As arak spread, it could have influenced the development of Chinese shaojiu both in the form of its production technology and in its name. Thanks to post-Yuan literature like Li Shizhen's *Bencao gangmu*, we know that the form of distilled liquor that continued to develop during later periods in Chinese history was called both *shaojiu* and arak. While people in East Asian countries like China, Korea, and Japan used the term *shaojiu/soju/shōchū* more frequently, the other term *arak* became universal. It remains so today, disguised in a number of linguistic variants, throughout much of Eurasia.

Conclusion

This chapter has comparatively examined the history of distilled liquors in Eurasia as a means to detect the possible origins and developmental history of the idea and practice of distillation in different Eurasian societies. At the center of this examination of Eurasian roots lies the question, how did arak, a universally known and available distilled beverage, with which Korean soju shares a common origin and tradition, appear and proliferate during the Mongol period? It seems highly likely that the distillation technology needed to produce arak appeared and developed independently in different areas of Eurasia, such as the Mediterranean region, South and Southeast Asia, and China. Distillation technology was inspired by a variety of ancient technological practices like alchemy, as the cases of the Islamic world and China show. Yet the fact that many regions share the same name – *arak* – suggests that the term, along with its related concepts and technology, migrated from the Islamic Middle East, the home of arak, thanks to the flourishing cross-cultural trade that spanned Mongol-era Eurasia. Indeed, a Mongol-era Chinese encyclopedia introduces arak as a foreign shaojiu (distinguished from the native shaojiu) made using a South, Southeast, or West Asian distillation method.

In order to validate the credibility of this transfer theory of a southwest trajectory, let us summarize a possible scenario of distilled-liquor transfer that we can consider based on discussions offered above. Distillation technology had been developing in China since ancient times. It is also possible that the Chinese made distilled alcohol based on grain-fermented liquors called *shaojiu* to a limited degree during the Song dynasty based on distillation technologies that included both its native and foreign forms transferred to China from Southeast, South, or West Asia during the Song era. As seen in the first documentary evidence, in Chinese sources written in the fourteenth century, the Mongols adopted Chinese distillation technology in order to create the kind of milk-fermented liquor they drank. They called this new liquor arak*i*/arajhi/arkhi, appropriating a foreign term already adopted by the Chinese, and thereafter popularized it wherever they expanded their political power across Eurasia. Yet this Mongol boom of distilled liquors stimulated further an earlier tradition of distillation in China. After conquering southern China, the Mongols encountered Chinese who manufactured distilled alcohols called *shaojiu* based on grain fermentation, a method that had evolved thanks to China's largely agricultural environment. Then the Mongols appropriated this, after which large-scale grain-based distillation flourished throughout the Yuan and Ming eras of Chinese history. At this time, the documentation of still-based alcohol distillation began, and large-scale distilleries of the period were excavated. The fact that Chinese shaojiu began to include foreign shaojiu, called arak, suggests the prevalence and persistence of ongoing, large-scale interaction and acculturation; indeed, in the Ming period, shaojiu and arak came to be identified as the same form of distilled alcohol.

Tasks that still remain in our search for distillation's Eurasian origins include the examination of the development of distilled alcohols after the Mongol period in places like Europe, Russia, and West, South, and Southeast Asia, as well as Mexico, where their popularity has long been widespread. The history of distilled liquors in Europe, which now predominate globally, only begins in the fourteenth or fifteenth centuries. It is quite possible that the European pot still with worm tubs was developed by adopted Islamic alembic prototypes that had been transferred gradually from the Islamic world through Spain in the medieval period. While stills in Europe developed differently from those in Asia, Europe's connection to the Middle East ultimately links them to distillation in the "Old World." However, we should investigate the possible influences of Mongolian portable traditions on some forms of European still technology. Signs of enduring influence are there. Recently in Iceland, for example, much "moonshine" was made using stills that resemble the Mongolian style; the same is true in Russia.[114] Future research may establish that the Mongol influence through distilled alcohol extended much further than we had ever thought, perhaps even globally.

This chapter has confirmed that distilled liquor and the technology to make it did not develop from a single point of origin but rather progressed from diverse sources of innovation in a variety of cultures spread across vast territory at different times. Therefore one must observe Eurasia as a whole in order to grasp the history of distilled liquor's beginnings. Once this continental perspective is appreciated, the role that the Mongols played in the sudden spread of distilled liquor throughout China and beyond during medieval times becomes clear. It is possible that the shared patterns of Mongol-facilitated technology transfer and transformation that shaped the development of distilled liquors like arak and shaojiu might also have applied to Korea, which fell into the Mongol sphere of influence at the same time as China and much of the Eurasian landmass. The cases examined in this chapter thus establish a clear context within which to observe the development of soju in Korea as a well-documented case of transfer in global history.

---

[114] When Paul Buell was in Hrafnarfjordur, Iceland, he went to a local history museum where he saw Mongolian-style stills identified as Icelandic moonshine stills. His host, who worked for the big local-studies center in the same city, wrote his dissertation on Icelandic moonshine, and his well-founded conclusion confirmed my impression.

2   The Mongols and the Rise of Soju in Koryŏ
     Korea

In order to understand the rise of distilled liquors in Korean society, it is imperative to examine the prior development of alcoholic drinks in Korea generally. Like many countries, Korea possesses only a very limited quantity of extant resources for studying its ancient period; political history comprises most of these topics, outnumbering cultural history. Nevertheless, a close comparative examination of scattered sources, including documents of various genres written in China and later in Korea, as well as archaeological findings, helps us to reconstruct the various aspects of historical society on the Korean peninsula at micro levels (see Figure 2.1).[1] This offers insights into the basic lifestyle of people living there, or into specific elements, such as the alcoholic beverages they consumed. Throughout their history, the kingdoms that controlled the Korean peninsula enjoyed close contacts with dynastic courts that ruled China, whose literati produced accounts of the Middle Kingdom's neighboring societies that survive to this day.[2] The considerable influence of Chinese literature on its closest neighbors prompted Koreans to develop their own means of documentation by borrowing Chinese scripts and adapting them, to create a native system for representing Korean words.[3]

As a consequence, the earliest extant Korean accounts of its ancient history date back to the era of the Koryŏ dynasty (918–1392), which lasted four centuries and whose demise overlapped with the Mongol downfall. Subsequently, written sources – mainly literature produced in Korea,

---

[1] The dates for the foundation of Silla and Paekche given in this timeline are "traditional" dates based on the chronological tables in *Samguk sagi* 三國史記 (History of the Three Kingdoms) by Kim Pusik (1075–1151) and not historically accurate dates, as has been demonstrated in more recent scholarship. For detailed chronologies and historical maps of the history of Korea, see, for example, Michael D. Shin, ed., *Korean History in Maps: From Prehistory to the Twenty-First Century* (Cambridge: Cambridge University Press, 2014).

[2] On the sections of foreign countries in Chinese official histories, starting from Sima Qian's *Shiji* (Records of the Historian), see Endymion Porter Wilkinson, *Chinese History: A Manual*, 2nd edition (Cambridge, MA: Harvard University Press, 2000), 736–737.

[3] Peter H. Lee, *A History of Korean Literature* (Cambridge: Cambridge University Press, 2003), 88–95.

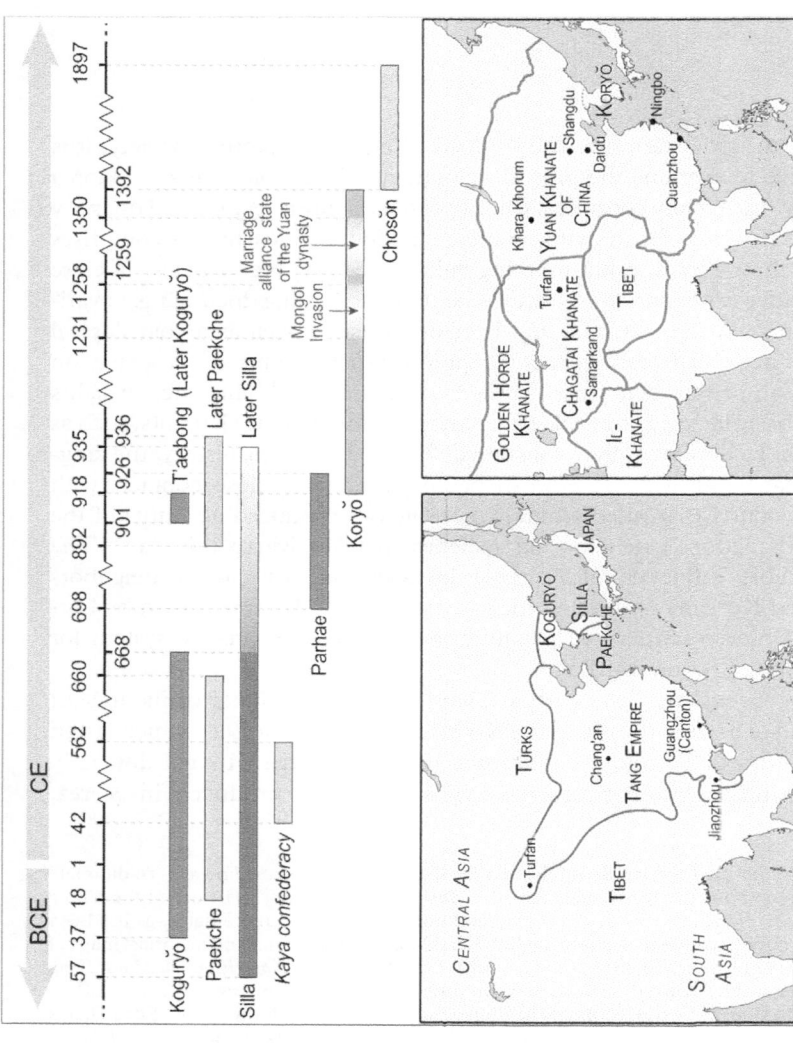

Figure 2.1 Timeline of the kingdoms and dynasties in Korea (top); the Three Kingdoms of Korea in broader East Asia (bottom left); the Yuan dynasty in China and Koryŏ dynasty in Korea as part of the greater Mongol Empire (bottom right).

supplemented by contemporaneous Chinese writings – provide the repository of data that makes it possible to trace the overall development of
alcoholic drinks on the Korean peninsula during premodern times. At the
same time, we can locate the oldest extant documents that verify the
existence of specific spirits like soju, and its cousin arak, during the years
of Mongol domination in Korea during the thirteenth and fourteenth
centuries. This provides us with a sufficient source base that allows
the necessary analyses of historical evidence that in aggregate reveals the
evolution of distilled liquors in Korean society (and, in time, beyond).

This chapter explores the development of alcoholic drinks on the
Korean peninsula from antiquity to Mongol times, based on the sources
mentioned above, in order to trace soju and arak's possible origins on the
Korean peninsula. These documents demonstrate that, while Koreans
developed their own styles of alcoholic drink over time, they did so thanks
to the influence of varieties introduced into their society via China, mainly
through the vectors of various political, economic, and cultural relations,
such as diplomatic gifts, trade, cultural or religious influences that crossed
borders, and other forms of contact. Judged against the broad panorama
of Korean history, soju's rise happened in a very short time – more
precisely, within the last century of the Koryŏ era, which was defined
largely by Mongol interference in and domination of Korean society. This
happened at a time when shaojiu, soju's Chinese ancestor, and arak began
to grow widespread in their consumption throughout China.

In order to identify soju's first appearance, we first analyze extant documents in their proper historical context in order to detect evidence that
points to those origins. Next, we review both old and ongoing debates
about the subsequent rise of soju and arak in Korea and how it compares
to the rise of shaojiu and arak in China. These debates have lasted centuries,
at least since the drink grew widespread across the Korean peninsula during
the late eighteenth century. Juxtaposing this secondary literature against
properly contextualized primary-source evidence reveals that current historiographical works on the origin of soju on the Korean peninsula fail to
explain what is revealed in sources because scholars have not fully
considered the sociocultural context of their sources from global historical
perspectives. Doing so widens our scope sufficiently to pursue a comparative
approach to the critical examination of relevant theories and debates that
have circulated in Korea, which acts as a bridge between the early founda-
tional scholarship and the new sources and methods now available to us
worldwide. As noted earlier, thanks to a richer body of documentary
evidence, the case of shaojiu and arak's transfer from China to Korea serves
as a useful illustration of how the distilled liquors that developed in China
during the Mongol period began to influence other societies through cross-
cultural contacts and the transfer of knowledge they enabled, which makes
this task of global history worth doing.

Traditions and Transfers of Alcoholic Drinks in Premodern Korea

Various pieces of scattered information hint that ancient people of the Korean peninsula, much like the inhabitants of any agrarian society, enjoyed drinking alcoholic beverages on special occasions such as heaven-worshipping events and folk festivals. It is clear that alcohol and food were important elements of the feasts central to these events.[4] After comparing the habits of different societies in the ancient period, Gina Hames argues that, for most ancient agricultural societies, consumption of alcohol proved to be particularly important to human life and culture in all aspects, including religion, class, and gender.[5] While Hames's study does not discuss the case of Korea, it is no exception, as seen in several episodes related to mythology and history. For example, at some folk festivals, women together played weaving games, and those who were defeated in the game had to offer the winners alcoholic beverages and foods, which they all enjoyed while drinking, singing, and dancing together. As Chong Dae Song points out, this probably developed from the practice of enjoying moon-watching during the mid-Autumn festival.[6] Indeed, it would have been impressive to see so many women gathered in merriment on such an occasion and drinking together. While we have to interpret these folkloristic episodes critically, we cannot deny that some of them reflect historical and cultural elements.[7]

What kinds of alcoholic beverage were they enjoying? Scholars have assumed that there were rice wine that had not been through a refining process and contained only about 6 to 9 percent alcohol; this could have been similar to modern-day *makkŏlli* (often spelled *makgeolli* in English), a milky, off-white, lightly sparkling rice wine that contains a slightly sweet, tangy, bitter, and astringent taste and is easier to make than other kinds of alcoholic drink. Today, it is a traditional and popular drink in Korea, having evolved into more than 800 varieties produced by different breweries.[8] It seems that its ancient form would have been simpler, so

---

[4] Chong Dae Song, *Chosen no sake* (Korean Alcoholic Beverage) (Tokyo: Tsukigi shokan, 1987), 23.

[5] Hames, *Alcohol in World History*, 23–32.   [6] Chong Dae Song, *Chosen no sake*, 24.

[7] For more episodes related to alcohol in the Three Kingdoms period and Great Silla, see Yi Sŏngu, *Han'guk sikp'um sahoesa* (Social History of Korean Foods) (Seoul: Kyomunsa, 1984), 197–203; and Chong Dae Song, *Chosen no sake*, 23–228. As in the histories of other societies, the oldest episodes of Korean history include mythological elements. Many stories about alcoholic drinks in the ancient period of Korea are related to myths, too.

[8] Hŏ Simyŏng, *Makkŏlli, nŏn nugunya? Saekkal innŭn sul, makkŏlli ŭi mŏdŭn kŏt* (Makkŏlli, Who Are You? Alcoholic Drink with Colors, Everything about Makkŏlli) (Koyang: Yedam, 2010).

that it could have been easily made and consumed; however, Chong Dae Song argues, a careful examination of records that originated in Korea's Three Kingdoms period (57 BCE–668 CE) shows that the variety of alcoholic beverages produced on the Korean peninsula at that time was great. The two oldest extant documents of Korean history written in the later Koryŏ period – *Samguk sagi* 三國史記 (History of the Three Kingdoms) by Kim Pusik (1075–1151) and *Samguk yusa* 三國遺事 (Memorabilia of the Three Kingdoms) by Iryŏn (1206–1289) – contain several terms for alcoholic drinks that are categorized under the rubric *ju* – a Korean name written with the Chinese character, pronounced *jiu* in Mandarin Chinese – which means an alcoholic beverage. Such varieties include *chiju* 旨酒, *mion* 美醞, and *yorye* 醪醴.[9] Chong argues that *chiju* and *mion* refer to high-quality drinks, while *yorye* signals a kind of turbid rice wine.[10] While the insufficiency of sources has made it difficult to trace how winemaking developed on the Korean peninsula over time, we can nonetheless assume that the people who lived there developed these varieties by improving flavors as they accumulated experience and know-how. Like many other items, important opportunities that stimulated the urge to improve varieties probably derived, in part, from new contacts with people in other societies, especially Chinese, who had developed many varieties of liquor during ancient times.

Koreans' close contacts with their neighboring societies from the ancient period, ancient and extensive due to its broader connections with Eurasian societies via the Silk Roads, are verified by documents created by contemporaneous Chinese and some Japanese writers. These documents further hint, from more objective perspectives, at the existence of an alcohol culture on the Korean peninsula. Particularly, close political and diplomatic relations between Chinese and Korean royal courts are described in detail in Chinese official histories.[11] However, these historical materials do not explicitly identify such forms of

---

[9] Mentions of *chiju*, *mion*, and *yorye* are found in *Samguk sagi* and *Samguk yusa*. For example, see Kim Pusik (1075–1151), *The Silla Annals of the Samguk Sagi*, trans. Edward J. Shultz and Hugh H. W. Kang (Sŏngnam: The Academy of Korean Studies Press), 49; Iryŏn, *Samguk Yusa: Legends and History of the Three Kingdoms of Ancient Korea*, trans. Tae-Hung Ha and Grafton K. Mintz (Seoul: Yonsei University Press, 1972), 165.

[10] Chong Dae Song, *Chosen no sake*, 23. Chong Dae Song mistakenly inscribed *yorye* 醪醴 as *yorju* 醪酒. As the word *yorye* appears in a passage in *Samguk yusa* as wine offered by a king for a memorial ritual, it is doubtful that it referred to turbid rice wine.

[11] See the sections about the kingdoms in the Korean peninsula described in Chinese official histories, starting with the section on the Eastern Barbarians in *Weishu* 魏書 (The Book of Wei) in *Sanguozhi* 三國志 (The Record of the Three Kingdoms) compiled by Chen Shou 陳壽 (233–297) and the same section in *Hou Han shu* 後漢書 (History of the Later Han) compiled by Fan Ye 范曄 (398–445).

grassroots cross-cultural contact and influence as clothing or food that endured over extended periods of time. We need to imaginatively reconstruct these connections by scrutinizing scattered sources of various literary genres in addition to archaeological reports.[12] For example, much contextual evidence shows that Chinese foodways such as cooking methods, materials, and chopstick culture, which had long-term influence on Korean food culture overall (as well as that of other East Asian societies, such as Japan and Vietnam), disseminated from China to neighboring lands like Korea since ancient times (see Figure 2.1, bottom left).[13] Books reinforced this transmission of culinary concepts as merchants and other people transported them from China to neighboring societies in East Asia. It seems plausible that any pattern in the cross-cultural transfer of foodways would be repeated in the transference of alcohol and its surrounding culture.

From this context, we can guess that, while we lack specific evidence, earlier scholars validly assumed that different kinds of Chinese alcoholic drink also migrated to Korea. For example, the best-preserved Chinese agricultural text of antiquity, *Qimin yaoshu*, written by Jia Sixie 賈思勰, an official in the court of China's Northern Wei dynasty (386–534), probably found its way to the Korean peninsula through interstate contact between the two courts, which over time might have influenced Korean traditional winemaking in addition to culinary techniques.[14] *Qimin yaoshu* contains detailed recipes for making fermented grain-based alcoholic beverages by using different kinds of yeast as fermentation starters.[15] In 668, Silla, one of three major kingdoms governing the Korean peninsula at that time, unified the southern and middle parts of the region by allying with China's Tang dynasty, which helped it to overcome Parhae 渤海, a rival Koguryŏ–Mohe kingdom to the north. After this unification occurred, the court of Great Silla (668–935 CE) increased its scholarly and cultural contacts with the Tang court in China. This had an effect on Silla, because the Tang dynasty promoted a cosmopolitan, open society that welcomed a large community of foreigners, who often formed their own communities, such as the famous Muslim community in Guangzhou, where members followed their own Islamic customs with the court's blessing. Many Koreans also actively participated in the maritime commerce that linked East Asian countries

---

[12] For a detailed overview of the development of Korean writings up to the Koryŏ period, see Lee, *A History of Korean Literature*, 1–147.

[13] A prominent example is the case of the transmission of chopsticks in the fifth century from China to other parts of Asia, including Korea and Japan. See Q. Edward Wang, Chopsticks: A Cultural and Culinary History (Cambridge: Cambridge University Press, 2015).

[14] Yi Sŏngu, *Han'guk sikp'um sahoesa*, 198.   [15] Ibid., 189–194.

like China, where Korean communities called *Silla bang* 新羅坊, which spread in Shandong and Jiangsu provinces, thrived as hubs of cross-cultural contact between the Tang and Silla under the watchful eye of self-governing, court-sanctioned administrative institutions called *Silla so* 新羅 所.[16] Some Silla elites studied the Chinese classics in order to succeed in the Chinese civil service examination. A few succeeded, advancing to become officials in Chinese government: a famous example is Ch'oe Ch'iwŏn (857–tenth century), who earned the coveted *jinshi* 進士 degree by passing China's highest-level civil service exam and afterward won his first appointment in a prefectural office in the country's south.[17] It is highly likely, then, that simply through personal contact with Chinese in China, people from Silla could have transmitted and received ideas about cultures of food and wine. In this way, China's rich food culture, which included alcoholic beverages, could have stimulated and influenced Korean wine-making traditions with new ideas and practices.

Some Chinese sources suggest that the Koreans developed distinctively good liquors that could compete with the best that China had to offer. For example, the section about foods and wines in a miscellany of Chinese and foreign legends and wondrous tales, *Youyang zazu* 酉陽雜俎 (The Miscellaneous Morsels from Youyang), from the ninth century, introduces a special brewing method that may have differed from the Chinese norm, called *Lelang jiufa* 樂浪酒法, a reference to the Han empire's Lelang commandery 樂浪郡, which existed in the northern part of the Korean peninsula from 108 BCE to 313 CE.[18] The *Taiping Yulan* 太平御覽 (The Imperial Reader of the Taiping Era), a massive Chinese reference encyclopedia compiled between 977 and 983 by court officers under the supervision of Li Fang, introduces a mythological episode relating to Koguryŏ, an ancient kingdom (37 BCE–668 CE) in the north of the peninsula, citing an earlier source. According to the story, the god of the Eastern Sea, angered when a lady from Koguryŏ who had come to Dantu 丹徒, a region in today's Zhejiang and Jiangsu provinces, rebuffed his efforts to meet her, overturned a liquor jug, causing its contents to flow into Lake Qu'e (曲阿; today Danyang in Jiangsu province). The wine in question is called *qu'e* 曲 阿 and reportedly was very tasty; Chong Dae Song asserts that its flavor probably results from its special brewing technique, which originated in

---

[16] Especially in what today are China's Shandong and Jiangsu provinces. For more episodes related to alcoholic beverages in ancient Korean sources, see Chong Dae Song, *Chosen no sake*, 23–28.

[17] Lee Ki-baik, *A New History of Korea*, 94.

[18] Chong Dae Song, *Chosen no sake*, 25; Duan Chengshi 段成式 (d. 863), *Youyang zazu* 酉 陽雜俎 (The Miscellaneous Morsels from Youyang) (ninth century), trans. Yoshio Imamura 今村与志雄, vol. 2 (Tokyo: Heibon sha, 1980), 44.

Koguryŏ, though the *qu'e* wine story sounds too mystical to credit and does not explicitly mention a brewing technique.[19] Another piece of evidence suggests that Korean spirits influenced Chinese by impressing them with their tastes. Two later Korean accounts, *Chibong Yusŏl* 芝峰類說 (Topical Discourses of Chibong) (1614) and *Haedong yŏksa* 海東繹史 (Unraveling the History of Korea) (1823), introduce a poem by Tang dynasty poet Li Shangyin 李商隱 (*c.* 813–858) that includes the following lines: "I am afraid that the aroma of a glass of Silla wine will go away with the wind at dawn."[20] This same line appears in Li's poem *Gongzi* 公子 (Young Nobleman).[21] The Tang poet's familiarity with Silla wines seems to have been obtained through exchanges with members of the kingdom's literati elite. In short, Silla enjoyed regular cross-cultural exchanges maintained by close contacts across borders, which probably affected the development of Korea's unique liquor-making methods due to the synthesis of borrowed brewing techniques and their traditional ones. Unfortunately, descriptions of Tang liquor by Silla literati no longer exist, and Tang dynasty literature contains little evidence of distilled liquors, so the presence of locally produced distilled liquors in Korea at that time seems unlikely.[22]

Under the Koryŏ dynasty (918–1392), which replaced Great Silla and lasted for almost four centuries, change came to the cultures of the Korean peninsula in ways that altered the course of wine's evolution there. From this period on, there appears more varied and detailed documentation of Korean liquors regarding the varieties of alcoholic beverages and foods consumed by Koreans at the time.[23] The sources that most effectively testify to the existence and influence of alcohol varieties in existence during the second half of the Koryŏ era are the writings of literati who became sharp, critical observers of their society. There was even a special genre of literature, called *kajŏnch'e sosŏl* 假傳體小說, "anthropomorphizing short stories," that personified wines in order to satirize the times and offer life lessons.[24] The

---

[19] Yi Sŏngu, *Han'guk sikp'um sahoesa*, 197. Li Fang, *Taiping yulan* (Imperial Reader of the Taiping Era), vol. 12 (Shanghai: Shanghai shudian, 1985), juan 46.

[20] Yi Sŏngu, *Han'guk sikp'um sahoesa*, 197–198.

[21] Wang Qixing, ed., *Jiaobian quan Tangshi* (Complete Tang Poems, with Annotations), vol. 2 (Wuhan: Hubei renmin chubanshe, 2001), 2818; Li Shangyin (*c.* 813–858) was a politician and famous Chinese poet who, along with Du Mu 杜牧 (803–852), represents late Tang literature.

[22] A thing to note here is that the kind of Silla wine described by Li Shangyin did not have a high alcohol content, so there is no chance that it was a distilled form. While a few documents about alcoholic drinks that could have been distilled are quite rare, it therefore seems that the people of Silla did not possess strong alcoholic spirits, to say nothing of distilled forms.

[23] See Yi Sŏngu, *Han'guk sikp'um sahoesa*, 210–226.

[24] See "Kajŏnch'e sosŏl" 假傳體小說 (Anthropomorphizing Novel), in *Han'guk minjok munhwa tae paekkwasajŏn* (Encyclopedia of Korean cultures), http://encykorea.aks.ac.kr/Contents/Index?contents_id=E0000328 (accessed December 2, 2018).

earliest form of such a genre, *Kuk Sun chŏn* 麴醇傳 (Story of Kuk Sun) by Im Ch'un 林椿 (fl. late twelfth century), presents both their positive effect of invigorating people's energy, and their negative aspects, such as overdrinking, which renders people corrupt and shamed. It ultimately criticizes the kings and nobles who fell into drinking and did not work properly for their state and its people. *Kuk Sun chŏn* then influenced Yi Kyubo 李奎報 (1168–1241), the most famous literary man of the Koryŏ period, to write his *Kuk Sŏnsaeng chŏn* 麴先生傳 (Story of Mr. Kuk).[25] Although he did not experience a life as miserable as that of Im Ch'un, Yi Kyubo writes of difficult conditions suffered in his society, and he approached this critically in his writings. All the characters in these two Mr. Kuk stories, in addition to story components, represent liquors of different varieties, even the yeast used for grain-based fermentation. Indeed, most of the references to liquors in literature, like these anthropomorphizing novels, involve fermented grains. Therefore we can assume that grains provided the primary ingredients for the fermented liquors commonly consumed.[26]

While mid-Koryŏ sources attest to the strong preference for consuming grain-fermented wines among people of that time, other literary sources reveal the variety of liquor types that existed during the mid- to late Koryŏ period. For example, Yi Kyubo, a civil official and famous poet of the thirteenth century who particularly enjoyed drinking liquor, not only authored the above-mentioned *Kuk Sun chŏn* (Story of Kuk Sun) but also wrote about different kinds of liquor in his poetry.[27] These varieties include *ihwaju* 梨花酒 (pear-blossom liquor), *chaju* 煮酒 (boiled or cooked liquor), *hwaju* 花酒 (flower liquor), *ch'ohwaju* 椒花酒 (Sichuan pepper liquor), *pip'aju* 波把酒 (liquor flavored with pipa fruit), *paekchu* 白酒 (white liquor), *pangmunju* 方文酒 (liquor brewed according to a special recipe), *ch'unju* 春酒 (spring liquor), *ch'ŏnilchu* 千日酒 (thousand-day liquor), *ch'ŏn'gŭmju* 千金酒 (*ch'ŏn'gŭm* tree brew liquor), and *nokp'aju* 綠波酒 (clear-blue wave liquor). There are many other names of liquors mentioned in other literary works as well. Altogether, Koryŏ society consumed more than twenty-five different kinds of liquor. Later sources show that they continued to enjoy most of these varieties until as late as the Chosŏn period (1392–1897).[28] Here, note that many sources hint at various kinds of alcoholic drink

---

[25] Yi Sŏngu, *Han'guk sikp'um sahoesa*, 224–225.
[26] Chang Chihyŏn, *Hanguk oeraeju yuipsa yŏngu*, 16.
[27] Yi Sŏngu, *Han'guk sikp'um sahoesa*, 220.
[28] We learn that the Korean alcoholic-drink system was established at that time. Pae Kyŏnghwa (Bae Kyung-Hwa), "Andong sojuŭi chŏllae kwajŏnge kwanhan munhŏnjŏk koch'al" (Literature Review for Transmission of the Andong Soju), MA thesis, Andong National University, 1999, 6.

besides grain-fermented liquors, such as those incorporating flowers and tree bark.[29] In any case, the basic ingredients of alcohol like rice grew only more precious as a commodity once popularity increased demand beyond staple food consumption. Much of this demand at this time came from the nobility, whose privilege it was to drink such varieties of alcoholic beverage. Im Ch'un and Yi Kyubo, for example, clearly reflect this elite trend, as they used the ingredients of these precious grain-fermented wines to compose their popular satirical works. Nonetheless, common people consumed alcoholic beverages too, albeit of a different variety. This can be verified in a source that describes the basic features of alcoholic beverages consumed in Koryŏ, an account written by a Chinese named Xu Jing 徐兢 (1091–1153) who visited Koryŏ as part of a diplomatic mission and wrote about his observations there.

Xu Jing's *Xuanhe fengshi Gaoli tujing* 宣和奉使高麗圖經 (Illustrated Account of an Official Mission to Koryŏ during the Xuanhe Reign [-1119–1125]), the earliest surviving source of information about Koryŏ liquors, drinking habits, and cooking ways, pre-dates the work of Im Ch'un and Yi Kyubo. Xu was a Chinese envoy whom the Song emperor Huizong (r. 1100–1126) once dispatched to Koryŏ on a special diplomatic mission to assess the situation there and request the kingdom's military support for the Song war against the Jurchens threatening northern China. His exceptional account, full of details about the political system and culture of Koryŏ society from the perspective of an outsider, includes a brief discussion of the alcoholic drinks he saw people drink in Korea.[30] Xu reports that, because the people of Koryŏ loved their liquor, they drank it regularly, some even going so far as to attend several drinking parties in a night. It seems that, unlike Chinese of the earlier Tang dynasty who praised Silla's distinctive liquor, Xu disliked the Koryŏ wine he tasted, an opinion he emphasized several times in his writings. Because they did not have sticky rice, he wrote, Koreans used standard non-glutinous rice and its malt to brew alcoholic beverages instead. In his opinion, this explains why Korean liquors were not as good as contemporaneous Chinese liquors made with sticky rice. For our purposes, Xu

---

[29] These sources include literary collections discussed above and other sources, such as episodes contained in *Koryŏsa* (History of Koryŏ, completed 1454).

[30] The account's full title is *Xuanhe fengshi Gaoli tujing* 宣和奉使高麗圖經 (Illustrated Account of an Official Mission to Koryŏ during the Xuanhe Reign (1119–1125)). For more information about the work, see Xu Jing 徐兢 (1091–1153), *A Chinese Traveler in Medieval Korea: Xu Jing's Illustrated Account of the Xuanhe Embassy to Koryŏ*, translated, annotated, and with an introduction by Sem Vermeersch (Honolulu: University of Hawai'i Press, 2016), 1–55.

leaves a helpful clue to the reliance on fermentation using a form of malted rice to create brews. Scholars assume from such passages that the yeast they would have used for fermentation was a traditional Korean fermentation starter known as *nuruk*, whose documentation dates from the Chosŏn period (late fourteenth to nineteenth centuries); however, the original method probably traces back to China during its Warring States period (471–220 BCE) and reached Korea centuries later, perhaps as early as the third century CE. The fact that they used a grain-based fermentation starter shows that common Korean alcoholic drinks were probably all based on grains.[31]

Another important characteristic of Koryŏ alcoholic beverages, Xu explains, is highlighted in the varieties consumed by kings and high nobles. These included a clear, strained wine called *ch'ŏngju* 清酒 and another called *pŏpchu* 法酒 – literally "recipe" liquor, a kind of ch'ŏngju. Both wines were brewed by the Yangonsŏ 良醞署 (Yangon Bureau, i.e., Bureau of Alcohol), a government office that provided alcoholic drinks to the court of Koryŏ. In the Koryŏ period when Buddhism was the state religion, Buddhist temples associated with privileged classes such as royal and noble families also produced alcoholic beverages.[32] By contrast, ordinary people could not obtain these kinds of fine liquor and instead drank cruder forms of spirit that were thickly colored and tasted turbid.[33] Xu adds that by his time, Korean alcoholic drinks were already divided into two classes: those consumed by nobles and those consumed by ordinary people. This big division of liquor types during the Koryŏ period attested by Xu Jing also finds expression in the *Koryŏsa* 高麗史 (History of Koryŏ) (1454), the principal history of the Koryŏ dynasty. Composed during the reign of King Sejong (r. 1418–1450) of the Chosŏn dynasty that succeeded Koryŏ, the text reports that the court established the Yangon Bureau during the reign of King Munjong 文宗 (r. 1046–1083) with the intention of brewing high-quality alcoholic beverages – namely the elite drinks ch'ŏngju and pŏpchu – for use in state ceremonies.[34] We learn from Chinese and Korean sources that ch'ŏngju was made by condensing fermented yet unstrained base liquor. Pŏpchu, a kind of clear strained wine, was brewed using a rich base composed of certain proportions of raw ingredients.[35] It was also called *ŏju* 御酒 ("royal

[31] About *nuruk* and grain fermentation, see Yi Sŏngu, *Han'guk sikp'um sahoesa*, 179–180; Pae Kyŏnghwa, "Andong sojuŭi chŏllae kwajŏnge kwanhan munhŏnjŏk koch'al," 7–17.

[32] Chong Dae Song, *Chosen no sake*, 32–33.

[33] Xu Jing, *A Chinese Traveler in Medieval Korea*, 199. Yi Sŏngu, *Han'guk sikp'um sahoesa*, 210–211.

[34] Yi Sŏngu, *Han'guk sikp'um sahoesa*, 211. *Koryŏsa* (History of Koryŏ, 1454), vol. 85, section II, on criminal law.

[35] Chong Dae Song, *Chosen no sake*, 29–30.

liquor") or *kwanju* 官酒 ("official liquor"); the court used them, for example, whenever they bestowed gifts upon their officials. Brewers used special compression-brewing methods to make high-grade liquors for use by the nobility in ancestral memorial ceremonies at the Royal Ancestors' Shrine. The turbid unstrained liquor called *t'akchu* 濁酒 and identified by Xu Jing as a wine for ordinary people also appears in other contemporaneous sources of the Koryŏ period. For example, scholarly poems mention white liquor, such as t'akchu and paekchu (white liquor), as well as a *pakchu* 薄酒 (light wine), which farmers consumed in their fields and travelers drank on the road.[36] All of these had a weaker taste (lower alcohol content) and were darker in color. They were easy to make, too. The Chinese envoy that recorded his observation of Korean culture during his brief visit there aptly described the essential characteristics of the major liquor types commonly consumed by different classes in Korea in the twelfth century. However, sources of the mid-Koryŏ period have yet to provide any clue to the existence of distilled liquors at this time.

The import and transfer of foreign alcoholic drinks offers an important topic through which to examine Koryŏ sources for evidence of distilled liquors in Koryŏ society. To this end, Chang Chi-Hyun (Chang Chihyŏn, 1928–) conducted the most complete analysis of available documentary sources. He argues that Koreans of the time imported alcohol varieties, directly and otherwise, whether as diplomatic gifts or commercial goods, from places outside Korea, including the lands of the Khitans, Jurchens, and Mongols of northern Asia and numerous countries in South and Southeast Asia.[37] Based on his analyses of passages in collected discourses on diplomatic relations recorded in the *Koryŏsa* and other period documents, Chang suggested possible dates for the importation of certain liquors from China to Koryŏ. The list below describes the names of liquors of foreign origin and their possible dates of transfer to Koryŏ based on Chang's analysis:[38]

• *haenginja pŏpchu* 杏仁煮法酒, literally "stewed apricot wine," a liquor made from apricot kernels boiled according to method, recorded in the 6th month in the 32nd year of the reign of King Munjong 文宗 (r. 1046–1083);
• *yangju* 羊酒, "mutton-steeped wine," recorded during or before the reign of King Yejong 睿宗 (r. 1105–1122);
• *kyehyang ŏju* 桂香御酒, "cinnamon-flavored court liquor," recorded in the 12th month in the 12th year of the reign of King Yejong 睿宗;

---

[36] Yi Sŏngu, *Han'guk sikp'um sahoesa*, 212; Chong Dae Song, *Chosen no sake*, 30–32.
[37] Chang Chihyŏn, *Hanguk oeraeju yuipsa yŏngu*, 10–36.   [38] Ibid., 36.

- *hwaju* 花酒, "flower liquor," recorded during the reign of King Sukchong 肅宗 (r. 1095–1105);
- *mayuju* 馬乳酒, "mare's milk liquor," or *tongnak* 湩酪, "milk liquor," aka kumiss, recorded in the 12th month during the 18th year of the reign of King Kojong 高宗 (r. 1213–1259);
- *p'odoju* 葡萄酒, "grape wine," recorded in the 2nd month in the 28th year of the reign of King Ch'ungnyŏl 忠烈 (r. 1274–1308);
- *sangjonju* 上尊酒, "supreme liquor," recorded in the 12th month during the first year of the reign of King Ch'ungsŏn 忠宣 (r. 1213–1259);
- paekchu 白酒, "white liquor," recorded in August in the 28th year of the reign of King Ch'ungnyŏl;
- *chungsanju* 中山酒, "Zhongshan liquor," noted during the mid-Koryŏ period;
- *chŭngnyuju* 蒸溜酒, "distilled liquor," implied during the late Koryŏ period.

Among the distilled liquors listed above, Koreans consumed liquors based on ingredients other than grains, such as *hwaju* (flower liquor), *p'odoju* (grape wine), *yangju* (sheep's milk liquor), and mayuju (mare's milk liquor). Evidence for the consumption of sheep's milk liquor appears in a reference to King Yejong 睿宗 (r. 1105–1122), who presented the drink to a Koryŏ general in honor of his achievement in defeating the Jurchens in 1107. The *Koryŏsa* does not explicitly discuss its origins; however, Chang Chi-Hyun argues that the Koreans probably received it from the Khitans or Jurchens through state-sponsored trade before this particular gift-giving event, because sheep's liquor was typically consumed among nomads and typically was not produced in agricultural societies. Sources also document mare's milk wine and grape wine given to the Koryŏ court by the Mongols. According to Chang, *tonglao* 湩酪 (milk liquor) was a nickname applied to a liquor based on mare's milk and commonly drunk by northern nomads, which was also called *ma tonglao* 馬湩酪 or "mare's milk liquor."[39] It appears in the *Koryŏsa chŏryo* 高麗史 節要 (Essentials of Koryŏ History) as part of a 1231 offering given by a Mongol general to Koryŏ's King Kojong 高宗 (r. 1213–1259) following the completion of diplomatic negotiations between their respective governments.[40] At that time, the Mongols had invaded and were devastating virtually all of the Korean peninsula while the Koryŏ government resisted them from the small island of Kanghwa 江華 off the peninsula's west coast. The Yuan emperor also gave grape wine to Koryŏ's King

[39] Ibid., 23–24.
[40] Kim Chongsŏ 金宗瑞 (1383–1453), ed., *Koryŏsa chŏryo* 高麗史節要 (Essentials of Koryŏ History) (Seoul: Asea Munhwasa, 1972), *kwŏn* 16.

Ch'ungnyŏl (r. 1274–1308) in 1302 and 1308.[41] As the sources show, most of these foreign liquors were used exclusively by kings and nobles as special royal gifts and not shared with commoners.

Among the foreign alcoholic drinks identified in Koryŏ documents, one type spread through the ranks of the ordinary people rapidly: distilled liquors including soju and arak, which the Koreans called *aralgil* 阿剌吉 and the Chinese alaji, clearly reflecting the Turkic form pronounced arajhi and Mongolian form arkhi. While the documentary sources clearly indicate the transfer routes of many foreign wines used as diplomatic gifts, as seen in the example of *tonglao* recorded by the Mongol nomads, they provide no official record of *distilled* liquors among them.[42] Several sources of late Koryŏ provenance hint, however, that by this time such liquor had become quite popular locally. Later sources demonstrate that it had become one of the most important spirits in Korean society by the time the Chosŏn overthrew the Koryŏ in 1392. Given that the liquor cannot be found in Yi Kyubo's mid-Koryŏ writings, it most likely was not popular until the mid-thirteenth century. It is only in the fourteenth century, when Koryŏ enjoyed unprecedentedly close relations with the Mongol-run Yuan dynasty governing China, that documentation of distilled liquors in Korea arises.

Popularizing Soju and Arak

Three passages from Koryŏ-era sources suggest that the popularization of distilled alcohol, specifically soju and arak, occurred during the dynasty's later years. Two of them, both identified in the *Koryŏsa*, explicitly mention soju. The first of these passages, in a biography of Ch'oe Yŏng 崔瑩 (1316–1388), introduces a general under Ch'oe's command named Kim Chin 金縝 (fl. 1360s), who loved soju excessively, failed to do his duty, and suffered punishment:[43]

Before this event, when Kim Chin was the head of Kyŏngsang province, he drank wines and played day and night along with officers under his command calling on many famous *kisaeng* [female entertainers]. Because Kim Chin enjoyed drinking soju, people in the army called him and his men the "soju group." And because he assaulted and insulted his soldiers and assistants if they displeased him, they all possessed resentments and grudges against him. When Japanese enemies burned

[41] *Koryŏsa chŏryo*, *kwŏn* 22; Chang Chihyŏn, *Hanguk oeraeju yuipsa yŏngu*, 23–27.
[42] While the means by which foreign wines were transferred can be traced quite clearly through documentary sources (as seen in the example of mayuju discussed above), relevant sources on distilled liquors provide inconsistent information about their origins, resulting in several divergent theories.
[43] Chang Chihyŏn, *Hanguk oeraeju yuipsa yŏngu*, 42–43.

and looted the barracks in Happ'o 合浦, soldiers said: "Have the soju group defeat the enemy. How can we fight?" They then retreated and made no effort to go and fight. Yet Kim Chin fled alone on horseback, and the army was defeated in the end. Then he [Ch'oe Yŏng] demoted Kim Chin to a commoner, condemned him to exile in Ch'angnyŏng 昌寧 county, and then moved him to the island of Kadŏk 嘉德.

Once he finished with Kim, Ch'oe "then executed Yi Tongbu 李東樆 and Kim Wŏngok 金元穀 of the Mongol regiment in Happ'o."[44]

Ch'oe Yŏng was one of the most important generals in late Koryŏ history. He supported the last kings of Koryŏ against the Yuan, even when facing the rise of a new power on the Korean peninsula that would inevitably vanquish him and establish a new dynasty, the Chosŏn.[45] Ch'oe had long been familiar with officers like Kim, if not with Kim himself. Perhaps he knew about an incident that had occurred during the reign of King U (r. 1374–1388), one of the last kings of Koryŏ. Clearly, the compilers of the *Koryŏsa* included Kim's story in Ch'oe Yŏng's biography in order to demonstrate that the general was upset with King U for not executing bad officers like Kim, who violated the military code of conduct in critical situations. (In the end, the general seized the initiative and punished them appropriately.)[46] The fact that a group of people who enjoyed soju to excess were called the "soju group" does not reveal whether the drink was strong, like distilled liquor; however, it does suggest that soju was known as a special kind of alcoholic beverage, different from other types of alcoholic drink, and that Koryŏ soldiers had already begun to consume it, often to excess, leading to improper behavior due to intoxication.

Another piece of evidence in the section of the *Koryŏsa* on prohibition during the reign of the same King U shows that soju was broadly popular in many sectors of society – on occasion to an extreme, it seems. An edict issued by the Koryŏ court in 1375 prohibits the purchase of soju and other luxury goods such as silks, gold, and jade wares in order to suppress their consumption. This piece of evidence strongly suggests that, because so many people squandered their fortunes on soju as a new kind of expensive indulgence, the court had to ban it and other luxury goods. The paragraphs above demonstrate that soju does not appear in any earlier piece of

---

[44] *Koryŏsa* (1451), *kwŏn* 113, biography, 26.
[45] Ch'oe Yŏng attempted to maintain the Koryŏ dynasty; however, a pro-Ming faction eventually executed him. After the last king of Koryŏ was forced to abdicate, the faction declared the new dynasty, Chosŏn. Lee Ki-baik, *A New History of Korea*, 162–165.
[46] For more on the last years of the Koryŏ dynasty, including the reign of King U with all its political corruptions and diplomatic tensions, see Lee Ki-baik, *A New History of Korea*, 160–165.

Koryŏ-period literature written by alcohol aficionados; however, this evidence from the late Koryŏ period suddenly verifies that the liquor had become popular among nobles by then; most likely, rich merchants and ordinary people enjoyed it, too, as far as they could afford it, as with silks and other luxury goods.[47]

Whether the soju mentioned in these two passages of the *Koryŏsa* was in fact distilled is not clear, but one can only assume that it was so, because it was new and distinctive; because its name shared the same Chinese characters as shaojiu, its Chinese contemporary; and because it was a luxury rather than an ordinary alcohol. One need only consider the contexts of passages in which authors regarded soju as a powerful specialty alcohol distinct from older, traditional Korean liquors. No other Koryŏ source stipulates the characteristics of soju at that time. However, another piece of evidence from the same historical period verifies the existence of distillates at the time: a poem entitled "Sŏrin Cho p'ansa" 西隣趙判事 (Minister Cho in the Western Neighborhood), in which Minister Cho, who lives in the western neighborhood, introduces "alaji" 阿剌吉 – read *aralgil* in Korean transliteration. The poet, Yi Saek 李穡 (1328–1396, also known by his pen name Mogŭn 牧隱), was an important Neo-Confucian scholar and a tutor to many major governmental officials in the courts of the late Koryŏ and early Chosŏn dynasties, most notably among them Chŏng Tojŏn 鄭道傳 (1342– 1398), who established the new dynasty's ideological, institutional, and legal frameworks. The poem's title explicitly declares that a certain Minister Cho in the western neighborhood (Sŏrin) introduced alaji. Joo Young-ha argues that Minister Cho is none other than Cho Unhŭl (趙云仡, 1332–1404), and that the name Sŏrin refers to a district in T'aep'yŏnggwan, a lodging quarter for envoys located in Kaegyŏng, Koryŏ's capital.[48] As Minister Cho had probably received the liquor from an envoy or merchant from the Yuan, the original term used by Yi Saek should be alaji, not *aralgil* as in its Korean form of pronunciation, even though the author was Korean. As we have seen in Chapter 1, alaji was used as a transliteration of *arak*, the name of a form of distilled liquor that began to rise in popularity during the Mongol Yuan period starting with *Yinshan zhengyao*. As Chapter 3 shows, Yi Saek traveled to and from Daidu (Chinese Dadu, nowadays Beijing) the Yuan dynasty

[47] *Koryŏsa*, *kwŏn* 85, criminal law, 39; Chang Chihyŏn, *Hanguk oeraeju yuipsa yŏngu*, 43–44. If soju had been a special product consumed only occasionally by kings and uppermost-level nobility, the government most likely would not have banned it.

[48] Chu Yŏngha, "Chŏng Dojŏn sŭsŭng Yi Saek ŭi soju sarang" (Chŏng Tojŏn's Teacher Yi Saek's Love for Soju), *Dong-A Ilbo* (East Asia Daily), June 22, 2015, http://news .donga.com/View?gid=72032136&date=20150622 (accessed May 2, 2017).

capital, and sojourned there as both a student and a diplomat, so it does not seem strange that he encountered a drink of foreign origin, nor that he knew it by its original name. Yi's passage on arak provides strong evidence that transfers of distilled liquors from China to Korea were taking place through scholarly contacts during the late Koryŏ and early Chosŏn eras.

In his poem, Yi Saek describes the liquor – which Chinese called *alaji* and Koreans *aralgil* – as "forming like autumn dewdrops, and dripping down at night ... after drinking half a cup of the liquor, a warm feeling spreads to the bone."[49] Apparently, the people who made the alaji liquor described in this poem did so by means of a distilling process, extracting a strong spirit from the fermented alcohol they used as its raw material. The fact that later Chosŏn-period sources refer to soju as *noju* (露酒, Chinese *lujiu*) – "dewdrop liquor" – hints at a relationship between the distilled alaji liquor described by Yi Saek. Most likely, the popular understanding among people of Koryŏ of distilled liquor as noju reflected in Yi's poem would independently influence people of the Chosŏn period because we cannot find the word *lujiu* as a nickname for the Chinese word *shaojiu*. These earliest pieces of documentation offer no new details about these liquors, such as their ingredients. Nor do Koryŏ-period sources offer new information that helps to explain how certain distilled-alcohol varieties began to increase in popularity throughout Korea during Koryŏ times.

In short, these three contemporaneous records suggest that, most likely, soju and arak suddenly appeared on the Korean peninsula during the era of Mongol domination in the thirteenth and fourteenth centuries. Considering the fact that soju and arak do not appear in any earlier record, it is evident that they transferred into the region from outside. The apparent time of this transfer was when Korea experienced the influence of a dynasty in China that was strongly connected to the larger Eurasian world as a part of the greater Mongol Empire (see Figure 2.1, bottom right). Understanding the conditions that exposed Korea to foreign relations in unprecedented ways usefully contextualizes the rapid rise of soju in Korea at this time. The fact that soju and arak in Mongol-era Korea were almost identical to shaojiu and arak in Mongol-run China, in terms both of the Chinese characters that compose their written names and also their characteristics as alcoholic beverages, strongly suggests that it developed from an original transference to

---

[49] Chang Chihyŏn, *Hanguk oeraeju yuipsa yŏngu*, 53; Yi Saek (1328–1396), *Kugyŏk Mokŭnjip* (Literary Collection of Mokŭn), trans. Im Chŏnggi (Seoul: Minjok munhwa ch'ujinhoe, 2008), vol. 9.

Korea from China. This is easy to imagine, given the close diplomatic relations that existed between Yuan China and Koryŏ.[50] The best way to corroborate this finding would be to retrace the transfer routes of distilled liquors from China to Korea during the Mongol period. This would be best achieved by examining the broader context of the unprecedented relations existing between the two countries and how that facilitated the alleged transfer of soju, which we do in the following chapter.

Before doing that, we must review the previous scholarship that addresses the origin of distilled liquors in Korea. Earlier works have utilized Korean sources from both the Koryŏ and succeeding Chosŏn dynasties, in addition to contemporaneous Chinese documents, to good effect.[51] For example, some scholars have argued explicitly that soju/shaojiu originated in Yuan-era China and then transferred to Korea during the latter Koryŏ era by utilizing several sources written during the subsequent Chosŏn era. This claim corroborates earlier arguments about the sudden and simultaneous rise of soju throughout East Asia. However, other scholars have used different sets of sources to refute the Yuan-period transfer theory and instead argue the possibility that distilled liquors appeared before the Yuan era began in China and subsequently transferred to Korea before the Yuan began. These discussions often demonstrate deep and early insights into Korea's interconnections with the wider world thanks to their careful analyses of the major Korean and Chinese sources available to them at the time. However, they often face limited understanding due to the shortcomings of the sources available to them. In particular, they largely lack recognition of the issue of distillation technology, including the development of stills, which were advanced in Chapter 1. It is worth investigating these debates because they have influenced our recent understanding of the origins and characteristics of soju. After reviewing the pros and cons of different theories raised in the course of these debates, we then aim to revise established origin theories based on new understandings of the entire historical context, which leads to new insights into the history of distillation and China–Korea relations under Mongol domination. Due to space limitations, let us first focus on recent major works that review earlier scholarship that has directly influenced recent debates on the origin of soju.

---

[50] For the classic works on Koryŏ–Mongol relations by Korean scholars, see Chang Tong'ik (Jang Dong-Ik), *Koryŏ hugi oegyosa yŏn'gu* (A Study of Diplomatic History in the Late Koryŏ Period) (Seoul: Ilchogak, 1994), and Ko Pyŏngik (Koh Byong-ik), *Tonga kyosŏpsa ŭi yŏn'gu* (Studies on East Asian International Relations) (Seoul: Sŏul taehakkyo ch'ulp'anbu (Seoul National University Press), 1970).

[51] These sources include different kinds of literature written by Korean scholars from the Chosŏn period, including informal essays, poems and writings for practical use like medical works.

Theories of the Origins and Transfer Routes of Distilled Liquors in Korea

In Korea, interest in the geography and history of soju's origins dates to before the rise of modern scholarship in the late twentieth century. Indeed, this interest apparently manifested first among Japanese, for no other reason perhaps than the fact of their colonial rule and the advantages it gave them over their Korean counterparts. In 1935, a group of Japanese scholars published a collection of analyses and discussions of soju's history in *Chōsen shuzōshi* 朝鮮酒造史, a "history of the production of alcoholic drinks in the Chosŏn period," as the title suggests.[52] The book provides unprecedentedly accurate and quantitatively driven information, so decades later, Pae Sangmyŏn, who had already advanced his field by establishing a liquor-research center in Korea, translated it into Korean in order to advance study of the country's history of alcoholic beverages.[53] Under its Korean title, *Chosŏn chujosa*, the book has influenced the study of traditional Korean liquors like soju by clarifying the spirit's origin. Despite the narrow time frame of their history, 1907–1935, the Japanese authors of *Chōsen shuzōshi* address soju's early history, arguing that it originated during the reign of Koryŏ's King Ch'ungnyŏl (r. 1274–1298). Their claim lacks proper documentation; nonetheless, many later studies, including encyclopedia articles, rely on this dating, a sign of the book's influence. However speculative, the claim helped to propel the field forward.

Research devoted specifically to the history of soju began to develop during the latter half of the twentieth century, as the academic foundations of Western-style scientific inquiry in Korea stabilized. Pioneers of this work include Chang Chi-Hyun (1928–) and Yi Seong-wu (1928–1992), specialists in the subjects of agricultural and food history, whose books and articles chronicle Korean dietary life and foodways from a cultural-historical perspective. Chang Chi-Hyun traced the origins and characters of soju/shaojiu and arak more rigorously, thanks to a thorough and systematic examination of documentary sources, in his 1979 monograph about the influx of foreign alcoholic liquors to Korea.[54] Though limited to Korean and Chinese documentary sources, Chang aptly argued that the influx of soju and arak opened a new chapter in the history of alcoholic drinks in Korea. Because his findings and argument

---

[52] Chōsen shuzō kyōkai (Committee of Alcoholic-Beverage Making in Korea), *Chōsen shuzōshi* (History of the Production of Alcoholic Drinks in the Chosŏn Period) (Keijō (Seoul): Chōsen shuzō kyōkai, 1935).

[53] Pae Sangmyŏn, trans., *Chosŏn chujosa: 1907–1935* (History of the Production of Alcoholic Drinks in the Chosŏn Period: 1907–1935) (Seoul: Kyujanggak, 1997).

[54] Chang Chihyŏn, *Hanguk oeraeju yuipsa yŏngu*, 38–100.

have influenced the historiography of Korean soju for several decades, including this book, we should review them carefully.

In his 1979 study, Chang Chi-Hyun summarizes the existing theories about the origins and transfer routes of distilled liquors raised by earlier generations of scholars. They are: (1) the Mongol Yuan-period origin theory; (2) the anti-Mongol Yuan-period origin theory (aka the pre-Mongol Yuan-period origin theory) proposed by Yi Kyugyŏng (1788–1856), a scholar of the late Chosŏn period; and (3) the theory of direct transfer from Persia (through Muslim merchants) suggested by Choe Nam-seon (1890–1957). Chang rightly points out that these Chosŏn scholars paid more attention to China than to Korea. Nonetheless, like the Japanese scholars, these early Korean theorists helped to build a sound foundation for a scholarship of Korea's distilled liquors, which Chang begins to build upon in his critique of their theories to provide his own insights into such questions as whether arak or soju first traveled to Korea.[55]

The theory of Mongol origins explained by Chang Chi-Hyun simply states this: that Yuan and Koryŏ documents demonstrate that shaojiu, a distilled liquor, was first created in Yuan China in 1277, nearly a century earlier than its first appearance in documentation in Koryŏ Korea, in about 1375. This narrows the window within which a transfer from China to Korea could have taken place – as early as 1314, the year when shaojiu consumption finally arose in Chinese society, and as late as 1367, the year before the Mongol Yuan dynasty collapsed in China and their influence in Korea waned.[56] Chang sometimes relies on a dictionary of world history – a secondary, not primary, source – to make his case for soju's source in Yuan China, which weakens the credibility of his argument.[57] However, as we have seen in Chapter 1, documentary and archaeological sources of the Yuan period seem to vindicate Chang's suggested time frame. Such evidence confirms, for example, the existence in China of a large-scale brewing industry under Mongol princes like Ananda and of large-scale distilleries. Moreover, the Yuan and Koryŏ courts enjoyed a close relationship at this time, so it makes sense that shaojiu found its way to Koryŏ rather quickly after its development in China, in the form of diplomatic gifts. Several Chosŏn-period sources found in Chang's own book also explicitly state that shaojiu appeared in China during the Yuan period.[58] This includes works on medicine. As he correctly points out, medical works written in Korea were surely

---

[55] Ibid., 82.   [56] Ibid., 46–48.   [57] Ibid., 48.
[58] Starting with a study by Hŏ Chun 許浚 (1539–1615), *Tongŭi pogam* 東醫寶鑑, which literally means "a precious mirror on the medicines of eastern countries"; Yi Sugwang's 李睟光 (1563–1628) *Chibong yusŏl* 芝峰類説 (Topical Discourses by Chibong) (1614);

influenced by their Chinese counterparts. When discussing the origin of Chinese shaojiu, Korean sources contain wording that resembles the prose in *Bencao gangmu*, written during the Ming period (1368–1644), even though copies of this classic were unavailable in Korea until the mid-Chosŏn period beginning in the seventeenth century.[59]

There is one issue, however, that we should approach critically in Chang Chi-Hyun's argument. In comparing shaojiu and arak, Chang claims that arak made from fermented milk had been consumed by Mongols before the Yuan dynasty began, while shaojiu was created by adopting arak's distillation technique in Yuan-era China. It suggests, therefore, that shaojiu transferred to Koryŏ at this time, while arak probably transferred to Korea through Koryŏ's relations with the Mongols before the latter established the Yuan dynasty in China. Chang then explores why the Chinese created shaojiu while they already had arak and concludes that, because the Chinese traditionally used grain as brewing material, they created shaojiu by distilling local materials; that is, fermented grains.[60]

The problem in Chang's argument is that, as we have seen in Chapter 1, shaojiu developed using traditional Chinese distillation technology and had been in China since the Song period, albeit on a smaller scale than during the Yuan period. This led to the conclusion, based on a thorough and comparative analysis of sources available, that shaojiu also had been considerably influenced by a new form of distilled liquor, arak, and that its distillation technology transferred from foreign countries through maritime connections. According to *Jujia biyong*, a Chinese household encyclopedia that was also introduced to Korea later in the eighteenth century along with other similar practical books, arak received the nickname "shaojiu of the Southern Barbarians," which brewers made by distilling many kinds of fermented wine.[61] Li Shizhen's *Bencao gangmu* identifies shaojiu and arak as the same kind of liquor, and cites *Yinshan zhengyao*, which documented arak for the first time, as its source.[62] This suggests that arak was made using different kinds of materials and grew similar to shaojiu after brewers began to use more readily available fermented grains.

In short, we can acknowledge the different origins of these two liquors, shaojiu and arak; however, we should pay attention to their mixing process in producing the two liquors and its rapid and widespread influence on the Mongols who fostered its growth in popularity further during

---

and Sŏ Yugu's 徐有榘 (1764–1845) *Imwŏn sibyuk chi* 林園十六志 (Sixteen Treatises of Imwŏn).
[59] Chang Chihyŏn, *Hanguk oeraeju yuipsa yŏngu*, 61.   [60] Ibid., 62   [61] Ibid., 62.
[62] See the discussion in Chapter 1 of this book.

the Yuan period. As we have no evidence for the Mongols' invention of distillation before the Chinese, it is safer to argue that the Mongols received influence from Chinese and foreign distillation methods to distill their own fermented mare's milk drinks that could preserve for a longer time before they facilitated further production of distilled liquors based on fermented grains once they came to rule all of China. As we have seen in Chapter 1, the liquor industry under Mongol princes like Ananda was most likely influenced by a Southeast or South Asian distillation technique and was labeled arak (Chinese alaji), a fact that serves as a good example of how shaojiu and arak were mixed.[63] Arak from Southeast Asia was probably made from the fermented sap of coconut flowers. The author of *Jujia biyong* identifies it by its distinct "Southern Barbarian" or Southeast Asian distillation method; thus arak was identified more by its particular distillation method than by any ingredient, a pattern that the Mongols later adopted, followed by other regions. As arak became synonymous with distilled liquor worldwide, it integrated with Chinese shaojiu and then subdivided into varieties. By the end of the Ming dynasty, when *Bencao gangmu* was written, spirits made with fermented grains had grown popular in China, and shaojiu and arak would have been perceived as the same liquor with different names. Therefore we can conclude that, when shaojiu was introduced to the Korean peninsula at the end of the Koryŏ dynasty, shaojiu and arak most likely were introduced together.

While Chang is convinced by some of the points made by Yuan-origin theorists, he is in fact more convinced by opponents of this theory, described by Yi Kyugyŏng in his long essay "Dialectical Argument for *Arigŏl* 阿里乞 [Soju] from Southwestern Foreigners," which begins with the following:[64]

Noju is *soju*, and some people call it *hwaju* 火酒 (fire wine). It became known in China from the Yuan period. In general, it was imported from foreigners of the maritime southeast (*Sŏnambŏn* 西南番). It was probably originally from the Tang dynasty period.

Among the primary sources he used, Yi Kyugyŏng cites *Jujia biyong*, which first discussed the Southeast Asian form of shaojiu written as 阿里乞, pronounced *aliqi* in China, the very *arigŏl* discussed by Yi above. As the title of Yi's essay shows, we can assume that he was primarily

[63] Chang Chihyŏn, *Hanguk oeraeju yuipsa yŏngu*, 73.
[64] Yi Kyugyŏng (1788–1856), *Oju yŏnmun changjŏn san'go* (Scattered Manuscripts of Glosses and Comments by Oju), ed. Kojŏn Kanhaenghoe (Seoul: Tongguk Munhwasa, 1959), vol. 1 (*kwŏn* 5), 151; Chang Chihyŏn, *Hanguk oeraeju yuipsa yŏngu*, 51.

influenced by this reference to aliqi in *Jujia biyong*.[65] He attempted to use the greater number of sources available to him in order to enhance his theory. First, he provided another name for soju, *noju* 露酒 (dewdrop liquor), which is only found in Korean sources (and not Chinese). (Note, too, that the Koryŏ-period poem composed by Yi Saek used this very term to describe arak.) Second, Yi cited other Chinese documentary sources to argue that soju/shaojiu could be traced to the Tang dynasty or perhaps even earlier.[66] In fact, Yi's grandfather, Yi Tŏkmu 李德懋 (1741–1793), another scholar of the late Chosŏn dynasty who wrote many books about practical matters, refuted the Yuan dynasty theory of shaojiu's origins supported by medical books, citing a Song dynasty account by Tian Xi 田錫 that discusses *Xianluo jiu* 暹羅酒 – "Thailand liquor," a term of possible Southeast Asian origin – a liquor double-distilled from shaojiu. By analyzing more information from Chinese sources and citing Yi Tŏkmu's work, Yi Kyugyŏng was able to argue that shaojiu already existed during the Song era, and that because shaochun (which some scholars have considered a nickname for shaojiu during the Tang dynasty) existed during the Tang, it is possible that shaojiu also existed before the Song.[67] Moreover, he even claimed that distillation technology may even have already existed as early as the Eastern Han dynasty (25–220 CE) by arguing that "cooked liquor" (*zhujiu* 煮酒), which appears in the literature of the time, could have referred to a form of shaojiu. Yi very likely expended great effort to collect sources, most of them earlier Chinese works like *Jujia biyong*, which could help to buttress a new argument about soju's origin. In any case, it is hardly strange that Yi's work impressed Chang.[68] It is more surprising that many of Yi's points have won support thanks to the newly found archaeological sources examined in Chapter 1.

While Yi Kyugyŏng offered many convincing arguments regarding the origin of distilled liquor in China, other arguments appear less credible. Yi discusses the spread of distilled liquors via maritime routes to places in East Asia, even places as obscure as Okinawa, by exploiting sources newly

---

[65] Chang Chihyŏn, *Hanguk oeraeju yuipsa yŏngu*, 54.   [66] Ibid., 64–75.

[67] Yi Sŏngu, *Han'guk sikp'um sahoesa*, 214; Yi Tŏkmu 李德懋 (1741–1793), *Ch'ŏngjanggwan chŏnsŏ* 青莊館全書 (Complete Works of Ch'ŏngjanggwan), ed. Kojŏn Kanhaenghoe, vol. 9 (Seoul: Sol, 1997), 12. As Yi Kyugyŏng had many more sources available to him than the *Jujia biyong*, he was able to ferociously argue that shaojiu originated before the Yuan period, even dating back to the Tang period, by citing Chinese works about distillation that developed from the ancient period.

[68] According to Chang, it was a pity that the creative term *noju* (dewdrop liquor) that the Koreans began to use as a synonym for distilled liquor did not gain use as a primary term rather than the imported term *soju* (*shaojiu*, "fire wine"). Chang Chihyŏn, *Hanguk oeraeju yuipsa yŏngu*, 64.

available to him at the time of his research. In other words, the situation regarding the transfer and spread of distilled liquors in Yi Kyugyŏng's lifetime had grown more complex. In his case, the arrival of Europeans in Asia might have functioned as a new variable in the economics of the region's liquor trade that meant that some of his evidence was outside the historical context of his discussion topic. Moreover, he received new information from Japanese contemporaries who had limited contact with Europeans such as the Dutch, a topic to which this book returns in the following chapters. For example, Yi Tŏkmu's work, which his grandson Yi Kyugyŏng cited, states, "Ananta's 阿難陀 soju (shaojiu) is called *aralgil ju* (alaji jiu 阿剌吉酒), and *awamori* 泡盛 liquor, the soju (shochu) of Ryukyu (Okinawa) and Satsuma, is called *p'osŏngju* (*awamori shu* 泡盛酒)."[69] As Chang clarifies, Yi cited a Japanese encyclopedic source, *Wakan sansai zue* 和漢三才図会 (Illustrated Sino-Japanese Encyclopedia), compiled in 1712, which was imported to Korea and circulated among scholars developing pragmatic scholarship in the eighteenth and nineteenth centuries, such as Yi Tŏkmu and Yi Kyugyŏng.[70] This encyclopedia originally says,

The liquor brought by the Dutch commercial fleet called Oranda 阿蘭陀 is alaji liquor, and the liquor of Ryukyu and Satsuma is awamori liquor, all of which are the shaojiu of their countries. Their taste is very strong, and they eliminate heart-burning, prevent chest pain, and maintain body moisture. These are all freshly distilled liquors, becoming shaojiu.[71]

We should note here that Yi Tŏkmu changed one Chinese character in the middle of the original term in the Japanese book, "Oranda 阿蘭陀," meaning Holland (the Netherlands), to "Ananta 阿難陀," the name of a Buddhist saint, in order to link this particular version of shaojiu with South Asia. The name could refer to another Ananda, with whom we are familiar from Chapter 1 – his Chinese name transliterated as Ananta 阿難答 – the Mongol prince who monopolized alcohol brewing and distilling within the Mongol court. It is also possible that Yi Tŏkmu was familiar with this princely arak distiller and changed the name of the original Japanese source as some form of acknowledgment. Whether done intentionally or serendipitously, he obviously misused the original Japanese source for the sake of his own argument.[72] Therefore, rather than taking the arguments of Yi Tŏkmu and Yi Kyugyŏng at face value, it is important to consider that the historical context surrounding the transfer of distilled alcohols that the two scholars describe differs from the context that

[69] Ibid., 66–67.
[70] These scholars are discussed in Chapter 4 below, and the Japanese book *Wakan sansai zue* in Chapter 6 below.
[71] Ibid., 66n39.   [72] Ibid., 66.

surrounded the original Yuan dynasty sources. The Dutch, to illustrate, did not even appear in Japan (indeed, anywhere in Asia for that matter) before 1500.[73]

While we put aside Yi Kyugyŏng's nineteenth-century arguments, it is important to pay attention to his reference to Song–Yuan-period sources, which can be taken as an indication of arak's origin in the southern maritime routes, even though the passage in question only marginally supports such an interpretation. For sure, many goods, including spices, came to China during and prior to the Mongol period through the maritime shipping and trade networks in which merchants from West, South, and Southeast Asia actively participated. This makes sense; the Mongol Yuan era witnessed an unprecedented boom in maritime trade, yet the sea routes driving this growing maritime international trade had actually been flourishing since even before the Mongol Empire arose. There is a possibility, therefore, that distilled liquors were transferred to China and even to Korea before the Mongol period, although, as we have seen earlier, no documentary evidence supports the existence of distilled liquors in Korea before the late Koryŏ period.[74]

It is necessary to review another soju propagation theory, proposed by Choe Nam-seon and critically reviewed by Chang Chi-Hyun, which compares to that of Yi Kyugyŏng. Like Yi, Choe Nam-seon, a pro-modernization scholar active from the late Chosŏn era to the early years of the Korean republic, also created passionate discourse about the origin of shaojiu/soju by using Chinese and Korean documents in the section called "Customs" of his book *Chosŏn sangsik* 朝鮮常識 (Common Knowledge of Chosŏn) (1948), a book that deals with issues of history and traditional culture in Korea that need to be understood through common sense. While Yi Kyugyŏng argues that arak traveled to Korea by way of maritime routes, Choe instead emphasizes overland routes, focusing on the arrival of Persian Muslims in Yuan China, Koryŏ Korea, and even Manchuria during the Mongol period.[75] He also proposed the possibility of a transfer of arak from West to South and Southeast Asia by sea; however, he did not consider the further extension of maritime routes to China and Korea. Choe's theory does not possess a concrete piece of evidence that suggests Muslims in West Asia widely consumed arak before Mongol times, as we have seen in Chapter 1. It would be safer to think instead that the Mongols promoted arak as their presence expanded

---

[73] A form of *arakhi/araq* is used in *Essential Things for Living at Home*, rather than the *alajhi* 阿剌吉 used by Yi Saek and Li Shizhen, or the *yalaiji* 軋賴機 used by Zhu Derun (ibid., 50).

[74] Pae Kyŏnghwa, "Andong sojuŭi chŏllae kwajŏnge kwanhan munhŏnjŏk koch'al," 61.

[75] Chang Chihyŏn, *Hanguk oeraeju yuipsa yŏngu*, 82–88.

further west *and* east via overland connections sometime during the thirteenth and fourteenth centuries. Documentary evidence supports this.

Considering these early scholarly discussions in his conclusion, Chang Chi-Hyun proposes his own complimentary theories. For example, he supplements Choe Nam-seon's theory by proposing that arak – or indeed any form of distilled liquor – could have been transferred to Korea by Muslims who sailed to Korea directly without ever passing through Mongol territory, including Yuan China.[76] Yet, as Chang himself admits, few Koryŏ-period sources written before Mongol connections developed testify to the existence of distilled liquors.[77] Such a view also neglects the important historical fact of the explosive growth in the volume of spirits consumed during the Koryŏ period.

As shown above, such a systematic discussion of soju's origin and transfers has grown since the 1980s; however, little further progress in research on this topic has been made in recent years. At the same time or shortly after Chang wrote, Yi Seong-wu argued that several Mongol army bases on the Korean peninsula subsequently became modern centers of soju development in Korea. This claim has begun to raise critical problems, as articulated in recent research by this book's author and Lee Sang-Hoon, because it lacks sufficient evidence to fill the multi-century gap after the production of distilled liquors by locals revived in the late twentieth century.[78] In general, it seems that later studies have merely recycled previous discussions.

Recently, the claim that shaojiu/soju originated before the Mongol Yuan period, as did the possible subsequent transfer of its technology to Korea, which Chang introduced as one of his soju origin theories, came to attract the attention of scholars. These discussions simply rely on new findings that prove Muslims traveled to the Korean peninsula before the Mongol era, even as early as the Tang dynasty (seventh to tenth centuries) when Silla was open to maritime contacts with outsiders. However, to date, it is unclear whether the Tang variety of shaochun was in fact a form of distilled liquor; indeed, many Chinese scholars would be skeptical about such consumption of distilled liquors during the Tang era. In

---

[76] Ibid., 88.

[77] Ibid., 46. On the maritime trade commodities between Song China and Koryŏ Korea, see Kim Yŏngje (Kim Young-jae), *Koryŏ sangin kwa Tong Asia muyŏksa* (Koryŏ Merchants and the History of East Asian Trade) (Seoul: P'urŭn yŏksa, 2019), 190–216; and "Songdae Chungguk kwa Koryŏ saiŭi haesang kyoyŏkp'um: Tongnam Asia chiyŏk kwaŭi pigyo nŭl t'ongan kŏmt'o" (Marine Trade Commodities between China and Korea in the Song Dynasty: Review by Comparing with Areas of Southeast Asia), *Yŏksa Munhwa Yŏn'gu* (Journal of History and Culture) 60 (2016): 151–189.

[78] See Chapter 4 of this book.

such a situation, it is too much to insist on the possibility that distilled liquor could have traveled directly from China or from West Asia during the Tang dynasty. As noted above, the lack of evidence for the widespread consumption of spirits in Islamic society proves to be the bigger problem. Recently, Lee Hwa Seon has argued that, as a distilled "flower liquor" (*hwaju* 花酒) developed in Okinawa from the early modern period, documentation from early Koryŏ times suggests the possibility that distilled liquors developed in the Korean kingdom before the widespread rise of Mongol power. However, as the author herself admits, it is unclear what elements constituted flower liquor at the time its existence was recorded, so further evidence is needed to support Lee's assumption.[79]

Rather, the time has come to revise the standard approaches to the history of distillation technology in Korea in favor of cross-disciplinary ways, in order to exploit both sources and research – both long-accumulated materials and newly available ones – comparatively and on a global scale. Chapter 1 depicted the Yuan explosion of distilled spirits thanks to their promotion by the Mongols. None of the various arguments about existing theories deny that the distilled spirits under discussion suddenly became popular in Korea during the Mongol period. Despite this fact, few early studies have drawn a connection between the popularization of soju as a "new" beverage and the unprecedented level of cross-cultural contact that flourished throughout thirteenth- and fourteenth-century Asia.

Here is a possible scenario: as the Mongols occupied Korea in the course of their military expansion, they probably brought stills with them to Korea to serve a variety of military, diplomatic, commercial, and societal purposes. The locals in question then moved quickly to appropriate distillation technology for themselves and subsequently adapt it to new ferments – as in the case of Korean rice wine, for example. Entirely new types of liquor soon developed (for example, the ancestral drink for modern Korean soju), not just based upon milk or rice, but in an abundant variety. Moreover, the Mongols promoted this distillation technology along with a host of other kinds of technologies, ideas, and goods. It is important, then, that new studies reflect the broadening historical context that these new contributions create.

In special consideration of this, the next chapter traces some of the likely routes by which distilled liquors like soju and arak traveled in their transfer to Korea thanks to the country's close political connections with members of Mongol society at this time. Based on our new and rich body

---

[79] Yi Hwasŏn (Lee Hwa Seon), *Tongasia sul munhwasa* (A History of East Asian Wine Culture) (Seoul: Hyangŭm, 2018), 59–60.

of sources, we can reconstruct Yuan–Koryŏ relations, particularly the aspect focused on the circulation of soju into and throughout Korea. This serves as strong circumstantial evidence providing clues to the origin of soju and the routes it traveled during the end of the Koryŏ period when soju most likely was introduced to Korea.

Conclusion

In this chapter, we examined the medieval documentary sources comparatively to explore the situation of the rise of Korean spirits. Similar to the case of China, which abounds in extant classical literature, Korea possesses a rich supply of documentary sources that raise the chance of success of tracing the origins of distilled liquors like soju and arak in Korea since ancient times. As with other societies, Korean society has attached great social significance to alcohol since ancient times. A cup of alcoholic drink consumed during a restful break in an agricultural society would have played an important role in the lives of the peninsula's people of different classes. Gradually, alcohol varieties increased as interactions with communities in China and elsewhere grew. Nonetheless, no singular piece of evidence, documentary or archaeological, exists to verify the existence of distilled liquors before the end of the Koryŏ period.

Evidence of such spirits in Korean sources began to appear conspicuously during the Mongol period, similar to the case of China at the same time. These three cases of documentation do not explicitly explain the transfers of distilled liquors from the Yuan to Koryŏ realms, or its rapid propagation. However, they strongly hint at the situation that defined the distilled spirit's rise in political, military, economic and cultural arenas in Koryŏ society under Mongol influence. The effect of these transfers probably spread rapidly once Mongol and Koryŏ soldiers valued distilled liquors with greater preservative power through the Mongol army camps stationed in Koryŏ, and once nobles and literati made use of their easier access to new spirits from China. Whatever the trajectory, the literature reviewed in this chapter establishes a sufficient body of historical sources with which to proceed in our global history of Korean soju.

As we have seen so far, these pieces of historical evidence strongly suggest the fact that distilled liquors spread during the late period of the Koryŏ dynasty under the aegis of Mongol influence. After the mid-Chosŏn period (seventeenth century), however, when soju grew widespread on the Korean peninsula, the debate about soju's origins grew active and has continuously influenced the topic ever since. Debates about the general origin and transfer of distillation technology on a worldwide basis grew huge, and new findings and studies have fueled this even more, as we saw in Chapter 1. However, regarding the history of distilled liquors in Korea, there have been no new studies of major significance since the pioneering work was undertaken in the 1980s. Against this backdrop, we revisited the controversy over the origin of Korea's spirits and tried to find a more reasonable interpretation of it. We tried to re-examine the debate over the origin of soju given our updated understanding of the development of distillation and distilled liquors throughout Eurasia.

Korean scholars have generally developed a number of theories based on the study of Chinese sources used in the early debate over the origin of shaojiu in China and added some of these new sources to the work of developing more expanded views. Earlier scholars like Yi Kyugyŏng and modern scholars like Chang Chi-Hyun have provided systematic analyses of the multilingual sources available to them, but indeed some pitfalls exist, given the limited number of sources they were able to utilize. What they could have further pursued was an examination of the technological aspects of distillation and distilled-liquor transfers, which are not found in any historical or archaeological record. This is a difficult task to undertake, at least until new archaeological excavations yield new evidence. To that end, in Chapter 4 we trace the major part of technology transfer through the extant records of the Chosŏn era, in which soju was rapidly spreading and becoming an important alcoholic beverage in Korea.

Having confirmed a sufficient repository of historical evidence on which to build this global history of Korean soju, the book moves on to address the broader historical situation based on new secondary literature and new evidence related to soju found in that context. This discourse will reveal an important relationship between Mongolia and the Mongol-run Yuan dynasty in China and Koryŏ Korea that previous scholarship has overlooked and more recent scholarship has acknowledged. This will ease the work of pinpointing particular events in the history of soju and arak's cross-cultural transfer under the broader context of cross-cultural contacts. To this end, Chapter 3 advances key evidence of the early transfer process and establishes the historical implications of the transfer of distilled liquors from the Yuan to Koryŏ by charting the historical context of Koryŏ's relations with the Yuan and its political situation in the broader Mongol Empire.

3   Contextualizing Soju
Political Relations and Cultural Transfers between
the Mongol Empire and Koryŏ Korea

The transfer of a cultural element that changes an aspect of a society's culture does not occur without reason. The earliest extant documentary sources about soju and arak previously discussed suggest that the distilled liquors were transferred from China to the Korean peninsula during the Mongol period, a time of close relations between the two societies. This was also a critical time in the individual histories of both societies, like many other contemporaneous societies in Eurasia under Mongol sway.[1] For almost 150 years, the Mongol invasions, empire, and other forms of intervention influenced Chinese and Koreans in unprecedented ways. For China, the entire territory of their historical empire fell under foreign rule for the first time when the Southern Song dynasty succumbed to the Mongols in 1276 – an ironic situation, as the Chinese had once looked down on the government of their new sovereigns. While it lasted for only about a century, Mongol rule opened China to many new elements of the world at large, thanks in great part to the empire's vast territory, which stretched from Hungary to the Pacific. In recent decades, an increasing number of scholars have begun to pay attention to broader aspects of cross-cultural contact and exchange, long overlooked, revealing the empire's facilitation of movement by people, goods, and ideas across the Afro-Eurasian continent during this brief century.[2] This situation affected peoples living along the edges of this vast empire as well, like

---

[1] For a succinct overview of the Mongol Empire during the thirteenth and fourteenth centuries, see David Morgan, *The Mongols* (Oxford: Blackwell, 2007 [1986]). For the trends in Mongol studies since 1986 when the book's first edition was published, see the newly added chapter and updated bibliography in the 2007 edition.

[2] Despite its violent beginnings, the Mongol Empire subsequently ushered in policies that produced the Pax Mongolica or "Mongol Peace," which fostered travel and commerce across Eurasia that expanded and intensified cross-cultural interaction that circulated people, goods, ideas, and technologies through the diverse and dispersed societies they subjugated. This new perspective of the Mongols is still in its early stages of development. See Thomas T. Allsen, *Culture and Conquest in Mongol Eurasia* (Cambridge: Cambridge University Press, 2001); and Park, *Mapping the Chinese and Islamic Worlds*, 193.

Korea; which, as a consequence, found itself at the end of the Koryŏ dynasty (935–1392) exposed to the wider world beyond China, a new world that introduced an unprecedented variety of new cultural elements that Korean society assimilated on an unprecedented scale.

People living on the Korean peninsula had been connected indirectly to various societies in Afro-Eurasia during earlier periods through overland and sea routes. They were more directly connected to the wider world, however, when their country became part of the Mongol Empire in 1258, after the Koryŏ court made a peace agreement with the Mongols. Unlike the Chinese, the Koreans avoided direct rule, so the Koryŏ dynasty remained intact. Yet, as both a vassal state and a *quda* (marriage alliance) state of the Great Yuan State (or the Yuan dynasty, the suzerain state of the Mongol Empire), Koryŏ possessed many characteristics that were unique among the societies of Mongol-dominated Eurasia (see Map 3.1).

People in Koryŏ were exposed to many forms of direct interference by the Mongol regime in China and Mongolia; at the same time, however, they enjoyed opportunities for direct contact with Eurasia at large thanks to their kingdom's place within the larger regime. While Korea's subject relationship within the Mongol hierarchy produced both positive and negative forms of cross-cultural influence, local society soon took these newly assimilated cultural elements and modified them according to local conditions and needs. Soju is a good example of a commodity that was adopted into local society after its transfer from China and then developed further to suit its new Korean environment. As seen in Chapter 1, shaojiu began to spread within China during the Yuan period, yet its origin and further development involved broader Eurasian cross-cultural exchanges. So it was with Korea. The story of soju's introduction into Korea typifies the interconnected nature of Korean society during Mongol times, though it stands as only one among many examples of cultural elements that transferred between Korea and other societies through the frequent and large-scale interchange that occurred between people at all social levels traveling between Koryŏ Korea and Yuan China and places beyond.[3]

This chapter explores the broader and specific contexts in which soju and many other new cultural elements were transferred between the two societies, many of them moving from the Yuan dynasty in China and Mongolia (and the broader Mongol Empire) to the Koryŏ dynasty in Korea, but also some vice versa. It will present further evidence for the transfer of soju that has been overlooked by previous studies but made

---

[3] Yi Kae-Seok (Yi Kaesŏk) points this out in his book *Koryŏ Tae Wŏn kwan'gye yŏn'gu* (Studies on Koryŏ–Dayuan Relations) (Seoul: Chisik sanŏpsa, 2013), 16–17. He also argues that historians should study such topics of cultural exchange in order to fill gaps in current studies.

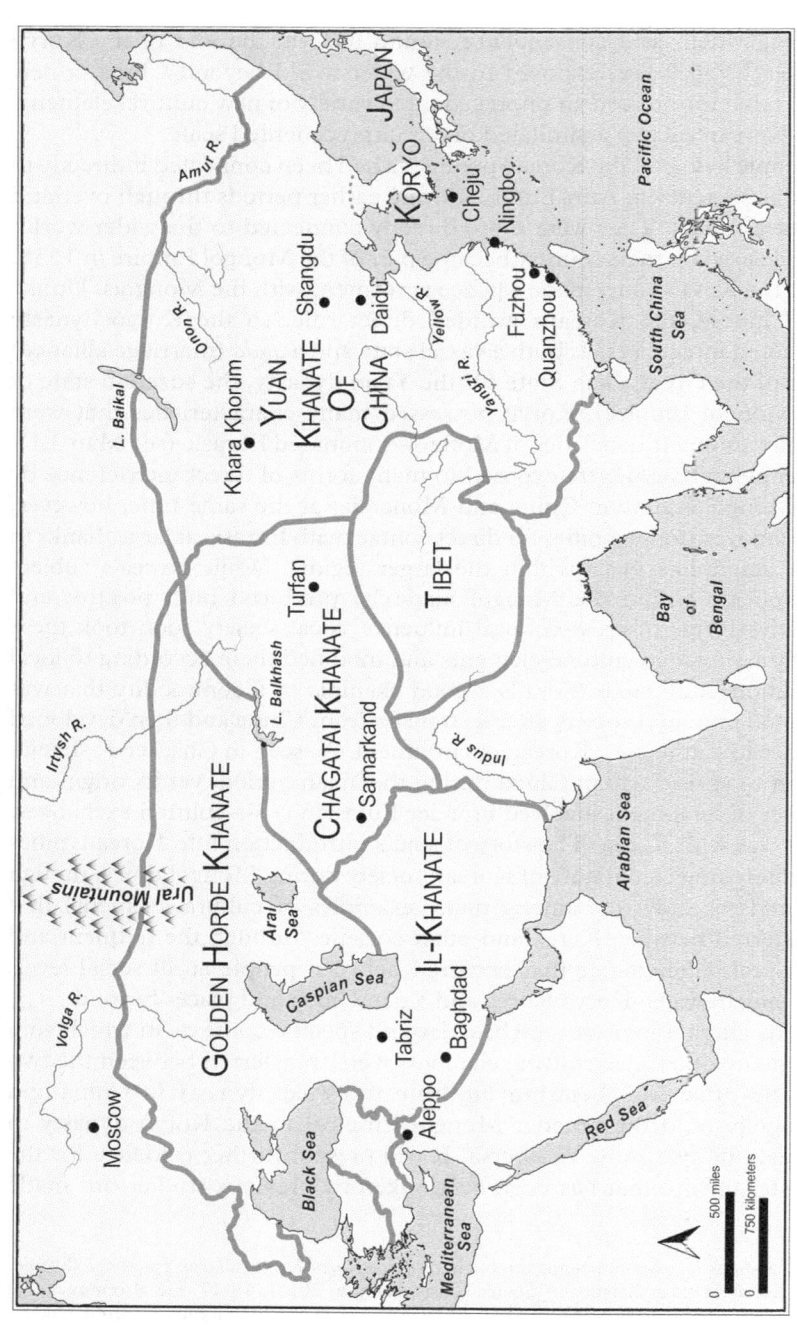

Map 3.1 The Mongol khanates, 1280, after Morris Rossabi, *The Mongols: A Very Short Introduction* (Oxford: Oxford University Press, 2012), Figure 7, 62–63. Some aspects of the map have been modified by the author.

apparent by this book's expanded perspective. The following questions drive this chapter: what political, social, and economic conditions facilitated movements of people, goods, and ideas that led to exchanges of cultural items between the two societies? How many of the movements of cultural elements happened at Eurasian scales, as in the case of soju? What further evidence does the broader historical context make clear? In order to tackle these questions properly, we must first create a grand overview of the Mongol world order and the place of Korea within it in order to gauge the scale of Korea's cultural connection to the broader Eurasian world under Mongol influence. The chapter then examines Yuan–Koryŏ political, economic, and social relations, focusing on contacts at different levels of society among Mongols, Chinese, Koreans, and sometimes other ethnic groups like Muslims (Huihui), which all facilitated various types of cultural exchange, including foods and alcoholic beverages such as soju. After that, we will look at several major items of cultural exchange and transfer that provide examples comparable to soju. Some of these elements, formed at this transnational period, would have lasting and profound influences on Korea's later history. This confirms the transfer of soju as a part of a larger exchange of goods and ideas that resulted from the integration of the peninsula into the greater long-distance exchange networks of Eurasia.

Korea in the Mongol World Order: Vassal State versus *Quda* (Marriage Alliance) State

As a nomadic group whose power suddenly rose on the Mongolian steppe in the early thirteenth century, the Mongols rapidly conquered the wide expanse of Eurasia in just a few decades and established the world's largest contiguous empire in history. No society in world history has had such considerable and critical influence on more various societies in premodern Afro-Eurasia during such a short period of time. Many significant studies have been made of societies that became major parts of the Mongol Empire (known to the Mongols as *Yeke Mongol Ulus*), such as China, Iran, and Russia, in whose histories the Mongols marked new eras.[4] Yet the empire's supranational *ulus* (joint patrimony)[5] also exerted significant influence over many other states all over its periphery, like Korea – places that were not under direct Mongol rule yet were significantly influenced by it in different aspects and on various levels. Examples of other societies that received such substantial Mongol influence include

---

[4] There have been many studies on this, which I cannot adequately summarize here.
[5] Paul D. Buell and Francesca Fiaschetti, *Historical Dictionary of the Mongol World Empire and Its Successor States*, 2nd revised and expanded edition (Lanham, MD, and Oxford: The Scarecrow Press, 2018), 299.

countries at the empire's western edge like Armenia and the Seljuks of Anatolia (in modern-day Turkey).[6] Indeed, as David Robinson shows in his pioneering study of late Yuan–Koryŏ relations from a Northeast Asian regional perspective, Korea's case resembles in many ways several other small- to mid-sized Eurasian polities under Mongol rule. This makes them useful as comparisons when analyzing cases of cross-cultural contact between Koryŏ and the Mongol Yuan courts. At the same time, we cannot simultaneously deny the uniqueness of Korea's case, because Koryŏ became a close ally – in terms of both its location and its place in the larger Mongol state system – while simultaneously functioning as a vassal and marriage-alliance state with the greater *ulus* of the Yeke Yuan, the suzerain state encompassing China and Mongolia that the Mongol Empire created after the establishment of its four khanates in 1260. To understand Korea's close and special relationship with the Mongols, it is best to view it from a holistic perspective that includes the entire Mongol Empire. Robinson has already argued for "the need to see the Koryŏ kingdom as part of the Great Yuan *Ulus*."[7] This book suggests the further need to see Koryŏ Korea from the perspective of cross-cultural contact as an integral part of a wider Eurasia influenced by the Mongols. A succinct review of the political relations that Koryŏ developed with the Mongol Empire is a good place to start.

Koryŏ encountered the Mongol state soon after Chinggis Khan unified the Mongol chiefdoms in 1206. Though their Eurasian conquests began with their destruction of the Jurchen Jin empire in 1234, the Mongols encountered Koryŏ militarily as early as 1218. However, in the wake of this initial encounter, both sides agreed to a negotiated settlement, and Koryŏ avoided further Mongol incursion into its territory by forging a brethren alliance with them.[8] This situation did not last long: once the Mongols began their rapid expansion into other regions of Eurasia, their ambitions inevitably turned eastward, which led to strikes against Koryŏ. While the house of Ch'oe, the military dictators of Koryŏ, resisted the Mongols for almost thirty years by moving the royal capital offshore to Kanghwa island, it could not hold out forever, and Koryŏ officially surrendered in 1259, when its Crown prince (and later King Wŏnjong, r. 1260–1274) met Khubilai and agreed to guardianship relations.[9] In

---

[6] For the case of the Seljuks, see Sara Nur Yildiz, "Mongol Rule in Thirteenth-Century Seljuk Anatolia: The Politics of Conquest and History Writing, 1243–1282," PhD dissertation, University of Chicago, 2006; A. C. S. Peacock and Sara Nur Yildiz, eds., *The Seljuks of Anatolia: Court and Society in the Medieval Middle East* (London: I. B. Tauris, 2013).

[7] Robinson, *Empire's Twilight*, 8–9.   [8] Yi Kaesŏk, *Koryŏ Tae Wŏn kwan'gye yŏn'gu*, 14.

[9] For the most thorough analysis of the Mongol invasions of Koryŏ in English to date, see William E. Henthorn, *Korea: The Mongol Invasions* (Leiden: E. J. Brill, 1963). For

other words, despite their successful military campaign, the Mongols allowed Koryŏ to continue as a kind of vassal state with its own king. The relationship appears to have been a fruitful one for Koryŏ. It became one of the few countries to maintain systematic relations of intermarriage with the Mongol imperial family, because Koryŏ kings became the sons-in-law of Mongol emperors through intermarriage.[10] This double status is reflected in the title *Chŏngdong haengsŏng sŭngsang puma Koryŏ kugwang* – "Koryŏ king as chancellor of the Eastern Expedition Field Headquarters and son-in-law of the Yuan emperor" – which King Ch'ungnyŏl received after suffering many upheavals during the early period of political relations between the Mongols and Koryŏ.[11] It was in this context – as the vassal of an already massive Eurasian empire trying to extend its domain further east into Japan and Sakhalin – that Korean society integrated into the development scheme of the entire Mongol Empire.

Scholars of Koryŏ history have paid attention to the very special and complicated relationship that the Koryŏ court forged with its Mongol imperial overlords. Earlier studies focused on its few features of traditional tributary relations, which resembled those between earlier Korean kingdoms and Chinese dynasties. More recent studies began to look closely at new characteristics, such as marriage alliances, that can be better explained in the larger context of the greater Mongol Empire.[12] Paul Buell discusses intermarriage as "a key part of the imperial governance system, the *yeke Tore*, the 'great imperial system,' just as was conquest, taxation and booty generation," which the Mongols traditionally used to expand their political power. He furthermore argues that, while Koryŏ was not the only country to be connected with the Mongols through state-level marriage alliances, it was the only state that sustained them in such a long-lasting and close way: for example, one Koryŏ king was a grandson of Khubilai, while the last Mongol emperor was the son of

---

information on military rule, including that of the Ch'oe family from 1170 to 1270 and the Ch'oe house's struggles with the Mongol invasions, see Edward J. Shultz, *Generals and Scholars: Military Rule in Medieval Korea* (Honolulu: University of Hawai'i Press, 2000).

[10] For the marriage alliances of the Mongol empire, see Paul D. Buell and Judy Kolbas, "The Ethos of Sate and Society in the Early Mongol Empire: Chingiz Khan to Güyük," in *The Mongols and Post-Mongol Asia: Studies in Honour of David O. Morgan*, ed. Timothy May, *Journal of the Royal Asiatic Society* 26, nos. 1–2 (January, 2016): 43–64.

[11] Koryŏ basically maintained such status to the end of the Mongol rule.

[12] For a comprehensive survey of the historiography of Koryŏ–Mongol relations, see Yi Kaesŏk, *Koryŏ Tae Wŏn kwan'gye yŏn'gu*, 14–58. See also Yi Myŏngmi (Lee Myung-mi), *13–14 segi Koryŏ–Monggol kwan'gye yŏn-gu: Chŏngdong haengsŏng sŭngsang puma Koryŏ kugwang, kŭ pokhapchŏk wisang e taehan t'amgu* (A Study of Relations between Koryŏ and the Mongols in the 13th and 14th Centuries: An Exploration of the Complex Status of the Koryŏ King as Chŏngdong haengsŏng sŭngsang and Imperial Son-in-Law) (Seoul: Hyean, 2016), 14–21.

Empress Ki from Koryŏ.[13] This demonstrates the special status that Koryŏ enjoyed at the heart of the Mongol Empire. Additionally, Buell points to a special kind of folklore that claims that Chinggis Khan once went to Koryŏ and then, after being indulged by Korean ladies, did not want to go back to Mongolia. While the folklore is not true – Chinggis Khan never went to Korea in his lifetime – this story highlights a close intimacy between the Mongols and Koreans since early Mongol times.

According to Kim Hodong, an important transition in Mongol imperial history – that is, the establishment of the four Mongol khanates in the four major areas of Eurasia – also affected the fate of Koryŏ within the empire. After the death of the fourth Grand Khan Möngke in 1259, two of his younger brothers, Khubilai (1215–1294), based in China, and Arigh Böke (d. 1266), based on the Mongolian steppe, fought to become the next Grand Khan. Khubilai finally defeated Arigh Böke and officially became the fifth Grand Khan of the Mongol Empire.[14] Earlier studies of the Mongol Empire tend to focus on the division of the Mongol Empire after this civil war. Recent scholarship, including that of Sugiyama Masaaki, Kim Hodong, and Paul Buell, argues, however, that although the new khans enjoyed more freedom in governing their individual territories than had their predecessors, the Mongol Empire was actually not all that divided, as the other khanates continued to acknowledge the Yuan in China as the suzerain state within the Mongol Empire.[15] According to their view, the Yuan state established by Khubilai in China and Mongolia was the Khan's *ulus* within the *Yeke Mongol ulus* or "Great Mongol Empire," rather than just the Yuan dynasty in Chinese dynastic tradition; the term Da Yuan ("Great Yuan") is simply a Chinese appellation for *Yeke Mongol ulus*.[16] Recent studies lend support to this idea, by highlighting the many nomadic structures and forms that the Yuan state possessed

[13] Paul D. Buell, "Korea as Part of the Mongolian World: Patterns and Differences," *International Journal of Eurasian Research* 5, no. 10 (Jan. 2017): 137–146.
[14] Morgan, *The Mongols*, 138–139. For another of Khubilai's rivals, Qaidu (d. 1301; a grandson of the second Mongol Khan, Ögedei), and the wars between different Mongol khanates leading to Qaidu, see Michal Biran, *Qaidu and the Rise of the Independent Mongol State in Central Asia* (Richmond: Curzon, 1997).
[15] See Sugiyama Masaaki, *Mongoru teikoku to daigen urusu* (The Mongol Empire and the Great Yuan Ulus) (Kyoto: Kyoto Daigaku Gakujutsu Shuppankai, 2004); Kim Hodong, "The Unity of the Mongol Empire and Continental Exchanges over Eurasia," *Journal of Central Eurasian Studies* 1 (2009): 15–42; Buell, "Korea as Part of the Mongolian World," 137–146.
[16] That is, Khubilai distributed power to other khanates because of convenience, yet the khanates connected to the Yuan closely in many ways, including through enduring diplomatic relations. See Kim Hodong, *Monggol cheguk kwa Koryŏ: k'ubillai chŏnggwŏn ŭi t'ansaeng kwa Koryŏ ŭi chŏngch'ijŏk wisang* (The Mongol Empire and Koryŏ: The Birth of the Khubilai Government and the Political Status of Koryŏ) (Seoul: Sŏul taehakkyo ch'ulp'an munhwawŏn, 2015 [2007]), 3.

as the heart of the Mongol Empire.[17] Through its political expansion on an unprecedented scale thanks to both nomadic and imperial institutions and concepts such as the Quriltai (Great Royal Assembly) and the formation of the single Mongol *ulus* incorporating others, which provided a more flexible means of connecting diverse societies than did sedentary societies, the Mongols successfully extended their influence both east and west. This was the large empire that Koryŏ had to deal with on its western frontier. It was not only Koryŏ that had to subject itself to the Mongols under disadvantageous terms. Kim Hodong argues that, in the course of bringing Koryŏ under its influence, Khubilai also had to negotiate privileges with the kingdom's ruling family to supplement the loss of its power in order to unify the empire after Khubilai became the Grand Khan.[18] Yoon Eun-sook also discusses the Otchigin royal family in Manchuria as an example of how Khubilai used marriage alliances with Koryŏ as a means of dealing with threats to his transition into power: facing difficulty in controlling the rising fortunes of the Otchigin family, Khubilai had to give his granddaughter to the crown prince of Koryŏ, making her a future queen and bringing the Koryŏ royal family into the "golden family" of the Mongols in order to check the rise of a potentially destabilizing power and maintain the balance in the region.[19] Thus a close and enduring political relationship formed between the Yuan and Koryŏ, forged through the *quda* marriage alliance.

In fact, Yuan–Koryŏ relations developed in more complicated ways over time, and many scholars have attempted to execute subtle analyses of different situations in different historical contexts, as summarized by Yi Kae-Seok in his monograph about Koryŏ–Yuan relations.[20] Lee Myung-mi focuses on three different forms of status that the king of Koryŏ held simultaneously within the Mongol state: the king of a vassal state, Koryŏ; the chancellor of the Eastern Expedition Field Headquarters (Chŏngdong haengsŏng); and the Mongol emperor's son-in-law. Lee confirms that the Koryŏ kings held all these three forms of status, each of which would be used interactively to produce political results. For example, some Koryŏ

---

[17] For example, see Sŏl Paehwan (Seol Paehwan), "Mongwŏn cheguk k'urilt'ai (*Quriltai*) yŏngu" (A Study of the *Quriltai* in the Mongol Empire), PhD dissertation, Seoul National University, 2016.

[18] Kim Hodong, *Monggol cheguk kwa Koryŏ*, 101–120.

[19] Yun Ŭnsuk (Yoon Eun-sook), *Monggol cheguk ŭi Manju chibaesa: Otch'igin wangga ŭi Manju kyŏngyŏng kwa Yi Sŏnggye ŭi Chosŏn kŏn'guk* (The History of the Mongol Empire's Rule in Manchuria: Rule of the Manchurian Management of the Otchigin Royal Family and the Founding of Chosŏn) (Seoul: Sonamu, 2010), 20.

[20] For a succinct discussion about the previous scholarship on Koryŏ–Yuan relations, starting with Japanese scholarship of the 1930s and ending with the most recent Korean, Chinese, and Japanese scholarship produced in the last few decades, see Yi Kaesŏk, *Koryŏ Tae Wŏn kwan'gye yŏn'gu*, 14–61.

monarchs like King Ch'ungsŏn took advantage of this complex status to ensure Koryŏ's continuing independence as a state and to promote its privileges within the Mongol Empire at large.[21]

Reflecting on these recent achievements of earlier studies, we should contemplate a few other key points. Many studies have discussed cross-cultural exchanges between China and other societies in the Pax Mongolica like Iran, India, and Europe; yet no other society in the empire at large could have had greater contact and more exchanges with the Great Yuan Ulus than Koryŏ, considering the close political relations between the two states and the geographical proximity of Korea to China.[22] We could assume that such a special situation facilitated transfers of cultural elements on a much larger scale to Korea than to any other society under Mongol influence. Therefore we should look at Mongol Yuan–Koryŏ cross-cultural exchanges more closely, because their study promises greater understanding of the effects of the Mongols on cross-cultural exchange than that of any other relationship within the Pax Mongolica. It is not only the character of relations between the Yuan and Koryŏ that we have to acknowledge. The fruits of continuous large-scale cross-cultural contacts across Eurasia connected Korea to the West, which resulted in transfers. We have seen in earlier chapters that the development of distilled alcoholic beverages was connected to similar ideas and technologies built in accord with Western traditions, and that the Mongols could have contributed to their spread to the peoples of the broad Eurasian expanse. This suggests that similar cases in political, economic, social, and cultural fields might be found. As a subordinate vassal state and a close brethren ally of the Yuan, in the broad scheme of international politics, various classes of Koryŏ's people began to forge unprecedented human contacts with counterparts in the Yuan, which naturally led to exceptional cross-cultural exchange.

Yuan–Koryŏ Interactions and the Evidence for Soju and Arak Transfers

In Chapter 2, we briefly reflected that people of the Korean peninsula enjoyed close contacts with people in neighboring countries like China in various ways and degrees, largely through diplomatic and commercial

---

[21] Lee Myung-mi (Yi Myŏngmi) argues that in the early stage of Mongol overlordship, the Koryŏ kings were forced to endure changes to their conduct of kingship; yet, in order to rebuild the kingship that had collapsed under the previous military regime, they also voluntarily assented to the power of the Mongols, which Koryŏ not only acknowledged but used to their advantage. See Yi Myŏngmi, *13–14 segi Koryŏ·Monggol kwan'gye yŏn-gu*, 33.

[22] For example, see Yi Kanghan (Lee Kang Hahn), *Koryŏ ŭi chagi, Wŏn cheguk kwa mannada* (Koryŏ Porcelain and the Mongol Yuan Empire) (Sŏngnam: Han'gukhak Chungang Yŏn'guwŏn ch'ulp'anbu, 2016).

relations. Nonetheless, political and military conflicts did occur. For example, Silla, one of the three kingdoms of ancient Korea, allied with the Tang dynasty in the 660s to defeat rival states Paekche (18 BCE–660 CE) and Koguryŏ (37 BCE–668 CE) and unify the Korean peninsula under Great Silla (or Later Silla), which ruled Korea from 668 to 935.[23] The royal family of the fallen Koguryŏ kingdom founded Parhae (called "Bohai" in Chinese) in 698 in present-day Manchuria and the northern part of the Korean peninsula. It consisted of not only Koguryŏ refugees but also a Tungusic people called the Mohe (Malgal), later called Jurchen, the name we associate with the Jurchen Jin dynasty that ruled northern China (1115–1234). Parhae flourished until its fall to the Khitan Liao dynasty in 926. Both Great Silla and Parhae were connected to wider Eurasian trade networks through the Silk Road trade that connected Central Asia and Japan. After the Koryŏ dynasty replaced Great Silla in 935, Korea experienced particularly severe internal upheavals, yet nothing had greater impact on medieval Korean society at all levels than the unprecedented foreign interference of the Mongols in the final century of the Koryŏ dynasty.[24] As seen above, when Khubilai made Koryŏ a vassal state of the Mongol Empire, he needed a close ally and therefore provided several privileges to Koryŏ, including his promise to allow the kingdom to keep its traditional system and customs and making Koryŏ's Crown prince his son-in-law. Actual human contacts, however, were complicated, and many traditional systems collapsed, while many Koreans had to endure undesirable cross-cultural exchange in addition to commerce – in the form, for example, of military recruitment and traffic in humans. Accordingly, many people in Koryŏ were exposed to gradual cultural influence directed from the Mongol Empire (though it also traveled in the other direction).

First of all, we should look at human interactions through large-scale military clashes in the history of Sino-Korean relations, however rare they may be. During the period when the Koryŏ court resisted Mongol conquest, invading armies devastated large areas in the Korean peninsula.[25] Once the Koryŏ court agreed to become a vassal of the Mongols, their new rulers

---

[23] These dates for the ancient kingdoms of Korea are traditional dates based on earlier sources and therefore are not historically accurate. See Chapter 2, footnote 1.

[24] Once the dynasty strengthened and centralized its royal power, and after fighting with powerful aristocratic families, the court was put under the control of a military regime, which lasted for a century in the mid-Koryŏ period. Lee Ki-baik, *A New History of Korea*, trans. Edward W. Wagner and Edward J. Shultz (Cambridge, MA: Harvard University Press, 1984), 110–154.

[25] Edward J. Shultz, *Generals and Scholars: Military Rule in Medieval Korea* (Honolulu: University of Hawai'i Press, 2000).

forcefully conscripted Koryŏ soldiers to crush the Sambyŏlch'o, a prestigious military unit within the Koryŏ military regime that continued to fight – to the end – by moving to Cheju Island. Soon after defeating the Sambyŏlch'o and occupying the entire territory of Koryŏ, the Mongols established the Eastern Expedition Field Headquarters in order to project their military expansion further east to Japan, and organized Mongol–Koryŏ allied forces for the expedition (which typhoons famously destroyed in 1274 and 1281).[26] The presence of the Mongol army, which consisted of not only Mongols but also recruits of Chinese, Korean, and other ethnicities, expanded the types of cross-cultural contact that people of the Korean peninsula engaged in.[27] The Mongols established army camps in several major cities of Koryŏ, such as Kaesŏng (the capital) and Chindo, as well as Cheju Island, and stationed their soldiers in certain military areas for several decades until they were expelled from Koryŏ at the end of the dynastic period (see Map 3.2). While there, these soldiers could have brought with them foods that they enjoyed and then introduced them to locals.

According to Yi Seong-wu, from this early time on, these places of Korean–Mongol contact became renowned for their production of high-quality soju.[28] Many works, both scholarly and general, followed Yi's argument here to explain the origin of soju in Korea. As we have seen in Chapter 2, some recent works criticize this view. The most detailed argument to efficiently disprove this earlier view can be found in Lee Sang-Hoon's recent study of Korean fortified fermented wine made by adding soju. In this book, a review of the origin of distilled liquor in Korea, Lee argues that it is difficult to connect Mongol influence to the development of soju in these locations because the period of direct Mongol military influence in these regions was brief and because many other regions of Korea that were also crucial to Mongol military campaigns did not see a similar development of local soju.[29]

Agreeing with Lee Sang-Hoon's convincing argument, we can modify the previous view to look at the general influence of the Mongol military rather than specifying places and periods. In the official history of the Koryŏ dynasty, one of the extant records describing soju (examined in

---

[26] For detailed studies on the Eastern Expedition Field Headquarters, see Chang Tong'ik, *Koryŏ hugi oegyosa yŏn'gu*, 14–109; and Ko Pyŏngik, *Tonga kyosŏpsa ŭi yŏn'gu*, 184–292.

[27] See Yi Kaesŏk, *Koryŏ Tae Wŏn kwan'gye yŏn'gu*, 128–197.

[28] See Yi Sŏngu, *Han'guk sikp'um sahoesa*, 216.

[29] Yi Sanghun, "Urinara kanghwa parhyoju ŭi chŏn'gae wa t'ŭkching" (The Development and Characteristics of Korean Fortified Fermented Wine), MA thesis, Seoul Venture University, 2014, 77. As Chapter 5 below shows, many local specialty brands from these regions began to revive during the late 1980s with promotional help from the South Korean government, and the gap between these two periods is too great to make a meaningful connection.

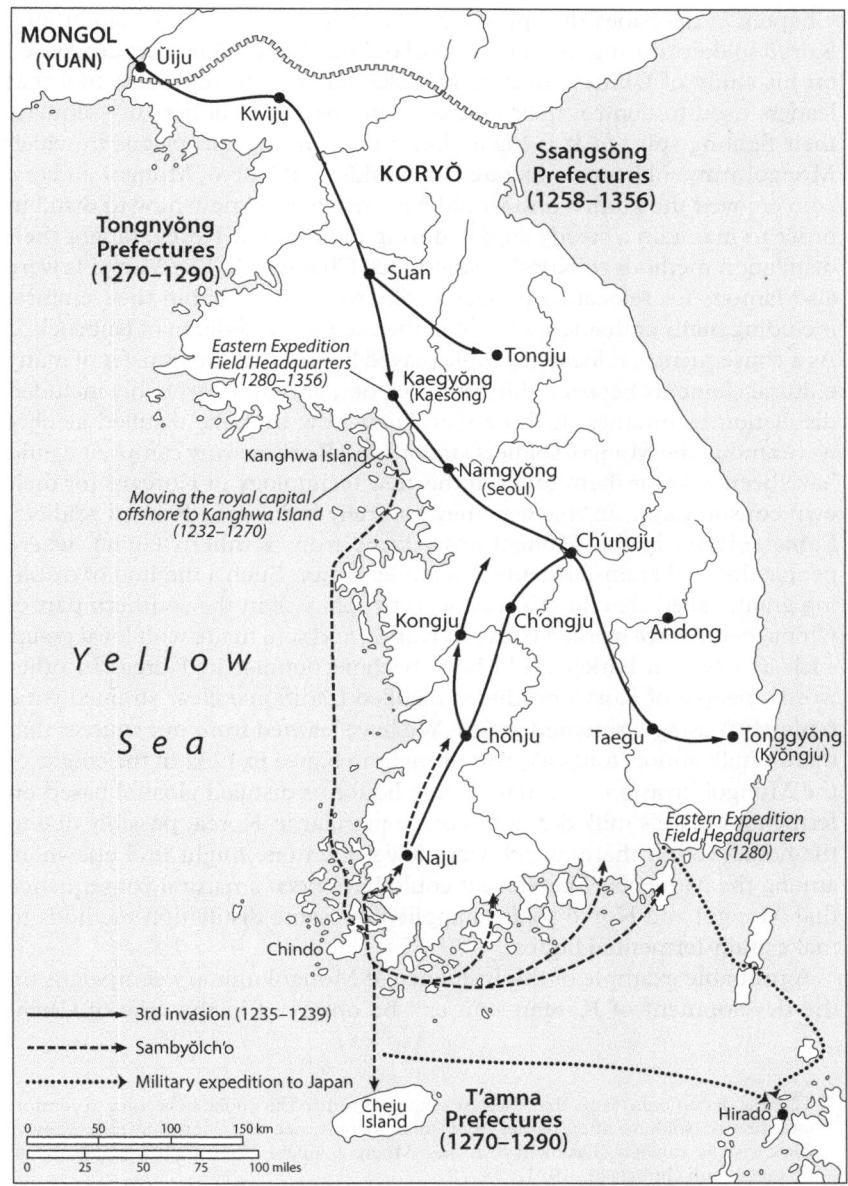

Map 3.2 The Mongol army's advance on the Korean peninsula, after
Michael D. Shin, ed., *Korean History in Maps: From Prehistory to the
Twenty-First Century* (Cambridge: Cambridge University Press, 2014),
68–73. Some aspects of this map have been modified.

Chapter 2) describes the spread and heavy consumption of soju among Koryŏ soldiers during the late Mongol period. As He Manzi argues, based on his study of Chinese documents, alcohol was an important tool that leaders used to control their armies to inspire tired soldiers or stimulate their fighting spirit.[30] It is highly likely that, during campaigns in which Mongol army soldiers co-operated with soldiers of Koryŏ, Mongol soldiers, who enjoyed the kumiss and would have needed to know how to distill in order to maintain a steady supply during their stay in Korea, taught their distillation methods to Koryŏ soldiers (see Chapter 1). The Mongols were also famous for relocating people to different places within their empire, including many craftsmen, as seen in the account of William of Rubruck.[31] As a consequence, relocated people played key roles in the transfer of many cultural elements between different societies, and these probably included distillation techniques. If those who knew how to make distilled alcohol were among the Mongol soldiers stationed in Korean army camps, it would have been easy for them to teach this new technology to Koreans for their own consumption. In this way they naturally influenced Korean soldiers. Some soldiers in the Mongol army came from southern China, where people distilled grain-fermented alcoholic liquor. Such a method of distilling grain, called shaojiu (Korean *soju*) by Chinese in the southern part of China, could have worked better in Koryŏ, and soju made with local grains such as rice and barley could have become popular in Korea. In other words, people of Koryŏ produced distilled traditional clear strained wine (ch'ŏngju) using fermented grains. We have learned from our sources that mare's milk liquor (tongnak) was brought to Korea in 1231 in the course of the Mongol invasion. Yet mare's milk liquor or distilled alcohol based on fermented mare's milk did not become popular in Korea, possibly due to the horses found there, which were fewer than one might find elsewhere among the Mongols. Therefore it could have been a natural consequence that Mongol and Koryŏ soldiers applied the same distillation methods to make grain-fermented liquor.

A probable example of this influence of Mongol military campaigns on the development of Korean soju can be observed in the case of Cheju

---

[30] Chinese documents testify that Chinese emperors utilized alcoholic beverages to comfort and reward soldiers after battles. For more details, see He Manzi, *Zuixiang riyue: Zhongguo jiu wenhua* (Drunken Sun and Moon: Chinese Wine Culture) (Shanghai: Shanghai guji chubanshe, 1991), 78–79.

[31] For example, see an episode that took place in Khara Khorum involving William the craftsman from Paris in William of Rubruck's account. William of Rubruck, *Rubruck's Report to King Louis IX of France*, trans. Peter Jackson, in Peter Jackson, *The Mission of Friar William of Rubruck: His Journey to the Court of the Great Khan Möngke 1253–1255* (London: Hakluyt Society, 1990), 183, 209–211.

Island. After resisting the Mongol campaign to conquer the island, the Sambyŏlch'o army was finally crushed there by an allied Mongol–Koryŏ force in 1273. The Mongols remained to rule the island until the Yuan dynasty in China fell in 1368. During Mongol rule, which lasted for a century, the people of Cheju Island were influenced by the Mongols in many ways, including ethnicity, language, architecture, clothes, and foods.[32] Scholars have demonstrated many words of Cheju dialects that most likely were influenced by the Mongol language, such as *poktak* (hat), from *bortu* (a Mongolian word originally from Turkish), and *hŏbŏk* (jar) from *qobura* (a Mongolian literary word).[33] Documentary sources demonstrate that many families of Cheju Island with surnames such as Ch'o 肖, Yang 梁, An 安, Kang 姜, and Tae 對 are descendants of the Mongols, an indication that the Mongols influenced the Cheju through intermarriage and other modes of contact.[34] Recent archaeological excavations, including the "refuge palace" site for the Mongol emperor, Shunti, show Mongol influences on artistic elements in buildings, such as the remarkably delicate stones and Mongolian roof tiles in Pŏphwa 法華 Temple that began to be rebuilt in 1269 and was completed in 1279.[35] The Mongols even built mountain ranches on Cheju Island to grow the five livestock breeds of Mongolia, including horses and sheep, which also affected the islanders' dietary customs.[36] While we have no concrete evidence from

---

[32] Yi Kaesŏk, *Koryŏ Tae Wŏn kwan'gye yŏn'gu*, 304–328.

[33] Pae Yŏnghwan (Bae Young-hwan), "Cheju pangŏn sok ŭi Monggol ch'ayongŏ e taehan yŏn'gusa chŏk kŏmt'o" (A Historical Review of Mongolian Loanwords in Cheju Dialect), *Ŏmun nonjip* (Journal of Language and Literature) 68 (December 2016): 24–25. For more details on Mongolian loanwords found in Cheju dialects, as well as in the entire Korean language, see ibid., 7–36; and Kwŏn Sŏnghun, "Cheju pangŏn sok ŭi Monggol ŏ ch'ayongŏ" (Mongolian Loanwords in Cheju Dialect), *Tongak ŏmunhak* (Dong-ak Society of Language and Literature) 70 (February 2017): 53–67.

[34] Yi Haeng 李荇, *Sinjŭng Tongguk yŏji sŭngnam* 新增東國輿地勝覽 (Newly Enlarged Geographical Survey of the Eastern Country [Korea]) (1531), (Seoul: Tongguk munhwasa 東國文化社, 1958).

[35] Kang Ch'anghwa (Kang Changhwa). "Cheju Pŏphwasaji ŭi kogohak chŏk yŏn'gu" (Archaeological Study of Pŏphwa Temple Ruins in Cheju). *Chejudosa yŏn'gu* (Journal of Cheju History Studies) 9 (2000): 28–33; Kim Kyŏngju (Kim Gyeong-ju), "Kogo charyo ro salp'yŏ pon Wŏn kwa T'amna" (Study of the Yuan Dynasty and Tamla [T'amna] Based on Archeological Materials), *T'amna munhwa* (Tamna Culture) 52 (2016): 129–160. Scholars have hypothesized that further excavation and investigation of the ruins of the "palace site" in Kangjŏng tong would confirm that the Ming and Koryŏ regimes planned construction of a large-scale housing complex within the precinct of this temple where the refugees of the Mongol royal families who fled to Jeju after the fall of the Yuan could reside and settle. Kim Ilu (Kim Il-woo), "Monggol hwangje Sunje (順帝) ŭi Cheju p'inan kungjŏn t'ŏ t'amsaek" (In Search of the Refuge Palace Site for the Mongol Emperor, Sunje), *Monggol hak* (Mongolian Studies) 46 (2016): 27–61.

[36] Yi Chongsu, "13 segi T'amna wa Wŏn cheguk ŭi ŭmsik munhwa pyŏndong punsŏk" (An Analysis of the Food Culture Acculturation of T'amna in the 13th century), *Asea yŏn'gu* (Journal of Asiatic Studies) 59 (March 2016): 157–160.

the period to prove such exchanges, later sources suggest that soju also began to spread there during that period; not only that, it even became the most important alcoholic beverage there because it can be preserved for longer periods of time than any other local fermented beverage.[37] As there were more horses in Cheju, the Mongols could have distilled more fermented mare's milk there, though we have no evidence yet for that. It seems that Cheju locals used grain-based fermented wine to make their distilled soju. Because rice is not one of their main grain products, the people of Cheju developed a form of soju based on fermented foxtail millet and barley, which grow better than rice, in contrast to other parts of the Korean peninsula where soju based on rice wine was more dominant, and rice was the most important grain for making soju.

In this way, it is highly likely that the Mongol military presence that continued for several decades, not only on Cheju Island but also on the entire Korean peninsula, was an important vector for the introduction and spread of distilled liquor in Koryŏ, yet this was part of a larger scale of cross-cultural contact between the people of the Yuan and those of Koryŏ. After Koryŏ sued for peace in 1259 and became a *quda* (marriage alliance) state of the Yuan, many economic and cultural exchanges took place through the diplomatic relations of that time, including the transfer of soju. Even after the failure of the Japan expedition, the Mongols maintained the Eastern Expedition Field Headquarters in order to put Koryŏ under its supervision and exploit resources there systematically. The hierarchical political relationship this helped to produce influenced various aspects of Koryŏ society in both positive and negative ways. Such a political environment allowed people in various sectors of society to exchange good and ideas.

The most active adoption of Yuan culture was made by the royal family of Koryŏ, who raised their status in the Great Yuan Ulus through the *quda* relationship in which the Koryŏ monarchs became imperial sons-in-law (*khuregen*). The first Yuan princess to become queen of Koryŏ was a granddaughter of Khubilai, a fact that symbolizes the strategic importance of Koryŏ to the Yuan. Although Khubilai provided Koryŏ with several privileges and promised to allow it to keep its traditional systems and customs, the Koryŏ kings nonetheless actively accepted Mongol culture in the course of preserving their kingdom amidst the uncertainty of early Mongol interference. Because Koryŏ's Crown princes were raised in the Yuan capital of Daidu, they were naturally accustomed to Mongol culture. When they returned to Koryŏ with their Mongol princess brides, they brought Mongol culture

---

[37] See Chapter 4 of this book.

to Koryŏ with them. Conversely, many Mongols traveled to Koryŏ as political relations normalized, and similarly influenced Koryŏ society, both directly and indirectly. As we will see in more detail below, people at the Koryŏ court adopted various cultural features such as dress and food through political rituals and parties, features which then later spread more broadly to commoners.

When we discuss Yuan–Koryŏ relations, however, we should not forget that the Koryŏ people also experienced the influence of the Yuan by traveling to Yuan China. As Koryŏ subordinated to the Mongols after its military defeat, the Mongols also brought many things from Koryŏ with them as tribute, a situation that led to direct cultural exchanges through human contact in both Koryŏ and the Yuan. The Yuan–Koryŏ political relationship has some features similar to those of earlier tributary relations between dynasties in China and Korea; however, the Mongols requested far more as tribute than had any earlier dynasty in China. They asked the Koryŏ court to supply many young women as *kongnyŏ* (young women selected as tribute) and men as eunuchs, as well as a large amount of various goods and special local products.[38] The privileges enjoyed by the Koryŏ royal family within the empire were thus gained through the exploitation and sacrifice of many commoners. The frequent supply of young women as kongnyŏ led to changes in Koryŏ social customs, indicated by such things as the rise of early marriage. Such harsh exploitation of Koryŏ resources by the Mongols resembled the hardships endured by Chinese, who were placed at the lowest level of the class hierarchy in their own country under the Mongols' direct rule (although it was not identical). As many Chinese co-operated with the Mongols in China, many Koryŏ people cultivated connections with the Mongols and the Koryŏ court, including powerful families such as Kwŏn Kyŏm of the Andong Kwŏn and Ki Chao of the Haengju Ki.[39] Some kongnyŏ, eunuchs, or other officials traveled to China and achieved power at the Yuan court. The most famous example is Empress Ki (1315–1369), who was brought to the Yuan court as a kongnyŏ yet became the consort of Emperor Shunti (Toghan Temür, r. 1333–1367). She and her family used her power to exploit their kingdom and ensure wealth and prestige in both Yuan and

[38] For the case of kongnyŏ, see Pak Kyŏngja (Park Kyung-Ja), "Kongnyŏ ch'ulsin Koryŏ yŏindŭl ŭi sam" (Life of Korean Women as Kongnyŏ), *Yŏksa wa tamnon* 55 (April 2010): 33–64. For the case of eunuchs, see Yi Kaesŏk, *Koryŏ Tae Wŏn kwan'gye yŏn'gu*, 248–274; Chang Tong'ik, *Koryŏ hugi oegyosa yŏn'gu*, 178–181.

[39] For example, several members of the Haengju Ki clan received honorary titles as kings (*wang*) from the Mongol emperor and "were able to parlay their connections with the imperial house into wealth and prestige that rivaled those of the Koryŏ kings." John B. Duncan, *The Origins of the Chosŏn Dynasty* (Seattle: University of Washington Press, 2000), 165–166.

Koryŏ courts.[40] Many people from Koryŏ lived among the Yuan in such ways.[41] A Koryŏ official document even testifies to the fact that a woman from Koryŏ married a prince from India.[42] Such a story demonstrates the exposure of some Koryŏ people to a broader Eurasian culture beyond China and Mongolia where they could wield power and influence people in similar ways. Some of them traveled frequently between the Yuan domain and Koryŏ and played an important role in cultural exchange. They could have brought to Koryŏ Yuan cultural items such as shaojiu (soju) – or vice versa.

Such forms of political and social relations also had considerable influence on economic exchange between the two societies. As Koryŏ became part of the Mongol Empire, it automatically entered into the Eurasian-scale economic system that the Mongol Empire had created. The Koryŏ court collected goods, including human tribute, which the Mongols requested and sent to their Yuan superiors. This tribute included grains, silver, horses, ginseng, ramie cloth, and celadon porcelain. In their requests for tribute, the Yuan court even gave specific numbers, so the pressure borne by the Koryŏ court to deliver was heavy. Yet the court also took advantage of their unprecedented situation to facilitate private trade between the two societies. As Lee Kang Hahn argues, the last century of the Koryŏ era witnessed an unprecedented increase in the outward activities of Koreans, who participated in foreign trade networks and also became more open to the outside world than ever before thanks to the Pax Mongolica.[43] Documents testify to trade in a variety of goods, including local Koryŏ specialties like ginseng and ramie cloth and Yuan Chinese products like high-quality silks and books.[44] Many new things traveled to Koryŏ, just as Koryŏ goods were introduced to the broader Eurasian market.[45] Even traders from Central and Western Asia, including the famous *ortaq* merchants, who came to China on a larger scale than ever before through the Mongols' expanded Eurasian trade, traveled all the way to Koryŏ to conduct business.[46] Many Central and Western

[40] See Chŏng Kusŏn (Chung Koo-sun), "Koryŏ mal Ki hwanghu ilchok ŭi tŭkse wa mollak" (The Rise and Fall of the Family of Empress Ki at the End of the Koryŏ Dynasty), *Dong Gook Sa Hak* 40 (2004): 167–185.
[41] See Yi Kaesŏk, *Koryŏ Tae Wŏn kwan'gye yŏn'gu*, 275–298; Robinson, 118–129.
[42] Yi Kaesŏk, *Koryŏ Tae Wŏn kwan'gye yŏn'gu*, 294.
[43] Yi Kanghan, *Koryŏ wa Wŏn cheguk ŭi kyoyŏk ŭi yŏksa*, 21–25.
[44] Yi Chinhan (Lee Jin Han), *Koryŏ sidae muyŏk kwa pada* (Trade and the Sea during the Koryŏ) (Seoul: Kyŏngin munhwasa, 2014), 200–209.
[45] On the transfer of a large quantity of silver from Koryŏ to the Yuan and then further to Central Asia and Southeast Asia, see Kim Yŏngje, *Koryŏ sangin kwa Tong Asia muyŏksa*, 227–245.
[46] Lee Kang Hahn (Yi Kanghan) argues that an important trade item was slaves. Yi Kanghan, *Koryŏ wa Wŏn cheguk ŭi kyoyŏk ŭi yŏksa*, 133–152.

Asians came to Koryŏ as merchants, and still others came in other career roles, living in Korea within both elite and commoner societies.[47]

Koryŏ merchants also traveled often to Daidu, where they exchanged many goods. The fact that a large number of Koryŏ merchants went there directly to trade is proven by the publication of language textbooks made specifically for Koryŏ people so they could learn and practice the Chinese language. Two extant language textbooks used at that time, *Nogŏltae* 老乞大 (Chinese *Lao Qida*, "Old Cathayan") and *Pakt'ongsa* 朴通事 (Interpreter Park), not only inform us of the language structure used at that time but also testify to the various aspects of people's daily lives in Daidu and the things that these people had to know to live and trade there.[48] As *Nogŏltae* was an elementary-level Chinese textbook for Koryŏ merchants, it provides scholars with content that describes their major concerns regarding basic information about travel, trade, and goods. On some pages, the book talks about alcoholic drinks; however, it does not specify types, probably because its authors tried to provide basic words for learners. However, *Pakt'ongsa*, the other textbook for more advanced-level Chinese-language learners, provides more details about a variety of things, including alcoholic drinks. *Milinqin shaojiu* 蜜林擒 燒酒, an alcoholic drink from southern China made by putting honey in shaojiu (soju), was popular in southern China.[49] Its mention in *Pakt'ongsa* hints that it became widespread in Daidu, too, and was possibly transferred to Koryŏ through merchants and Koryŏ people of other kinds who traveled to Daidu, such as governmental officials and scholars. The book also shows that Koryŏ merchants wanted to buy silks of high quality in Daidu to bring them back to Koryŏ to sell. Chapter 2 discusses a passage in the official history of Koryŏ that reports that, at the end of the dynasty, the government prohibited luxury goods such as soju and silks. Probably such luxurious items imported from the Yuan gradually grew popular among Koryŏ's wealthy, prompting the government's ban. When foreign merchants, including West

[47] Yi Kaesŏk, *Koryŏ Tae Wŏn kwan'gye yŏn'gu*, 334–344.

[48] For recent Korean translations of *Nogŏltae*, see Chŏng Kwang (Chung Kwang), trans. *(Yŏkchu) wŏnbon Nogŏltae* (A New Annotated Translation of the Original *Nogŏltae*) (Seoul: Pangmunsa, 2010); and Yi Yukhwa, trans., *Wŏnbon Nogŏltae shinju shinyŏk* (A New Annotated Translation of the Original *Nogŏltae*) (Seoul: Sinasa, 2015). For a Japanese translation, see Kim Bunkyō 金文京, Gen Yukiko 玄幸子, and Satō Haruhiko 佐藤晴彦, trans., *Rō kitsu dai: Chōsen chūsei no chūgokugo kaiwa dokuhon* 老乞 大―朝鮮中世の中国語会話読本 (*Rō kitsu dai:* Chinese Conversation Textbook in Medieval Chosŏn) (Tokyo: Heibon sha, 2002). For a Korean translation of *Pakt'ongsa* with detailed comments, see Chŏng Sŭnghye et al., *Pakt'ongsa Wŏn nara Taedo rŭl kŏnilta* (Pakt'ongsa Wanders in the Yuan Dynasty's Capital Daidu) (an annotated translation of *Pakt'ongsa*) (Seoul: Pangmunsa, 2011).

[49] Chŏng Sŭnghye, *Pakt'ongsa Wŏn nara Taedo rŭl kŏnilta*, 20–21. In general, the Chinese textbooks also do not provide many specific names of liquors, yet the fact that *Pakt'ongsa* talks about shaojiu (soju) suggests a gradual spread in the Yuan period of shaojiu (soju), which would be popular in the later dynasties of both China and Korea.

Asians, came to Koryŏ for trade, Koryŏ merchants must have benefited in turn, having met many foreign merchants of different ethnic groups and experienced and handled various kinds of goods from different parts of Eurasia.

As a high-level Chinese-language textbook, *Pak'ongsa* might also have been useful to Koryŏ scholars who were coming and going between China and Korea on government business. While more people in earlier dynasties were able to rely on professional interpreters who served official diplomatic missions, the increased contacts engaged many officials more directly and deeply in Yuan politics and brought Yuan and Koryŏ officials and scholars into direct contact on an unprecedented level. The new situation in which many Koryŏ people had to go to Yuan China and meet with Mongols and Chinese directly might have required the publication of such textbooks for such people. Many Koryŏ scholars had opportunities to advance in the Yuan dynasty government through the empire's civil service examination, which was available to Koryŏ scholars. The Mongols had once abolished the traditional Chinese civil service examination at the end of the Song dynasty, yet the exams were revived in 1315 by Emperor Renzong (r. 1311–1320), albeit with significant changes, including the placement of Neo-Confucianism at the center of the exam curriculum for the first time in history. These scholars, some of whom had opportunities to study in Daidu and interact with Yuan officials and literati in an unprecedentedly multinational environment, played an important role in transferring high culture to Koryŏ society. Not only did the Koryŏ scholars introduce their compatriots to Neo-Confucianism, which would become the most important ideology in Korea under the later Chosŏn dynasty, but they also transferred other cultural items. One of the most important literati of late Koryŏ was Yi Saek (1328–1396), who mentioned arak as a distilled liquor for the first time in his poem (see Chapter 2).

Yi Saek is a famous literatus and an important official who lived from the late Koryŏ to early Chosŏn periods; his case illustrates the life of Koryŏ literati who lived in this transitional period.[50] Yi Saek's special life experience was influenced considerably by his father, Yi Kok (1298–1351), who was from a middle-class family that lived in the countryside yet rose in society through successful study and civil service examinations in both Koryŏ and the Yuan empire.[51] Yi Kok's success in the Yuan civil

---

[50] For the importance of his literary works in the Koryŏ period, see Yi Saek (1328–1396), *Kukyŏk Mokŭnjip* (Literary Collection of Mokŭn), 1:1, and Yi Ikchu (Lee Ik Joo), *Yi Saek ŭi sam kwa saenggak* (The Life and Thoughts of Lee Saek (Yi Saek)) (Seoul: Ilchogak, 2013), 15–18.

[51] Yi Ikchu, *Yi Saek ŭi sam kwa saenggak*, 3.

service examination after several failures allowed him to advance to important posts in both Koryŏ and Yuan governments. His frequent stays and unusual governmental post in Daidu gave his son Yi Saek the opportunity to study in the Imperial Academy (Guozijian), the highest academic institute in the Yuan dynasty. While Yi Saek's case was particularly special and successful, it seems that many other scholars from Koryŏ tried to study there during the Yuan because that would ensure their success back home, and in doing so they shared common international experiences. King Ch'ungsŏn also facilitated contacts and exchanges between the Confucian literati of the Yuan and Koryŏ.[52] The object of study that consumed the lives of Yi Saek and other Koryŏ scholars preparing for the Yuan civil service examination was Neo-Confucianism. A rejuvenated form of Confucianism modified by a moral form of metaphysics that developed during the Song dynasty (960–1279), Neo-Confucianism became orthodoxy in society and politics under the Yuan, not only in China but across East Asia.[53] Upon his return to Koryŏ, Yi Saek taught the new philosophy to his disciples, who then contributed to the establishment of Chosŏn and the adoption of Neo-Confucianism as a state ideology for the new dynasty. Neo-Confucianism, therefore, offers an example of an important cultural element transferred from Yuan China to Koryŏ Korea that would have a lasting influence on Korea in the long run. While many studies of Yi Saek focus on his literary pieces and political role in the Koryŏ–Chosŏn transition, less attention is paid to his experiences in the international environment of China and Korea under the Mongols and his role in cross-cultural exchange. Let us look at these aspects in order to trace possible routes to Koryŏ of cross-cultural traffic in new ideas such as Neo-Confucianism, new commodities such as arak liquor, and other possible new cultural elements.

In his youth, Yi Saek spent many years in Daidu, where he studied at the Imperial Academy. According to his literary works, which reflect on his life and times, he got along more closely with certain Koryŏ people than with the many Chinese and Mongols who dominated the capital.[54] It seems that he grew lonely because his oral Chinese was not proficient enough to allow him to communicate with his Chinese colleagues. He was

---

[52] Kim Namil, *Koryŏ mal Chosŏn ch'ogi ŭi segyegwan kwa yŏksa ŭisik: Yi Saek* (李穡) *kwa Kwŏn Kŭn* (權近) *ŭl chungsimŭro* (World Views and Historical Consciousness in the Late Koryŏ and Early Chosŏn Periods: Focusing on Yi Saek and Kwŏn Kŭn) (Seoul: Kyŏngin munhwasa), 11.

[53] Peter K. Bol, *Neo-Confucianism in History* (Cambridge, MA: Harvard University Asia Center, distributed by Harvard University Press, 2008).

[54] Yi Ikchu, *Yi Saek ŭi sam kwa saenggak*, 56–57.

successful in the civil service examination because his written Chinese was good, as was the case with many other Koreans at that time and earlier.[55] He very likely studied spoken Chinese and possibly some Mongolian, because he gradually expanded his social networks after he was raised to an important governmental post that connected the Yuan and Koryŏ governments. He communicated with famous contemporaneous Chinese literati such as Ouyang Xuan (1283–1357) and Zhao Mengfu (1254–1322).[56] We can learn from his literary writings that he accepted the Yuan as the authority of the world order.[57] In Daidu, a genuinely cosmopolitan city with people of many different ethnicities, it would have been natural for him and other Koryŏ literati to accept the new culture of the Mongol Empire that by Yi's time had been flourishing for a century. When the Yuan government established the Imperial Academy (Guozijian), as a parallel institute it also created the Islamic Imperial Academy (Huihui Guozijian 回回國子監), where many Western scholars worked with books from their regions. The extant works of Yi Saek do not reveal many details about the cosmopolitan characteristics of his social circles; however, new philosophical and cultural elements like Neo-Confucianism and arak liquor appear sporadically.

Many people in Koryŏ other than Yi Saek experienced the new culture under the Yuan through contacts made through the cosmopolitan Mongol Empire. Many literati and people in various fields traveled to China during Mongol rule, interchanged with various peoples in the cities there, and contributed to cultural exchange. In his pioneering study of the Uighurs in Koryŏ and Chosŏn Korea, Michael Brose argues that even non-Chinese Uighurs wound up in Korea and served the Korean kings with their Neo-Confucian literary and administrative skills.[58] From ancient times, East Asian societies like Korea and Japan received considerable influence from Chinese and broader Eurasian culture, and many forms of culture such as thought, religion, and foodways were introduced by intellectuals and through the books that they brought with them. Thanks to this new scale of contact, Koryŏ society assimilated more cultural items, many of which had enormous influence on their society. Looking at another important Chinese city for international contact, the major port of Ningbo, sources indicate that many scholars from different

[55] For more cases of Koryŏ literati who were successful in the civil service examination in Yuan China, see Chang Tong'ik, *Koryŏ hugi oegyosa yŏn'gu*, 173–178.
[56] Yi Saek, *Kukyŏk Mokŭnjip*, 1:3.
[57] Kim Namil, *Koryŏ mal Chosŏn ch'ogi ŭi segyegwan*, 7.
[58] Michael C. Brose, "Neo-Confucian Uyghur Semuren in Koryŏ and Chosŏn Korean Society and Politics," in *Eurasian Influences on Yuan China*, ed. Morris Rossabi (Singapore: Institute of Southeast Asia Studies, 2013), 125–158.

countries, including Koryŏ, exchanged ideas and other cultural items there, including books.[59] Those cultural items that transferred through human contact had derived from political and economic relations and as a consequence influenced people's lives on a grand scale for a long time. Alcoholic drinking culture, including soju, is one of these items. Some items began to exert more influence in the new society after some time had passed. There are many such cases from the Koryŏ period to the Chosŏn period, each of which is supported by rich documentation. As an initial attempt that surveys major cultural exchanges between the Yuan and Koryŏ from the context of a broad Afro-Eurasian perspective, the next section examines several examples of items in these exchanges, such as science, foods, and dress. This will help us to understand the broader context for the transfer of soju and the richness of the cultural exchange that occurred at that time.

Examples of Cultural Transfer and Influence

Science, Including Astronomy and Cartography

From ancient times, various fields of advanced science that developed in China influenced those in East Asian societies, including Korea and Japan. Astronomy had a particularly strong influence. It played the most important role in premodern society for state rituals and agricultural life. From ancient times, people in Korea tried to interpret celestial objects and phenomena in order to improve their agricultural lives by developing unique astronomical practices.[60] However, they received considerable influence in the study of astronomy from Chinese dynasties, whose specialists, persistent and accurate observers of celestial phenomena, were among the most advanced in the world in developing the field.[61] By the time of the Koryŏ dynasty, the Koreans had adopted a calendar that was brought to them from China during the Tang dynasty. Since then, however, the Chinese underwent major changes in their calendrical system, particularly during the Yuan period.[62] This happened because the Yuan court under the Mongols launched path-breaking innovations

[59] Miya Noriko, *Mongoru jidai no shuppan bunka* (The Publishing Culture of the Mongol Period) (Nagoya: Nagoya University Press, 2006), 575–580.
[60] See Chŏn Yonghun (Jun Yong Hoon), *Han'guk ch'ŏnmunhak sa* (History of Astronomy in Korea) (P'aju: Tŭllyŏk, 2017), 61–124.
[61] Joseph Needham and Wang Ling, *Science and Civilisation in China*, vol. 3: *Mathematics and the Sciences of the Heavens and Earth* (Cambridge: Cambridge University Press, 1959), 171.
[62] Han Yŏngo (Hahn Young-Ho), and Yi Ŭnhŭi (Lee Eun-Hee), "Ryŏmal Sŏnch'o Pon'gungnyŏk wansŏng ŭi tojŏng" (Accomplishment of a Domestic Calendar System

in the field of astronomy, in particular by hiring many western scholars, like Jamāl al-Dīn (Zhamaluding, fl. end of the thirteenth century), who in 1267 brought seven astronomical instruments, including an astrolabe, with him from Iran to the Yuan court, which, in combination with his knowledge, contributed greatly to innovations in astronomy and cartography by combining Islamic and Chinese elements. Working at special offices for foreign scholars created by the Mongol government, like the Islamic Imperial Academy (Huihui Guozijian 回回國子監) and the Islamic Astronomical Bureau and Observatory (Huihui Sitiantai 回回司天臺), parallel institutions to the traditional Chinese ones, these western scholars played significant roles in developing the Huihui calendar (*Huihui li* 回回曆, or Muslim calendar) that Chinese officials came to use along with the official Chinese Shoushi calendar (*Shoushi li* 授時曆). Guo Shoujing, who developed the Shoushi calendar, most likely had academic exchanges with Jamāl al-Dīn, who was the supervisor of the Palace Library (*xing mishujianshi*), where he supervised both astronomical bureaus as subordinates and contributed to exchanges between Chinese and western scholars.[63]

Koryŏ acquired Guo Shoujing's Shoushi calendar, which they regarded as a novel and remarkably advanced system, later in the reign of King Ch'ungnyŏl and King Ch'ungsŏn.[64] Yet it was not easy for Koryŏ scholars to immediately utilize it, because they had been using a considerably outdated calendar system. Scholars in the late Koryŏ and early Chosŏn period, particularly during the reign of King Sejong, made desperate long-term efforts to remeasure and correct errors in order to implement this new system in Korea. Once they had assimilated the advanced Chinese system, resolved its mathematical problems, and finally produced a suitable calendrical system, which they called "Ch'ilchŏngsan Naep'yŏn" (Inner Section of Ch'ilchŏngsan), they imported the Huihui calendar, the Muslim calendar made along with the Yuan Shoushi calendar, and applied it to making a supplementary system, which they called "Ch'ilchŏngsan Oep'yŏn" (Outer Section of Ch'ilchŏngsan).[65] We should note that both new calendar systems produced in Yuan China and assimilated into the Korean system were fruits of Eurasian cross-cultural contacts under the Mongols.

during the Late Koryŏ and the Early Chosŏn Period), *Tongbang Hakchi* (Journal of Korean Studies) 155 (2011): 32–33.
[63] Park, *Mapping the Chinese and Islamic Worlds*, 99–100.
[64] Han and Yi, "Ryŏmal Sŏnch'o Pon'gungnyŏk wansŏng ŭi tojŏng," 3ϵ.
[65] Han Yŏngo, Yi Ŭnhŭi, and Kang Minjŏng (Kang Min-Jeong), trans., *Ch'ilchŏngsan Naep'yŏn: hae wa tal, tasŏt haengsŏng ŭi ch'ŏnmunhak* (Ch'ilchŏngsan Naep'yŏn: Astronomy of the Sun, Moon, and Five Planets), vol. 1 (Seoul: Han'guk kojŏn pŏnyŏgwŏn, 2016), 38–39.

Two other academic fields that made new developments in the Mongol era and transferred from China to Korea in the late Koryŏ–early Chosŏn period were geography and cartography. Through the cross-cultural exchanges of the Pax Mongolica, people in different societies, including China and Europe, acquired geographic knowledge that had been less familiar to them in previous periods and expanded their worldviews of Afro-Eurasia. In particular, the Chinese significantly expanded their geographic knowledge and worldview with the help of geographic and cartographic works that scholars from Central and Western Asia brought to the Yuan court. We know, for example, that *Honil kangni yŏktae kukto chi to* 混一疆理歷代國都之圖 (The Map of Integrated Lands and Regions of Historical Countries and Capitals) (hereafter *Kangnido*, following Gari Ledyard's abbreviation) was created in Korea thanks to a Korean scholar who was active during the transitional years between the late Koryŏ and early Chosŏn. We can guess that many of the Koryŏ scholars who had been to the Yuan Empire were able to conceptualize the larger world of Afro-Eurasia through the opportunities they had to view Chinese world maps, which constituted the base maps for the production of the *Kangnido* in 1402.

We can speculate about influences on other academic fields, such as linguistics. Some scholars convincingly suggest the possibility that the ʽPhags-pa script used by the Mongols influenced the creation of the Korean Hangul alphabet during the early Chosŏn period. Although the style of the ʽPhags-pa script is quite different from that of Hangul, because the Chinese characters that the Koreans had been using are logograms, nonetheless, having seen the ʽPhags-pa script as an alphabet system used by the Mongols could have given some inspiration.[66]

As scholars working on scientific fields in premodern East Asian societies were less respected than Confucian literati, we do not know how further in-person meetings and exchanges among scientists were made under the auspices of state patronage, although we can assume that such contacts existed at the private level to a limited degree. Still, books could have been an important medium for transferring scientific knowledge to Korea, as it was easier for people to transport them through the open commercial environment within the Mongol Empire. Distillation is not such a high-level technology, so the Koreans probably learned how to make soju with a portable device by way of direct human contact and guides rather than books. Yet we see some examples of major books about medicine and foods written in the late Yuan and early Ming eras that were

---

[66] See Chŏng Kwang, *Hunminjŏngŭm kwa Pʼasŭpʼa munja* (Hunminjŏngŭm and the ʽPhags-pa Script) (Seoul:Yŏngnak, 2012).

transferred to Korea and began to generate more direct influence on scholarly understanding of soju and its distillation. These books discuss many topics related to medicine and foods, which deserve close examination.

Medicine, Foods, and Drinks

Medicine is another important scientific field that experienced unprecedented exchange among different societies in Eurasia during the Pax Mongolica. Like the case of astronomy, Chinese and West Asian doctors developed medical knowledge and practices in their traditions over many centuries, and they were each able to produce medical works that include knowledge of the other tradition thanks to scholarly exchanges. This is attested in an encyclopedia of Chinese medicine entitled *Tānksūqnāma* that was produced by Rashīd al-Dīn (1247–1318) in the Il-khanate of Iran and the *Huihui yaofang* (Muslim Medicinal Recipes) produced during the early Ming dynasty based on the knowledge available in the Mongol Yuan period through a Syrian medical family resident in China.[67]

Unlike the Huihui calendar that was transferred to the Korean peninsula, where it had significant influence on astronomy under both Koryŏ and Chosŏn dynasties, *Huihui yaofang* seems to have been less widely read and probably exerted a more limited influence in Korea as a book. No evidence shows that it was introduced to Koryŏ or Chosŏn in this medium. Yet it is highly likely that some elements and content of western medical knowledge circulating throughout the Mongol Empire were incorporated into Chinese medical books circulating in Yuan China, which were then brought to the Korean peninsula during the Koryŏ–Chosŏn transition. In the history of Korean medicine, as Shin Dongwon demonstrates, no significant medical works were produced before the Chosŏn dynasty, and therefore the influx of many Chinese medical works in the transitional period would have greatly contributed to an innovation in the history of Korean medicine. As Chapter 4 below shows, several medical books commissioned by the court in the early and mid-Chosŏn era, including *Ŭibang yuch'wi* 醫方類聚 (Collection of Various Kinds of Medical Arts) (compiled 1445) and the *Tongŭi pogam* 東醫寶鑑 (Precious Mirror of Eastern Medicine) (compiled 1596–1610), tried to combine existing knowledge in many extant medical books collected at court. These books cite many Chinese works written in the Yuan and Ming periods and earlier, introducing the use of soju to cure diseases.

[67] Joseph Needham, *Clerks and Craftsmen in China and the West: Lectures and Addresses on the History of Science and Technology* (Cambridge: Cambridge University Press, 1970), 17.

Medical books include much information about foods. Such books introduced a certain quantity of information about new food ingredients, as in the case of alcoholic drinks, including soju. As later Chosŏn books like Hong Mansŏn's (1643–1715) *Sallim Kyŏngje* cite Yuan-period food books like Hu Sihui's *Yinshan zhengyao* (Proper and Essential Things for the Emperor's Food and Drink) (1330) and *Jujia biyong* (fourteenth century), these books were surely brought to Korea by scholars from the mid- to later Chosŏn period (seventeenth to eighteenth centuries). Much more information about new foods and foodways, and about foods generally, however, was transferred from Yuan China to Koryŏ Korea through human contact on a large scale. We have seen in Chapter 2 that special alcoholic beverages other than soju, such as mare's milk liquor (mayuju, called tongnak; Chinese *marujiu, tonglao*) and grape wine, were brought to the Korean peninsula as gifts or commercial goods through diplomatic and commercial contacts, as clearly attested in official documents.

Close contacts with the Mongols at that time might have significant influence on the food culture of Koryŏ in general. Because Buddhism was a dominant religion, the people of Koryŏ did not eat much meat. Xu Jing, who during the mid-Koryŏ period talked about varieties of alcoholic drink in his eyewitness account based on his diplomatic mission from the Song to Koryŏ in the twelfth century, also testifies that the Koryŏ people avoided killing animals for meat and were not good at butchery because of their faith in Buddhism, and therefore it was difficult for people, except kings and officials, to eat sheep and pig meat.[68] Yet the Mongols' meat-eating culture was transferred to Koryŏ and began to spread across the Korean peninsula from the Mongol period, along with the revival of a hunting culture and an increase of livestock husbandry and the supply of meat even for commoners.[69] Even Buddhist monks in Koryŏ, who had many chances to travel to China and have contact with Buddhist monks there, were affected by such a cultural change and ate meat dishes in the course of enjoying their luxurious lives.[70] Such a change in culinary

[68] Cho Wŏn (Cho Won), "*Ŭmsŏnjŏngyo* wa Tae Wŏn cheguk ŭmsik munhwa ŭi Tongashia chŏnp'a" (*Yinshan Zhengyao* and the Influence of Yuan Culinary Culture in East Asia), *Yŏksa Hakbo* (Korean Historical Review) 233 (March 2017): 199–200; for Xu Jing's mission and account, see Chapter 2 of this book.

[69] Yu Aeryŏng (Yu Ahe-Ryung), "Monggo ka Koryŏ ŭi yungnyu sigyong e mich'in yŏnghyang" (Mongol Influences on Meat Eating in Koryŏ), *Kuksagwan nonch'ong* 87 (1999): 221–237.

[70] Hŏ Hŭngsik, *Koryŏ pulgyosa yŏn'gu* (A Study of the History of Koryŏ Buddhism) (Seoul: Ilchogak, 1986), 514–515. Hŏ Hŭngsik also argues that Buddhist monks in Koryŏ became involved in sexual moral hazards as a consequence of the unrestrained trends introduced by the Mongols.

culture could have boosted consumption of distilled drinks with a strong alcoholic content that worked better with meat.

Some specific kinds of foods with Mongol and Central Asian origins, along with these big changes in eating trends, were probably transferred from the Yuan empire to Koryŏ too. Among the prestigious court cuisine enjoyed by the Mongols is a soup called *shülen*, the banquet soup, basically a lamb broth with almost anything added to create a great variety of distinctive dishes. Such dishes are well documented in the *Yinshan zhengyao* for China but are equally well known in Iran and elsewhere, an indication of its spread from China to Iran and other places like Korea. Some people have argued that shülen became *sŏllŏngt'ang*, a traditional soup dish in Korea. Of course, there are other theories about the food's origin. Yet recently accumulated investigations and ideas suggest that meat-soup dishes like sŏllŏngt'ang were influenced by similar ones from China during the Koryŏ period.[71] The most crucial evidence comes from *Mongŏ yuhae* 蒙語類解 (Categorization and Translation of Mongolian Language), a Mongolian–Korean dictionary compiled in the mid-eighteenth century under the Chosŏn dynasty, which interprets a dish called *kongt'ang* 空湯 as "soup that boiled meat" and identifies it as *shuru* in the Mongolian language.[72] Buell argues that shülen and other Mongol foods described in the *Yinshan zhengyao* show significant Turkic influence, from Uighurs in particular but also from smaller Turkic minorities in general, so he could also claim Turkic heritage for the book's author. The new foodways that people in Koryŏ received, therefore, are an example of the spread of Central Asian foodways throughout Eurasia thanks to the Mongol promotion of cultural exchange. These are clear examples that show that Korea enjoyed international cultural exchange by being part of the Mongol Empire. Like the case of soju seen in Chapter 2, these forms of Yuan cooking introduced to Koryŏ were localized to suit their new cultural environment, as seen in the example of beef instead of sheep as the meat of choice in Korea. As Cho Wŏn argues, more foods like *sarup*, a soft drink introduced in the *Yinshan Zhengyao* that was also introduced to Koryŏ by the Yuan, came from the Islamic world to the Mongol imperial court in the course of the fourteenth century.[73] Sources demonstrate a great influence of Koryŏ foods on Yuan cuisine, including ginseng, which was used as a major ingredient for healthy foods such as

---

[71] Pak Chŏngbae, *Hansik ŭi T'ansaeng* (Birth of Korean Food) (Seoul: Sejong sŏjŏk, 2016), 150–159.

[72] This source also defines soju 燒酒 as *arik'i*, a Korean transliteration of *arak*. *Mongŏ yuhae* 蒙語類解 (Categorization and Translation of Mongolian Language) (Seoul: Sŏul taehakkyo Kyujanggak Han'gukhak yŏn'guwŏn, 2006), 97.

[73] Cho Wŏn, "*Ŭmsŏnjŏngyo wa Tae Wŏn cheguk ŭmsik munhwa*," 202–204.

ginseng soup in the *Yinshan Zhengyao*.[74] This is a topic that cannot be discussed further here because of limited space, yet deserves thorough study later.

Dress

Another important cultural item that could be open to influence from other cultures is dress and clothing. Seen from earlier wall paintings in excavated tombs, Koreans shared a similar style of dress with the Chinese in ancient periods of history. Several studies have demonstrated that the new Yuan–Koryŏ relations led to some new important influences on Koryŏ clothing, some of which would continue to be worn in the Chosŏn dynasty. According to Kim Munsuk's study based on comparative analyses of books, paintings, and archaeological remains of costumes before and after the Mongol period, some specific aspects of Mongol costume, like braiding hairstyles, headdresses for women and men, and coats, were adopted into Koryŏ costume.[75] This was part of the phenomenon called *Monggol p'ung* 蒙古風 or "the Mongolian style." Kim Yunjung examines further the historical context for such influences.[76] For example, political protocol, including diplomatic rituals like the Jisun festival, initially brought such changes. With the start of subject relations, Koryŏ kings had to wear purple robes instead of their previous yellow royal robes because only the Yuan emperor could wear such a color. As King Ch'ungsŏn began to engage in Yuan politics more deeply, Koryŏ court society began to assimilate Yuan imperial attire, like the *kogo* (Chinese *gugu* 罟罟, a headdress worn by Mongol elite women) and *taejŏnga* (Chinese *dadingr* 大頂兒, a luxurious accessory to a garment).[77] This process of assimilation further intensified, as diplomatic relations between the two dynasties grew closer.

In his study of clothing and the food life of the people of the Koryŏ period, Pak Yongun argues that, as relations between the Yuan and Koryŏ courts grew closer, many people at the Koryŏ court, among the

[74] Ibid., 199.

[75] Kim Munsuk (Kim Moonsook), "13–14 segi Koryŏ poksik e suyong toen Monggo poksik e kwanhan yŏn'gu" (The Mongol Costume Adopted in the Koryŏ Costume from the 13th to 14th Centuries), *Monggorhak (Mongolian Studies)* 17 (2004): 223–246.

[76] Kim Yunjŏng (Kim Yunjung), "14 segi Koryŏ-Wŏn kwan'gye hwakchang kwa Koryŏ ŭi Wŏn poksik munhwa suyong" (The Expansion of the Koryŏ and Yuan Relationship and Koryŏ's Acceptance of Mongolian Clothing in the 14th Century), *Yŏksa Hakbo (Korean Historical Review)* 234 (June 2017): 63–112.

[77] Ibid., 76–83. For an image of *gugu guan* (*boqta*), see Eiren Shea, *Mongol Court Dress, Identity Formation, and Global Exchange* (New York: Routledge, 2020), Plate 20.

kingdom's upper class and starting with the king, accepted certain modes of Mongol attire like braided hairstyles and headdresses like the kogo, which were encouraged as well as regulated under the king's orders.[78] However, Kim Yunjung argues that fashion styles reflected in the two language textbooks *Pakt'ongsa* and *Nogŏltae*, as well as in artifacts from archaeological excavations, demonstrate that Mongolian clothing spread to commoners as well, and that this gradually changed their sense that the cultural forms of their country should not change.

Such a voluntary adoption of Yuan dress resulted from the gradual acknowledgment among Koryŏ people that Yuan culture was civilized, a phenomenon that is found in many other cultural aspects, such as Neo-Confucianism. Despite such huge cultural influences received by Koryŏ people from the Yuan, Kim Yunjung argues, through their unique organization of the imperial ancestral temple and efforts to write their own history, Koryŏ successfully maintained its name and national identity while assimilating new cultural elements. Indeed, this trend, in which Koreans expand their cultural boundaries after seeking in previous times to adhere closely to singular customs of the past, is apparent in earlier periods of Korean history. The biggest difference in the cultural influences Koreans received from the Mongol Empire was the unprecedented scale on which it occurred.

A variety of assumptions about the Mongol influence on Korean garments have been raised by scholars. Kim Moonsook argues that, while some view the kogo, the special upper-class Mongol ladies' headdress, as the origin of the *chokturi*, a bride's headpiece worn by ladies in the Chosŏn dynasty, we need further study to trace their possible connections, because the forms of these two headdresses differ from each other. Scholars argue that a Mongolian coat called *terlig* became popular among Koryŏ subjects and continued to be worn as a similar Korean coat called *t'ŏllik* or *ch'ŏmni* 帖裏 by people in the Chosŏn dynasty that followed.[79] Some scholars propose the possibility that the Mongols' request for many kongnyŏ or tribute women led to changes in the size of women's upper garments because, in order not to be selected as kongnyŏ, they wanted to pretend to be pregnant.[80] We need to study this further, open to various possibilities. Sources also testify to the influence of the Koryŏ fashion on that of Yuan society thanks to the many Koryŏ people brought to the Yuan empire, such as kongnyŏ. This phenomenon, which corresponds to *Monggol p'ung* or "Mongolian style," was called *Koryŏ yang* 高麗樣 or

---

[78] Pak Yongun, *Koryŏ sidae saramdŭl ŭi ŭiboksik saenghwal* (Clothing and Food Life of the People of the Koryŏ Period) (P'aju: Kyŏngin munhwasa, 2016), 24–29.

[79] Kim Munsuk, "13–14 segi Koryŏ poksik," 244.

[80] Based on personal conversation with Dashdondog Bayarsaikhan on May 12, 2015.

"Koryŏ style."[81] Like the Mongolian style, it is possible that this Koryŏ style continued to influence some Chinese in the Ming period after the Ming dynasty replaced the Yuan dynasty, a topic to investigate further.

Conclusion

This chapter has discussed how unprecedented political changes in the history of Eurasia during the Mongol period affected Korea's relations with China and the wider world. Despite being part of the Mongol Empire as a subject, vassal state, Koryŏ created a royal intermarriage relationship with the Mongol court of the Great Yuan Ulus in China and Mongolia, the suzerain state of the Mongol Empire, and through cultural channels it created actively received various cultural elements to promote the kingdom's status and ensure its sovereignty. Rich political, economic, and social relations between the Yuan and Koryŏ led to diverse cultural exchanges between the two societies, and people in Koryŏ were consequently exposed to various kinds of influence from the broader Eurasian world. Those involved in this contact were people of different sectors, ranging from members of the royal family to soldiers, officials, scholars, merchants, women, and eunuchs. Although we have only a few direct forms of documentation about the transfer of soju from the Yuan to Koryŏ, much richer documentation about various other cultural items in larger contexts provides indirect yet strong circumstantial evidence for the rise of soju in Koryŏ Korea. It was not only military contacts but also many other forms of contact in various sectors of society, including commercial and scholarly contacts, that facilitated soju's transfer. That is, soju was one among many items that were transferred, exchanged, and localized innovatively through Korea's connection to the Yuan and the wider Eurasian world. As this book's history of soju develops, it reveals a need to pursue more detailed studies of various other cases of cultural exchange, which historians of Koryŏ have gradually begun to advance in the recent years.

Careful attention shows that it was not only items from China but, more accurately, items transferred through a broader Eurasian sphere – and more precisely through Yuan–Koryŏ social relations – that influenced Korea thanks to the Yuan empire's central place within the Mongol Empire as a whole. In this larger context, soju developed in Korea, a possible product of a Chinese distillation tradition combined with West and South or Southeast Asian arak traditions. In this same framework, the same process can be seen in the development of other cultural

---

[81] Pak Yongun, *Koryŏ sidae saramdŭl*, 28.

forms, such as West Asian astronomy and Central Asian meat cuisine, to name just two. While the Mongols exerted a major cultural influence on Koreans through the suzerain–vassal relationship, as well as through marriage alliances, Koreans were not always passive in the process of cultural assimilation, and often actively adopted new forms of Eurasian-scale culture. Much documentary evidence demonstrates that Koreans across the social spectrum – Koryŏ kings such as King Ch'ungsŏn, literati like Yi Saek, and even commoners – gradually accepted the Yuan-centered world order and embraced the new Eurasian cultural elements they encountered through that order – luxury items such as soju/arak and silks, ideas such as a new form of Confucian philosophy, science, and food and dress of central Eurasian origin. Koryŏ received a bigger influence from its contact with the Yuan than was once believed, because it was connected to the broader social world of Afro-Eurasia. Through the same vector, Koryŏ also influenced Eurasian neighbors like China – a topic that reaches beyond the scope of this chapter, but which deserves scholarly attention.

It is intriguing that both Yuan China and Koryŏ Korea met similar fates because of the many problems common to the region, including the rise of the Red Turban Rebellion (1351–1366).[82] After the fall of the Yuan dynasty and of Koryŏ, the Mongol legacy survived in both societies in the new Ming and Chosŏn dynastic eras that followed (founded in 1368 and 1392 respectively). As scholars of Chinese history began to discuss the Mongol influence on Ming dynasty China, scholars of Korean history began to demonstrate that the cultural influences deriving from Mongol influence during the Koryŏ era continued to influence Korean society in the Chosŏn era, a significant fact given that the Chosŏn was not only the dynasty that succeeded Koryŏ in 1392 but also the final traditional dynasty of Korea, enduring until 1897. The next chapter will focus on continuing development of soju in Korea under the Chosŏn, which is well documented and provides important clues to understanding the development of distilled alcohol in other areas.

[82] See Robinson, *Empire's Twilight*, 130–159.

4 Distilling Soju at Court and Home in Chosŏn Korea

Soju became one of Korea's national drinks because it spread rapidly and became extensively popular during Korea's last dynasty, the Chosŏn. Founded in 1392 when it replaced Koryŏ's ruling house, long allied to the failing fortunes of China's Mongol Yuan dynasty, the Chosŏn dynasty lasted for five centuries, until it was replaced by the Great Korean Empire proclaimed in 1897.[1] During the Chosŏn era, people on the Korean peninsula developed various systems, ideas, and cultures, many of which were inherited from the late Koryŏ period when the people and government of the Korean peninsula lived under Mongol domination. This included soju. On Cheju, an island in the Korean Strait where military occupation created a strong Mongol legacy, soju gradually developed into a major alcoholic drink, because the humid natural environment there makes it difficult to preserve fermented drinks. Additionally, soju frequently appears in court chronicles, which date to the beginning of the Chosŏn dynasty, as a drink consumed by officials and also a ritual gift that circulated throughout the Korean peninsula. The exact scale of soju's consumption by Koreans cannot yet be determined as no source gives statistical information. However, frequent references to soju in official histories and literary works of other kinds help us to at least speculate about soju's Chosŏn-era growth in popularity throughout the Korean peninsula, along with other kinds of alcoholic drink like strained and turbid wines. Because it takes a larger volume of grain to produce high-alcohol soju than it does to make fermented drinks, it remained a luxurious recreational drink – so much so that it became a target for the drinking bans that the government frequently imposed on Korean society in order to prevent grain shortages as well as alcoholism. Still, many sources reveal that widespread consumption persisted among people at various levels of society because the spirit's high percentage of

---

[1] For a standard text on the history of Chosŏn in the broader context of Korean history, see Lee Ki-baik, *A New History of Korea*, trans. Edward W. Wagner with Edward J. Shultz (Cambridge, MA: Harvard University Press, 1984), 172–299.

alcohol and preservability suited it to a variety of social functions, most notably ritual and medicine. It is assumed that consumption grew as rice production in Korea increased, which explains why even more people consumed it during the eighteenth and nineteenth centuries. Soju intrigued many Chosŏn intellectuals, some so much so that they systematically examined its origins and characteristics, as seen in Chapter 2.

This chapter explores the rich and dynamic development of soju as a popular type of liquor in the Chosŏn period and the assorted roles it began to play in Korean society as its popularity grew. As previous chapters show, the scant extant documentation on distilled liquors from the Koryŏ period – a total of three references – pales in comparison to that from the Chosŏn, whose five-century history ended with long paper trails of both state and private discourse. The vast body of official chronicles, including the *Chosŏn wangjo sillok* 朝鮮王朝實錄 (Veritable Records of the Chosŏn Dynasty) and *Sŭngjŏngwŏn ilgi* 承政院日記 (Journal of the Royal Secretariat), not only reveals details about court events or larger political changes but also testifies to major cultural developments. Historians also benefit from the enriched literary tradition of the Chosŏn: this includes works by scholars influenced by Neo-Confucianism, the dynasty's ruling ideology; by the members of a broadening literary class, including women, after the creation of the Hangul alphabet in 1444; and by scholars who wrote about an expanded assortment of topics, including the livelihoods of everyday people and a host of other practical, quotidian pursuits.[2] The Chosŏn environment seems to have nurtured both the consumption and production of books about foods as well as liquors. In this vast sea of knowledge, we highlight those portions that together reconstruct the historical patterns of soju's development and functions in the context of Chosŏn society.

Observing these selected Chosŏn-era sources offers evidence that suggests key problems as well as solutions that help to contextualize soju's Chosŏn-era history within the book's larger theme. Most importantly: when does the word *soju* first appear in Chosŏn-era sources, and does it replace another brand called arak gradually or immediately? This question suggests not only a problem of identification – most of all, tracing terminological trajectories through Chosŏn-period documents – but also one of interpretation. Once we have located the first documentary references to the word *soju* in Korea, can we be certain that it continues to refer to the same distilled alcohol throughout the entire peninsula and the Chosŏn era? We certainly cannot assume that the stills and distillation methods referenced in Chosŏn sources

---

[2] For a brief overview and examples of the literature from the period, see Peter H. Lee, ed., *An Anthology of Traditional Korean Literature* (Honolulu: University of Hawai'i Press, 2017).

resemble those documented for the rest of premodern Eurasia; therefore, comparisons must be made with care. Next, we must solve soju's evolutionary paths – as producers created variants by hybridizing recipes with diverse applications in mind – which can only be traced by distinguishing changes in its recipes, materials, distillation technology and methods, and drinking ways. Last, while Korean soju assimilated much from China in the course of its development, local environmental characteristics peculiar to the peninsula encouraged the creation of distinctly Korean forms of soju, as in the case of *kayangju*, Korean home-brew culture. How did this happen? The discussion of these problems unfolds in three subsections: (1) the general development of soju as a type of distilled liquor during the Chosŏn period; (2) soju as a medicine, its side effects, and bans on its use; and (3) its position in the modern Eurasian world. A comprehensive understanding of soju's development during Chosŏn times provides a rich history of a distilled spirit in a key stage of its development, which involves its introduction into Korean society as a new cultural item and the subsequent processes of adaptation and innovation that ensured its localization.[3]

Changes in Kinds and Characteristics of Soju as Distilled Liquor

Given its earlier spread during the previous Koryŏ era, it is no surprise to find the Chosŏn court chronicles mentioning soju as an important gift item in Korean politics and diplomacy, or its frequent consumption by the nobility. To illustrate, *Chosŏn wangjo sillok*, the "Chosŏn Veritable Records," mentions soju more than 300 times.[4] The first son of King T'aejo (r. 1392–1398), who founded the Chosŏn dynasty, loved to drink alcohol so much that he died in 1393 after drinking too much soju.[5] During the reign of the third Chosŏn king, T'aejong (r. 1400–1418), the court used soju to conduct its affairs overseas, offering many bottles as a special present to the ruler of Tsushima island, a part of Japan (Muromachi period, *c.* 1336–1573) in the Korean Straits; during the same period of time, he bestowed many bottles on his own officials, in numbers ranging from ten to 100 bottles per occasion.[6] The official

---

[3] As we have seen in Chapter 3, not only distilled liquor but also many other cultural items transferred from Yuan China to Koryŏ Korea, where they continued to evolve under the Chosŏn.

[4] Many other official records, such as *Sŭngjŏngwŏn ilgi* (Journal of the Royal Secretariat) and private literary works also mention soju, corroborating official accounts.

[5] *Chosŏn wangjo sillok*, T'aejo 2 (1393): 02–12-13[01].

[6] *Chosŏn wangjo sillok*, T'aejong 4 (1404): 04–01-09[04], T'aejong 7 (1407): 07–10-19[03], Sejong 6 (1424): 06–01-09[02], Sejong 1 (1419): 01–05-19[07]. On the diplomatic relations between Chosŏn and Tsushima and other Japanese islands, see James

chronicles of the Chosŏn dynasty abound with such information: the numerous occasions on which the court gave multiple bottles of soju to various people on different occasions, soju's uses as a special medicine, and events in which some people died because of overdrinking. Such themes appear consistently throughout the Chosŏn chronicles. The evidence is clear: soju became widespread in Chosŏn Korea. But was the soju described here the distilled spirit we know today?

As in Koryŏ times, the earliest references made to the word *soju* in the Chosŏn court chronicles suggest that it constituted a variety of distilled spirits. The evidence is clear: a drink strong enough to cause sudden death, suggesting alcohol poisoning; its use as a medium of high-value gift exchange; its applications in medicine; and especially its listing parallel to other traditional alcoholic drinks like ch'ŏngju, a clear strained liquor, though not a distilled one.[7] A careful, comparative examination of documentary evidence reveals the various ingredients and distillation methods that producers used in different times and places. Passages in the court chronicles dated to the middle of the Chosŏn dynasty cite several kinds of soju: *paeksoju* 白燒酒 (white soju), *chasoju* 紫燒酒 (purple soju), *hongsoju* 紅燒酒 (red soju), and *hwansoju* 還燒酒 (double-distilled soju).[8] Other scholarly and culinary works written by Chosŏn literati explicate specific soju varieties and their recipes. Yi Kyugyŏng, who investigated the origin of soju during the late Chosŏn period, argues that single- and double-distillation methods already existed in Korea during the late Koryŏ period.[9] Other Chosŏn-era references to soju testify to the formulation of a special infusion method in which herbs like ginseng and cinnamon were used to create distinct varieties.[10] Indeed, such alterations should not seem strange, given the spirit's rapid spread in Chosŏn Korea. As Chapter 3 shows, Chinese of the Mongol Yuan period were already creating new kinds of soju in large distilleries, or bringing *milinqin shaojiu* 蜜林擒燒酒, a honeyed variety, to Koryŏ to trade with merchants there, as one passage in *Pakt'ongsa* reports. Given the extensive contact and trade between China and Korea during Mongol times, it

---

Bryant Lewis, *Frontier Contact between Chosŏn Korea and Tokugawa Japan* (London and New York: RoutledgeCurzon, 2003), 17–46.

[7] An official order for a new tax law written at the end of the Chosŏn dynasty clearly placed soju in the category of distilled liquor. *Sŭngjŏngwŏn ilgi*, Sunjong 3 (1909): 03–01-18[04].

[8] For example, see *Chosŏn wangjo sillok*, Sejo 3 (1457): 03–06-05[03], Yŏnsan'gun 6 (1500): 06–06-22[02], Chungjong 2 (1507): 02–05-28[05], 12–12-22[01] (1481), Sŏngjong 12 (1480): 12 11–08-19[01]; *Ilsŏngnok*, Chŏngjo 20 (1796): 20–02-11[11], Chŏngjo 23 (1799): 23–03-23[13].

[9] Chang Chihyŏn, *Hanguk oeraeju yuipsa yŏngu*, 89–90. For Yi Kyugyŏng's theory about the origin of soju, see Chapter 2 of this book.

[10] Chang Chihyŏn, *Hanguk oeraeju yuipsa yŏngu*, 90.

seems plausible that merchants were already shipping varieties of soju to Koryŏ and that this inspired locals to produce local alternatives. Sources suggest that both the transfer and the localization happened, but did so gradually, not at all as quickly as Yi Kyugyŏng supposed several centuries later.

While primary-source references remain too scarce to trace soju's exact chronological development during the Chosŏn period, they are enough to allow the determination of major long-term changes. A number of practical books – encyclopedic, medical, and culinary – contain several detailed descriptions of methods for making varieties of distilled liquor called *soju*. They show that society under the Chosŏn dynasty placed alcoholic beverages into two categories, fermented beverages and distilled spirits. In fact, authors identified all alcoholic beverages as belonging to three categories from the late Koryŏ period: ch'ŏngju (a clear refined wine), t'akchu (turbid wine), or soju (distilled liquor). Many cookbooks refer to an array of high-quality alcoholic drinks, most of them a form of ch'ŏngju, some of them t'akchu that could be made without special recipes as a home brew, and others made with added medicinals, herbs, or sweeteners. Soju consistently appears in these books as a specialty liquor whose varieties multiplied over time. Who made these liquors? And wrote about them and read about them?

It seems that Korean courts both produced and collected soju, which they used in medicines, at celebrations, and as official gifts. During the Koryŏ dynasty, the court most likely allocated much of this supply to the tribute they sent regularly to the Yuan court in China and reserved the rest for internal consumption by members of the royal family, nobility, and military, whom the court supplied to boost morale. According to *Koryŏsa*, a special bureau in the Koryŏ government called the Yangonsŏ 良醞署 made alcoholic drinks.[11] During the Chosŏn dynasty, its duties were assumed by another bureau, called Saonsŏ 司醞署, which fell under the authority of the Naeguk 內局, the government department in charge, among other things, of medical arts and medicines.[12] The bureau's name suggests that the Chosŏn court regarded alcohol as a medicine, and, like medicine, as something to be used very carefully. The Saonsŏ made, among other things, *hyangonju* 香醞酒, an aromatic alcoholic beverage produced especially for the court. Therefore, the term *hyangonju* does not always mean a form of distilled liquor. One entry in an official record cites a court doctor speaking before officials at the court's Naeguk as reporting that so much soju had been distilled that there was no need to brew hyangonju in advance of the

[11] See Chapter 2.   [12] Yi Sŏngu, *Han'guk sikp'um sahoesa*, 240–241.

sudden arrival of envoys. Clearly, soju was customarily used during any season on royal occasions both large and small.[13] In a similar way, many other records from the period confirm soju's frequent use by the Naeguk for medical purposes (see below). This passage demonstrates how important soju had become to the royal court by the middle of the Chosŏn period.

Whatever its actual productive capacity, the royal court could not have produced enough soju to meet its own demands. As a consequence, the court asked ordinary families to make their own soju and offer it as a form of tax. To illustrate, an entry in *Chosŏn wangjo sillok* dated to 1540 reports that a governor submitted to the court a proposal for the dismissal of a county magistrate because he imprisoned twenty members of a family that did not offer soju on demand.[14] Basically, the government discouraged both breweries and distilleries on a large scale; instead, it allowed individual households to produce varieties of fermented or distilled liquor. Indeed, they investigated households to discourage "excessive" alcohol production and banned distilling apparatus to avoid this perceived excess.[15] In China, in contrast, large-scale breweries and distilleries had been developing since the Yuan period; many of them survived beyond China's next two dynasties into the twentieth and twenty-first centuries.[16] The Chinese developed a liquor industry at larger scales for consumption at parties and festivals, not to mention for trade and taxation, as attested in many documentary and archaeological sources. However, except for an archaeological remnant of a Saonsŏ office discovered near the site of a dynastic palace, no archaeological evidence of large-scale breweries and distilleries in Chosŏn Korea has turned up.[17]

---

[13] Entered for the year 1639 in *Sŭngjŏngwŏn ilgi* (Journal of the Royal Secretariat), the court diary for the Sŭngjŏngwŏn, or Royal Secretariat, which recorded the Chosŏn king's public activities, including his daily interactions with court bureaucrats (1639: 17–06–18[07]).

[14] *Chosŏn wangjo sillok* 朝鮮王朝實錄 (The Veritable Records of the Chosŏn Dynasty), Chungjong 35 (1540): 35–12–02[03].

[15] As the case of Chŏng Yagyong, told below, illustrates, the government tried to control the scale of soju production by investigating soju kori in each household.

[16] The development of industrial and commercial enterprises and the increase in population during the Ming dynasty (1368–1644) encouraged the continuous increase in demand both for liquor and, in response, for its production and distribution. In time, liquor production left its agrarian base to become its own industry. As the liquor industry grew and matured, so did the distillation industry. During the Qing dynasty (1644–1911), both demand and supply, not to mention the number of varieties, grew even further. This laid the foundation for China's large national industry. Li Zhengping, *Zhongguo jiu* (Chinese Wines) (Beijing: Wuzhou chuanbo chubanshe, 2010), 8.

[17] See Wŏn Yuhan, *Han'guk minjok munhwa tae paekkwasajŏn* (Encyclopedia of Korean Cultures), s.v., "Saonsŏ 司醞署," Han'gukhak chungang yŏn'guwŏn, http://encykorea.aks.ac.kr/Contents/Index?contents_id=E0025829 (accessed November 12, 2018); "Saonsŏ t'ŏ (The Archelogical Site of Saonsŏ)," www.culturecontent.com/content/contentView.do?search_div=CP_THE&search_div_id=CP_THE010&cp_code=c

Nor has historical evidence, such as an official record of taxation levied against large-scale breweries. This suggests that all relevant parts of society – including the court, the bureaucracy, and many households – produced soju at relatively small scales in Chosŏn Korea. Such a situation is reflected well in a book completed in 1459 by a court doctor named Chŏn Sunŭi entitled *San'ga yorok* 山家要録 (An Essential Record of Farming Villages).

Chŏn Sunŭi's *San'ga yorok* is the oldest extant agricultural and culinary book in the history of Korea, and introduces the recipes of 229 foods and drinks, including rice and tofu dishes, porridges, soups, rice cakes, and, of course, alcoholic beverages.[18] While serving several kings from Sejong to Sejo as a court doctor of high repute, Chŏn Sunŭi accumulated a rich body of experience about both the arts of cooking and medicine practiced at court. He no doubt relied on this experience in authoring another major work, entitled *Ŭibang yuch'wi* 醫方類聚 (A Collection of Various Kinds of Medical Arts), the largest medical book in East Asia to date. Completed in 1445, *Ŭibang yuch'wi* was compiled as part of a government project using various domestic and foreign literature related to East Asian medicine.[19] In this medical book, Chŏn mentions soju seven times as a solution to which various kinds of medicine have been added to cure various kinds of disease. However, he does not explain how to make soju; this he does in his agricultural and culinary book *San'ga yorok*.[20]

In the course of writing this work (like other books such as *Ŭibang yuch'wi*), Chŏn gleaned much of his information from volumes he inherited from previous periods, which included books imported from China. For example, the first half of the book discusses forms of agriculture, including their know-how and technologies, relying in part on older works like *Nongsang jiyao* 農桑輯要 (A Collection of Essentials for Agriculture and Sericultural Industry) (complied in 1273), a Yuan dynasty text that was introduced to Korea sometime during the late Koryŏ period.[21] In another example, the section on dyeing probably derived

---

p1012&index_id=cp10120050&content_id=cp101200500001&print=Y (accessed November 12, 2018).

[18] Chŏn Sunŭi compiled the first dietary book in Korea, entitled *Singnyo ch'anyo* 食療纂要 (Compilation of Dietary Cure), in which he argued that the food we eat on a daily basis can act as a medicine able to cure our diseases. Chŏn Sunŭi, *San'ga yorok* (An Essential Record for Farming Villages), trans. Hong Kiyong and Yun T'aesun (Suwon: Nongch'on chinhŭngch'ŏng (Rural Development Administration), 2004), 5–15.

[19] Chŏn Sunŭi, *San'ga yorok*, 6–7.

[20] Chŏn Sunŭi, *San'ga yorok*, even cites a Sui–Tang medical book, *Qianjin yueling* 千金月令 by Sun Simiao 孫思邈, for soju (shaojiu). It suggests that shaojiu was used as a medicine in China during the Sui and Tang eras, yet as Chapter 1 shows, we cannot be sure if liquor called *shaojiu* at that time was indeed distilled liquor.

[21] Chŏn Sunŭi, *San'ga yorok*, 9–12. Chŏn wrote *San'ga yorok* using Chinese agricultural books, reorganizing content to accord with the peculiar features of Korean climate.

from *Jujia biyong*, a family encyclopedia from the Yuan–Ming period that contained instructions on such things as the distillation methods for foreign shaojiu called arak.[22] Some scholars have convincingly argued that many of the books on practical subjects published in China, like *Jujia biyong*, were imported to Korea only later in the Chosŏn era, so it is possible that Chŏn did not have the full encyclopedia when he consulted it. Indeed, large parts of the second half of his book, which relate to culinary culture, differ from the relevant content in *Jujia biyong*. This suggests that, when Chŏn compiled his agricultural and culinary opus, he incorporated new items of information drawn from local, contemporary sources instead of copying large quantities of content from imported Chinese books available to him. *San'ga yorok*, therefore, serves as a promising source of information about Korean food culture, including alcoholic beverages, in the early Chosŏn period.

The culinary portion of Chŏn Sunŭi's great work begins with recipes for alcoholic drinks. The first entry, "Chubang" 酒方 (How to Make Alcohol), is immediately followed by "Ch'wi soju pŏp" 取燒酒法 (How to Get Soju). This is the first concrete evidence of a local method for making soju in Korean sources, therefore it is worth examining it closely:

How to get soju:
    Pour 5 bowls of water into a large cauldron and simmer it. Then mix it with 1.5 *toe* of rice flour and make a porridge [with a consistency] like water made from washing rice. While still warm, put it in a jar and seal it, and after 3 to 4 days, it will smell sour and stinky. Steam one *mal* of glutinous rice and mix it with 3 *toe* of *qu* 麴 [yeast, Korean nuruk] powder, put it in the aforementioned jar, and wait it to mature. After that, divide the aged mixture into four pots. From each pot comes 4 *toe* of soju.[23]

In fact this passage focuses more on how to make a fermented wine that can be used to make soju. A fermented wine, of course, requires a starter, which when added to a mix of rice and glutinous rice begins the fermentation; to this end, the recipe calls for *qu*, a yeast powder, as the active ingredient. Called nuruk in Korea, this fermentation starter was imported from its Chinese counterpart by way of ancient Chinese cookbooks, such as the *Qimin yaoshu* and *Sishi zuanyao* 四時纂要 (A Compilation of Important [Activities] during the Four Seasons) (Tang dynasty, 618–907).[24] Many

---

[22] For more details about the distillation methods for foreign shaojiu called arak in *Jujia biyong*, see Chapter 1 of this book.

[23] Chŏn Sunŭi, *San'ga yorok*, 68, 170. A *toe* today equals 1.8 liters, a *mal* 18 liters.

[24] *Chŭngbo sallim kyŏngje* 增補山林經濟 (Expanded *sallim kyŏngje*) inscribes the character *qu* 麴 in a variant form as *qu* 麴; both characters actually have the same meaning. However, Japanese used the latter character 麴, pronounced *kōji* in Japanese. Yi Sŏngu, *Han'guk sikp'um sahoesa*, 227–228.

Korean cookbooks of a later era provide more detailed nuruk recipes made from a mixture of wheat, rice, and barley that is shaped into a large cake and hung to ferment for many days. These books all emphasize the importance of good-quality nuruk to the success of a wine. Once the mixture has matured for several days, the resulting base wine is then poured into four pots for distillation. Surprisingly, the entry on "How to Make Soju" in Chŏn's *San'ga yorok* does not explain how to use the pot to distill the fermented wine and thus make the final spirit. We must look elsewhere for that.

In fact, no document of the Chosŏn period provides a detailed description of stills used at that time. Only scattered documents describe different kinds of still made from different materials, such as bronze, pottery, and silver. The first document that depicts the actual shapes of these different kinds of still is *Chōsen shuzōshi*, written by Japanese scholars in 1935, during Japan's occupation of Korea in the early twentieth century.[25] These include contraptions such as the *nŭnji*, the earthen *to kori*, the bronze *tong kori*, and the metallic *soe kori* (see Figure 4.1), which were used in different areas of the Korean peninsula at that time.[26]

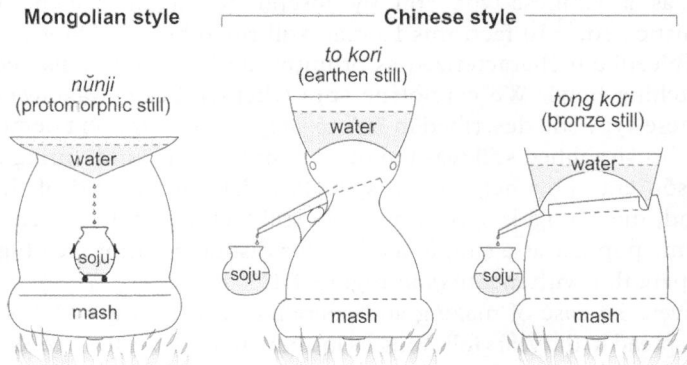

Figure 4.1 Types of traditional Korean still. Drawing by Matilde Grimaldi, after Chōsen shuzō kyōkai 朝鮮酒造協会 (Committee of Alcoholic-Beverage Making in Korea), *Chōsen shuzōshi* 朝鮮酒造史 (History of the Production of Alcoholic Drinks in the Chosŏn Period) (Keijō (Seoul) 京城: Chōsen shuzō kyōkai 朝鮮酒造協会, 1935), 159. Mongolian- or Chinese-style still types have been added to the original, following Needham's categorization. Compare Figure 1.1 above.

---

[25] See Chapter 5 for further discussion of this work.
[26] *Chōsen shuzōshi*, 159; *Chosŏn chujosa* (Korean translation), 218.

Koreans living under the early Chosŏn dynasty probably used large pots that look like protomorphic stills – called *nŭnji* in this Japanese report – inside which they would place a small cup, and over which they would place the pot's lid, upside down. After heating the fermented liquor inside the pot, they poured cold water on the lid so that the evaporated alcohol clinging to the inside of the pot and the inverted lid would rain down into the cup. That is why the Koreans called the distilling process "falling soju."[27] This method is not explicitly described in any cookbook; however, a diary-style record written in the late seventeenth century provides a wonderful hint. The author of the memoir, Yi Chihang, while adrift at sea, reportedly tried to distill seawater to drink as if he were distilling soju by this boiling-pot method. He describes how he "put the seawater in a pot, closed the pot's lid upside down, and received only half a bowl of distilled water, which dripped from the lid of the pot, and indeed, it tasted plain."[28]

The pot mentioned in *San'ga yorok* probably was part of the kind of still that might be found in any household. The distillation method that used such a pot differs considerably in its design from those used to distill related foreign spirits like shaojiu described in *Jujia biyong* and categorized as a Chinese-type still by Joseph Needham, which are more sophisticated.[29] In fact, this Korean still resembles the Mongolian type that Needham characterized as primitive and whose pot also contained a catching bowl. We cannot be sure whether Chŏn was aware of the Chinese-type still described in *Jujia biyong*, but Chŏn's text demonstrates that this simplified still was the most popular of its kind during early the Chosŏn era. This helps us to speculate that, at the end of the Koryŏ period, the Mongols spread portable-still technology to Korea, where it became popular as a simple method for distilling liquor (see Figure 4.2; compare this with the *nŭnji* in Figure 4.1).[30]

Given the ease of making and operating such a simple still, soju grew widespread in the years following its introduction. Recently, Lee Sang-Hoon has argued that the rapid and widespread popularization of soju consumption and production was actually facilitated by the speedy invention of more advanced forms of still, like the pottery-based soju kori (see Figure 4.1).[31] However, if this were so, then the discussion of distillation in early cookbooks like *San'ga yorok* would have referred to the more complicated soju

---

[27] Yi Sŏngu, *Han'guk sikp'um sahoesa*, 261.

[28] Yi Chihang 李志恒 (seventeenth century), *P'yojurok* 漂舟錄 (A Record of Drifting on a Ship), day 7 (probably May 6). This is a diary-style record that Yi Chihang wrote about his experience of drifting to Japan from 1696 to 1697.

[29] Compare Figure 1.6 above.

[30] Ishige, "Higashi yūrashia no jōryūshu," 84, 87. The Mongols probably introduced a portable distillation method similar to the one described in Chŏn Sunŭi's *San'ga yorok*.

[31] Yi Sanghun, "Urinara kanghwa parhyoju," 66–73.

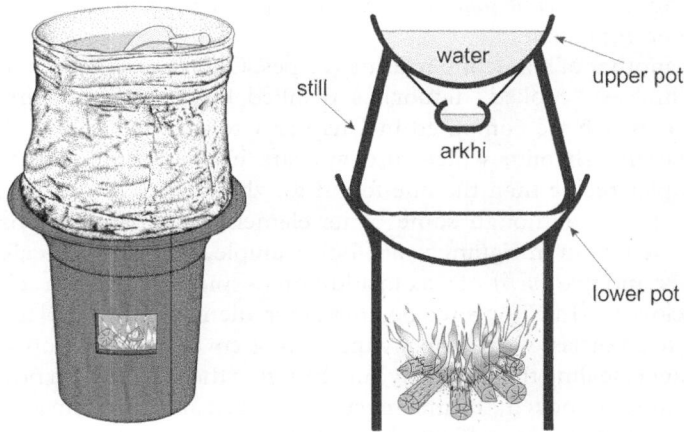

Figure 4.2 Mongolian stills for distilling arkhi (fermented mare's/cow's milk). Drawing by Matilde Grimaldi, after Ishige, "Higashi yūrashia no jōryūshu," 84, Figure 2 and Photo 2.

kori rather than a simple pot. Moreover, we can think in a logic different from that of Lee Sang-Hoon: because a simple Mongolian-style still was introduced to Korea by the Mongols, it was able to diffuse across the Korean peninsula more speedily. Indeed, recent archaeological excavations show that Koreans began to use pottery only as early as the eighteenth century, which also raises the possibility that more complicated stills developed only thereafter.

Although a court physician, Chŏn Sunŭi at least nominally aimed to help ordinary households, as the title *San'ga yorok* – "essential record of farming villages" – shows.[32] Therefore we can assume that many households – probably not all but possibly noble families and some ordinary families assigned to offer soju as tax – were able to distill soju using Chŏn's simple still when necessary. Following the first two entries about basic methods of brewing and distilling, the author continues to introduce recipes for various kinds of alcoholic drink. Most of the alcoholic beverages introduced here are strained wine, which could be made simply by fermenting the given ingredients in a jar for specified times. Yet the book

---

[32] Han Pongnyŏ argues that Chŏn probably was inspired by titles of similar Chinese books such as *Jujia biyong* and *Shanjia Qingjiao* 山家清供. Han Pongnyŏ, "*Sanga yorok* ŭi punsŏk koch'al ŭl t'onghaesŏ pon p'yŏnch'an yŏndae wa chŏja" (*Sanga yorok*: Analysis of Author and Published Dates), *Nongŏpsa yŏn'gu* (Korean Journal of Agricultural History) 2, no. 1 (2003): 14.

introduces a *memil soju* (buckwheat soju) recipe, using buckwheat as a major ingredient.

In another of his alcoholic drink recipes, Chŏn Sunŭi lists chaju (zhujiu in Chinese, "cooked" liquor), a distilled liquor that appears in *Jujia biyong*, which we compared in Chapter 1 to *nanfan* 南番 or "Southern Barbarian" shaojiu, which also appears in the Chinese text. This is a simpler recipe than the one found for zhujiu in *Jujia biyong*, as in the case of soju, although some of its elements compare to those in the Chinese text in intriguing ways. For example, *San'ga yorok* calls for two *qian* 錢 and five *fen* 分 of wax in addition to spices like pepper, cinnamon, and cloves.[33] In Chŏn's recipe, these ingredients should be placed in a jar with five bottles of medicinal wine, before covering the jar opening with oil paper, sealing the lid tightly, and bringing the mixture to a boil. Once it has properly cooked, the mix must be removed and placed in a small, cold place. After it has sufficiently cooled, the jar can be opened and its contents consumed.[34] A distillation step is conspicuously absent from *San'ga yorok*, in contrast to *Jujia biyong*, which calls for "heating to expel the white liquor to obtain the clear [distilled] liquor."[35] Yet we can assume that, as Chŏn Sunŭi appears to have abridged another soju recipe in his book, it is also possible that he drew from a Chinese book like *Jujia biyong* but modified the Chinese recipe because he wished to adapt it to Korean exigencies, such as different raw materials and different technologies. Quite possibly, he lacked the more sophisticated Chinese still or simply lacked the know-how to operate it.

In sum, *San'ga yorok* suggests that, although soju was less diverse in its varieties at this early Chosŏn phase in its evolution, it nonetheless became an important specialty alcohol not only for the court but also for ordinary households. In other words, soju had established a foothold in Korean society. Although we are not sure about how widely *San'ga yorok* circulated in ordinary society, it clearly reached enough people that later cookbooks include some of the book's major recipes.[36] During the Chosŏn period, soju gradually became a part of home-brew culture or kayangju, which served three important social functions: entertaining guests, practicing medicine, and conducting ancestral rituals.[37] This home-brew culture also provided the strongest motivation for the development of soju into different varieties during the Chosŏn period. Here, let us further discuss soju's characteristics, its varieties, and its distillation

---

[33] Compare its elements to those in the Chinese text in Chapter 1.
[34] Chŏn Sunŭi, *San'ga yorok*, 83.   [35] See Chapter 1.
[36] For example, see the case in *Suun chappang* (Various Methods of High-Class Food Culture), discussed below.
[37] See Chong Dae Song, *Chosen no sake*, 35–36.

technology as they developed over time by examining other culinary books written in the following periods.

After Chŏn Sunŭi published *San'ga yorok*, other local literati began to compile new culinary works, drawing upon this and other older texts as resources but also incorporating new information based on accumulated local knowledge. The two oldest extant books written by local literati that include soju recipes come from the Andong region in southeastern Korea. Intriguingly, the modern city of Andong enjoys fame for its Andong soju, which is considered the most famous traditional soju brand today. Chapter 3 argues that soju did not develop in Andong as a local specialty during the Mongol-era Koryŏ period, in contrast to the arguments of scholars like Yi Sŏngu.[38] In response, we must reconsider the Mongols' allegedly direct influence on the development of Andong soju. Attention ought to be paid to Andong's kayangju tradition during the Chosŏn period as a proper context for the rise of Andong soju since the beginning of the twentieth century.[39] It is assumed that soju developed conspicuously in Andong because the region was a center of culture and folk traditions. It is a place where many *yangban*, or aristocratic Chosŏn families, long lived and maintained Confucian traditions up to the present. Entertaining guests was a particularly important part of yangban culture in Chosŏn Korea, where Neo-Confucian principles ruled the day.[40] This defined advancements in the culinary arts and prompted literati to create the earliest culinary books.[41] Knowledgeable scholars or matriarchs associated with large regional landed estates compiled their accumulated knowledge of winemaking in a book that would ensure its preservation for their descendants, and in doing so assert their family status, as extant cookbooks from this early period attest.

[38] For Yi Sŏngu's argument about this, see Yi Sŏngu, *Han'guk sikp'um sahoesa*, 216.

[39] Andong soju has been a famous commercial brand since 1920, when a soju factory that adopted a stone Buddha statue of Chebiwŏn as its trademark was established in Andong. Since then, Andong soju was often called Chebiwŏn soju. Andong Chebiwŏn soju, like similar brands, declined after traditional soju made from rice was prohibited in the 1960s. Yi Sŏngu, *Han'guk sikp'um sahoesa*, 263. Andong soju revived again in the late 1980s, as we will see in Chapter 5 below.

[40] Pae Yŏngdong, "16–17 segi Andong munhwagwŏn ŭmsik chorisŏ ŭi tŭngjang paegyŏng kwa yŏksajŏk ŭiŭi – *Suun chappang* kwa *Ŭmsik timibang* ŭi sarye" (The Historical Significance and Sociocultural Basis of the 16th- and 17th-Century Andong Area's Culinary Manuscripts in Korea: Examples of *Suun chappang* and *Ŭmsik timibang* ŭi sarye), *Namdo minsok yŏn'gu* 29 (2014): 146. Pae also argues that alcoholic drinks became more important for social purposes in the patriarchal society of the Chosŏn period, where Neo-Confucianism ascended to the status of state ideology (ibid., 154–158). He also describes how environmental conditions, like the peninsula's regional watersheds, shaped the evolution of agriculture, from techniques of rice transplantation to the double cropping of rice and barley.

[41] Ibid., 163–168.

The earliest culinary book from Andong is *Suun chappang* 需雲雜方 (Various Methods of High-Class Food Culture) written in 1540 by Kim Yu 金綏 (1491–1555).[42] Consisting of two volumes, this work presents 121 recipes of the Andong region at that time. Some scholars point out some contents of *Suun chappang* that are similar to those in *San'ga yorok* and the earlier Chinese text *Jujia biyong*, and assume the possibility that the two earlier works influenced Kim.[43] Yet when we analyze specific details, it becomes clear that in some of his entries on alcoholic beverages he included new features. The first volume exclusively covers liquor making, including many kinds of fermented wines such as an array of clear, refined rice wines, but only one kind of soju, *chinmaek soju*, which was made from fermented wheat flour. This is different from *memil soju* (buckwheat soju), the only special soju recipe featured in *San'ga yorok*. We can only assume that soju had yet to win regional popularity, compared with the traditional fermented wines so abundantly profiled. Kim offers a simple recipe that notably mentions distillation but bypasses any instructions: "To ferment it, steam 10 *toe* of wheat flour, and pound it with 5 *toe* of yeast. After placing the starter inside a jar, fill it with water. After 5 days of fermentation, distill it."[44] No further direction follows. In any case, it seems safe to say that a lack of interest in such distilled spirits allowed the author to omit it.

The second culinary book from Andong, *Ŭmsik timibang* 飲食知味方 (Recipes for Tasty Food), which also includes detailed recipes of soju, appeared in 1670, a century after *Suun chappang*. Chang Kyehyang (- 1598–1680), often known as "Chang, Lady of Andong," compiled it. This is the first cookbook written by a female author in the history of East Asia, and the first cookbook written in Hangul, the Korean alphabet invented by King Sejong in 1444.[45] This unique feature of the book's authorship has led scholars to debate how best to use its contents to understand the culinary culture of the period it reflects. While many ladies in prominent clans probably developed and transmitted recipes of traditional foods and alcoholic beverages of the Chosŏn period, it is difficult to find many of them in published books. One reason for this could be the existence of a patriarchal Confucian culture, which discouraged women

---

[42] Ibid., 138. For a translation with original texts, see Kim Yu, *Suun chappang* (Various Methods of High-Class Food Culture), trans. Kim Ch'aesik (P'aju: Kŭrhangari, 2015).

[43] Kim Kwiyŏng, "Ŭmsik ŭro ponŭn Chosŏn sidae" (The Chosŏn Dynasty Viewed through Foods), in Kim Yu, *Suun chappang*, 15.

[44] Kim Yu, *Suun chappang*, 58.

[45] Han Pongnyŏ, "Ŭmsiksa esŏ pon *Ŭmsik timibang*" (*Ŭmsik timibang* Examined from the Perspective of the History of Korean Foods), in *Ŭmsik timibang* (Recipes for Tasty Food), by Chang, Lady of Andong (Taegu: Kyŏngbuk taehakkyo ch'ulp'anbu, 2003), 82.

from writing or publishing.[46] Paek Tuhyŏn argues in his linguistic study of Chang's cookbook that it lacked the author's signature because the book was written for private use as a way to transmit knowledge to younger generations of women in the family, rather than for public consumption.[47] This is plausible, given that the male literatus who wrote *Suun chappang* and assumed sole authorial credit probably wrote it with his mother. Debates have ensued about how to interpret these Chosŏn cookbooks. Yi Sŏngwu argues that while many cookbooks written in Chinese were actually copies of Chinese texts, *Ŭmsik timibang* stands out because a local noblewoman wrote it to explain her region's culinary ways.[48] Its content mixes information transmitted from earlier literary sources with that created by Lady Chang herself.

The situation changed during the late Chosŏn, when the number of books that include soju recipes soared. One such book was *Sallim kyŏngje* 山林經濟 (Forest and Economy), a book written by 1715 that drew considerable influence from *Jujia biyong*. Recently, Oh Young Kyun asserted the possibility that it was a convention in China and Korea for literati to write cookbooks and other practical books like encyclopedias for cultural cultivation rather than practical applications like cooking.[49] Joo Young-ha also suggests that we should look at these cookbooks and recipes more carefully because much of their content was probably copied from earlier ones and does not necessarily reflect the foods cooked in households at the time.[50] If so, then the existence of the two Andong books does not automatically prove that Koreans freely created soju as part of a growing home-brew culture up to the middle of the Chosŏn dynasty. While we are compelled to analyze these cookbooks critically and carefully as sources that reflect the realities of cooking at the times they were written, we can still use the books to understand popular recipes

[46] For the influence of the Chosŏn dynasty's founding ideology, based on Neo-Confucianism in 1392, on the transformation to a patriarchal society with regard to ancestor worship, the succession of family, inheritance, and marriage during the Chosŏn period, see Martina Deuchler, *The Confucian Transformation of Korea: A Study of Society and Ideology* (Cambridge, MA: Harvard University Press, 1992).

[47] Paek Tuhyŏn, "Kugŏsa esŏ pon *Ŭmsik timibang*" (*Ŭmsik timibang* Examined from the History of Korean Language), in *Ŭmsik timibang* 飲食知味方 (Recipes for Tasty Food), by Chang, Lady of Andong (Taegu: Kyŏngbuk taehakkyo ch'ulp'anbu, 2003), 56.

[48] Yi Sŏngu, *Chosŏn sidae chorisŏ ŭi punsŏkchŏk yŏn'gu* (An Analytical Study of Cookbooks of the Chosŏn Dynasty) (Seoul: Han'guk chŏngsin munhwa yŏn'guwŏn, 1982), 12.

[49] O Yŏnggyun (Oh Young Kyun), *Kŏga p'iryong saryu chŏnjip* ŭi chŏja wa p'yŏnch'an kŭrigo p'anbon – sadaebudŭl ŭi paekkwa sajŏn (Authorship, Compilation, and Editions of the *Jujia biyong shilei quanji* – An Encyclopedia for Literati), in *Chosŏn chisigin i ilgŭn yorich'aek* (Cookbooks that Chosŏn Intellectuals Read), by Chu Yŏngha, Young Kyun Oh, Ok Yŏngjŏng, and Kim Hyesuk (Seongnam: Academy of Korean Studies Press, 2018), 21–63.

[50] Based on personal conversation with Joo Young-ha (Chu Yŏngha).

in the mid-Chosŏn period, asking questions such as why some had to be modified – even simplified in the case of *San'ga yorok*.

Compared to *Suun chappang* and *San'ga yorok*, *Ŭmsik timibang* presents some new soju recipes that differ in some of their content from older books. *Ŭmsik timibang* provides detailed recipes that explain how to distill soju from rice and nuruk (yeast) using a pot as a still. After this general soju-making entry, the cookbook introduces, for the first time, two special kinds of soju made exclusively with wheat flour and glutinous rice – namely *mil soju* (wheat flour soju) and *ch'apssal soju* (glutinous-rice soju).[51] This shows that, as soju grew in popularity over time, people created more varieties using a growing diversity of local ingredients and distillation methods. All the soju recorded in these recipes uses fermented grains, such as rice, and flour, and nuruk as a fermentation starter. Also, for the first time, books in this late Chosŏn phase introduce terms like *nuruk*, using the Korean phonetic Hangul alphabet rather than the Chinese character *qu* 麴. *Ŭmsik timibang* introduces *t'arak* 駝酪, a fermented drink based on cow's milk that became a popular Korean food. Judging from this text, it appears that people did not distill t'arak to make soju; such a recipe does not appear in this or any subsequent text.

The recipes for special kinds of soju in *Ŭmsik timibang* do not mention the distillation process at all. Nevertheless, we know that a distillation process was undertaken to derive soju from fermented wine, which is implied in the recipe. In fact, the author was not good at presenting soju recipes in a complete and systematic way, as is also characteristic of other food recipes. For example, the cookbook's author created two entries for a nearly identical soju recipe by mistake. The two entries differ in a single way: one recipe calls for six days of fermentation, while the other calls for seven days. Both entries mention a method of replacing the water on top of the distillery pot. According to the descriptions found in other documents, this method involves pouring cold water on the lid, which was previously placed upside down over the vessel in order to accelerate the condensation of vapor, gained from the heated liquid, into liquid drops, the exact method of distillation for a portable still.[52] The first entry states only this: "change the water on top. Use this method to get three bottles of strong liquor."[53] The second entry expresses this in a slightly different way: "Once the water poured on the top becomes warm, replace it

---

[51] Chang, Lady of Andong, *Ŭmsik timibang* (Recipes for Tasty Food) (Taegu: Kyŏngbuk taehakkyo ch'ulp'anbu, 2003), 206.

[52] As a later picture of a still shows, the lid should be curved, so that a quantity of water can be poured into the inverted lid like a bowl. See Figure 4.1.

[53] Chang, Lady of Andong, *Ŭmsik timibang*, 205–206.

frequently. Prepare new water in a pot and pour it immediately, then you get a lot of nice soju."[54]

Based on these scattered descriptions, we can reconstruct a complete recipe for distillation: heat a fermented mixture of steamed rice and nuruk together in a pot and replace warmed water on the upside-down lid with new (cool) water. This is basically a more detailed version of the method described in *San'ga yorok*. It is thorough but incomplete, as it does not provide some crucial information about how to position the lid over the pot. One can assume two reasons for such incomplete descriptions, in addition to the redundancy of entries. First, people already possessed a basic knowledge of this pot method, so the author did not see a need for detailed explanation. Second, the author was not accustomed to writing a book and perhaps simply copied the recipe details from other books available at the time. At any rate, these recipes suggest that this pot method of soju making did not require high-level skill, a fact that, if true, explains how soju making spread so quickly on the Korean peninsula. Another piece of evidence showing widespread soju consumption at this time takes the form of a recipe for a fortified fermented wine called *kwahaju* 過夏酒 (lit. "summer-passing wine"). According to this recipe, soju is added to a regular grain-fermented wine in order to increase the latter's preservability, a useful thing to do during the summer when fermented wines turn bad easily.[55] By the time cookbooks like *Ŭmsik timibang* were written, people understood distillation and the uses of distilled spirits in enhancing the preservability of other kinds of alcoholic drink.

From a comparative perspective, it appears that household soju stills at this time were still small enough to be portable and to operate individually. The earliest culinary books examined above refer to the still as just a pot, and yet the vessel's name would soon appear as the technical term *chujŭng* 酒甑, also known colloquially as *kori* or soju kori. While Joseph Needham and his research team ignored Korean stills in their study of technology, Japanese and Korean scholars soon corrected the oversight and demonstrated the unique qualities of the kori still. According to Lee Sang-Hoon, this tool was used not to distill fermented solids, a method popular in China, but rather to distill fermented liquids, a process that would have been valuable to nomadic peoples like the Mongols who distilled fermented milks.[56] This supports the assertion that Mongols, more than Chinese, influenced distillation culture in Korea.

[54] Ibid., 207.
[55] For the most thorough study of kwahaju, see Yi Sanghun, "Urinara kanghwa parhyoju."
[56] Ibid., 66–68.

Once people grew accustomed to these distillation techniques, they probably began experimenting with more sophisticated methods and new materials, including the method Needham categorized as the Chinese type, in which two separate vessels are put together and a tube is positioned so that it draws the condensed spirit from the upper pot into a collection cup.[57] The Chosŏn court used a silver still, probably in order to detect poison in a drink. An official record reports that the court used silver pots to create a red soju with good taste and color because efforts undertaken with copper pots had failed.[58] This option was too expensive for ordinary households to consider. According to a Japanese report made in the early twentieth century, most people relied on a kori made by fitting two separate pots of bronze or clay together and sealing them using flour dough. A painting of the department of medicine at the royal court during the late Chosŏn period shows an object that looks like such a still consisting of two separate but interconnecting pots.[59]

On Cheju Island, people in more recent years developed a highly sophisticated still called a kosori, a one-piece porcelain still different from the two-piece kori used in other regions.[60] The exact date of this development is unclear; however, excavations of earthenware artifacts only date back to the eighteenth century, according to archaeologists. Indeed, this particular type of one-piece still could have been made quite recently.

A nineteenth-century scholarly text by Chŏng Yagyong (1762–1836), a scholar well known for his predilection for more pragmatic types of

---

[57] Court records mention a *pangguri* 方古里 as an earthenware vessel used to hold water or drink, and its shape is similar to that of an earthenware jar, though it is somewhat small. For example, a record in 1796 in *Ilsŏngnok* includes fifty pieces of *pangguri*, among many materials that should be prepared for use at court. I think that this is the source of the word *kori* 古里 in *soju kori*. In a previously published article of mine, I argue that *kori*, a Korean word meaning a ring or circular object, was probably adopted to signify the traditional bowl that was used to distill soju. A Korean dictionary defines it as being "made of copper or glazed earthenware as a folded pair, with one at the top and one at the bottom." Koryŏ daehakkyo minjok munhwa yŏn'guwŏn kugŏ sajŏn p'yŏnch'ansil, ed., *Han'gugŏ taesajŏn* (Grand Korean Dictionary), vol. 1 (Seoul: Koryŏ daehakkyo minjok munhwa yŏn'guwŏn, 2009), 445. However, this conclusion changed after I found the term *pangguri* in the official records. I still think the word *kori* is Korean, as we cannot find it in an extant corresponding Chinese source, and the author of the relevant historical records had to borrow Chinese characters to write it down because the Korean alphabet (Hangul) was invented only in 1444.

[58] *Chosŏn wangjo sillok*, Injo 16 (1638): 16–05-05[01].

[59] For more details about *Tongkwŏldo* 東闕圖 (*Donggwoldo*) (Painting of Eastern Palaces) held in the Korea University Museum and Donga University Museum, see "*Tongkwŏldo* 동궐도 (東闕圖)," *Han'guk minjok munhwa tae paekkwa sajŏn* (Encyclopedia of Korean Cultures), at https://encykorea.aks.ac.kr/Contents/Item/E0016317 (accessed August 1, 2020).

[60] Scholars do not know the name's source.

scholarship in the late Chosŏn,[61] testifies that many ordinary households made soju and kept two different kinds of kori – a pottery one and a bronze one. In his proposal to ban alcohol during bad droughts, he writes,[62]

Alcoholic bans [*chugŭm* 酒禁: prohibiting the production of spirits][63] during a famine year have now become a general practice. However, because local officials (*igyo* 吏校) exploit people under the pretext of carrying out the law, they are unable to stop the private production of spirits, and it is only the ordinary people who cannot bear their harassment. Moreover, turbid wine (*takju*) does not have to be strictly banned because it can be taken as a substitute for a meal and is also helpful to those who travel on the road. However, soju in the town, which causes the local officials' disorderly conduct, should be strictly prohibited. As a matter of course, stills (called *kori* in slang)[64] should be confiscated and stored in warehouses. There should also be a command that pottery manufacturers dare not produce them anew. Anyone who makes soju in secret will be compelled to pay fines that would add to the national treasury [famine-relief funds]. In supervision outside the town, it would be helpful to apply the example of the town only to both the village and the hamlet [where the state granary and markets are located]. It is easier to forbid soju in the coastal towns along the roads of the P'yŏngyang and Hwanghae regions (Sŏro 西路) and those including Tongnae 東萊 [on the east coast], because all of them use bronze stills (that produce twice the [normal] amount of soju).[65]

This text suggests that in each household, soju was popularly made using the soju kori. For home brewers, there were several kinds of soju kori to choose from, all of which could be basically categorized as pottery or bronze types; people could buy pottery and bronze kori at workshops. The original text calls kori *chujŭng*, although the text's original annotation by the author explains that *kori* was indeed its colloquial distinction. By the nineteenth century, after the initiation in most areas of Korea of a major 1708 tax reform called the Taedongbŏp 大同法 (Law of Taedong), which

[61] For more details about historical context and the philosophy and scholarship of the eighteenth-century reform-minded Korean thinker, see Mark Setton, *Chong Yagyong: Korea's Challenge to Orthodox Neo-Confucianism* (New York: State University of New York Press, 1997); Chong Yagyong, *Admonitions on Governing the People: Manual for All Administrators*, trans. Byonghyon Choi (Berkeley: University of California Press, 2010), xv–xxxiii.

[62] Chŏng Yagyong, *Yŏkchu Mongmin simsŏ* (An Annotated Translation of *Mongmin simsŏ* (The Mind of Governing the People)), trans. Tasan yŏn'guhoe 茶山研究會 (Seoul: Ch'angjak kwa pip'yŏngsa, 1985), vol. 6, 136, 282. I also consulted the existing translation by Chong Yagyong. Compare Chong Yagyong, *Admonitions on Governing the People*, 961–962.

[63] The court enacted the prohibition on brewing and drinking alcohol at times of disaster such as famine.

[64] According to the original annotation, this means a device that is used to promote distillation when making soju. It was made of pottery or bronze.

[65] The last sentence is an original annotation in the primary text.

lessened the peasantry's tax burden, the economy began to rise and people began to have more grain.[66] Despite the state's efforts, it seems that it was not easy to ban the production of alcohol, including soju, which uses more grain than other alcohol varieties. Thus more people gained an interest in soju, which motivated scholars like Yi Tŏkmu and Yi Kyugyŏng to explain the spirit's origins.[67] Other scholars offered detailed discussions of the varieties and characteristics of different kinds of soju in various kinds of relevant practical literature. These include books on practical use, such as *Imwŏn Kyŏngchechi* 林園經濟志 (Encyclopedia of Rural Life) by Sŏ Yugu 徐有榘 (1764–1845), and cookbooks, such as *Chubangmun* 酒方文 (The How-To Book of Alcohol) that contain more varieties of soju recipes.[68] These books exploded in the eighteenth and nineteenth centuries, testifying to the spread and popularization of soju along with the economic growth.[69] We can find an increase in soju varieties in the mid-Chosŏn period not only based on ingredients, but also based on distilling methods, which are verified in contemporaneous sources such as official chronicles and cooking and medical books. Let us, then, look at the evolution of soju, focusing on several special kinds that developed from the mid- to late Chosŏn period based on technological improvements.

A major soju of such specialty kinds is *hongju* (red liquor), which was also called hongsoju. As the name indicates, this soju is colored red, and it often appears alongside white soju in official records. In the nineteenth century, Yi Kyugyŏng argued that it had been developing since the Koryŏ period and later continued in the Chosŏn period. We cannot find a detailed recipe in the Koryŏ-period literature, and Yi Kyugyŏng does not support his argument by using a specific source. Yet in his *Kosa ch'waryo* 攷事撮要 (Selected Essentials on Verified Facts) (1554), Ŏ Sukkwŏn 魚叔權 introduces hongju in an entry called *hongnoju* 紅露酒, which is also called hongsoju. He provides the following recipe. Distill

---

[66] As for *Taedongbŏp*, which facilitated more active distribution of merchandise than ever before, see Chu Yŏngha, *Ŭmsik inmunhak*, 204.

[67] For more details, see Chapter 2 of this book.

[68] As the most comprehensive Chosŏn-period encyclopedia, with contents accumulated by the nineteenth century, entire volumes of Sŏ Yugu's (1764–1845) *Imwŏn Kyŏngchechi* have been translated and annotated by the Imwon Research Institute since 2003. The section on soju, included in the section *Chŏngjochi* (Records of Food and Cooking), presents recipes for more than twenty kinds of soju and for noju (dewdrop liquor) and kwahaju (fortified fermented wine) that used soju, and provides other related information, including how to make soju using flat wine. It cites many earlier Chinese and Korean sources available to the author. See Sŏ Yugu (1764–1845), *Imwŏn Kyŏngchechi: Chŏngjochi* (Encyclopedia of Rural Life: Records of Food and Cooking), trans. Imwon Research Institute (Chŏng Chŏnggi) (Seoul: Pungseok Cultural Foundation, 2020), vol. 4, 66–68, 127–152.

[69] Yi Sŏngu, *Chosŏn sidae chorisŏ ŭi punsŏkchŏk yŏn'gu*, 6–21.

three bottles and two *pokcha* (one *pokcha* = 450ml) of hyangonju, a kind of strained wine, and then you will get one bottle of liquor. In the course of distillation, cut one *nyang* of *chich'o* 芝草 (Chinese *zhicao*, *Ganoderma lucidum*) into thin pieces and put it in the bowl that receives the alcohol drops. Then you will receive hongnoju that is dyed a rich red color.[70] A similar recipe is also introduced in the most systematic medical work in East Asia at that time, *Tongŭi pogam*, by Hŏ Chun (1539–1615).[71] Chang Chi-Hyun points out that the only fundamental difference between Ŏ Sukkwŏn's recipe and that of Hŏ Chun is the ingredient for dyeing, as the former called for chich'o and the latter *chach'o* 紫草 (*Lithospermi radix* or lithospermun root). As Chang suggests, Ŏ Sukkwŏn probably committed a typographical error, using *chich'o* instead of *chach'o*. Nowadays, chich'o is identified with chach'o in East Asian herbal medicine. In sum, hongnoju in early Chosŏn recipes is basically soju colored with chach'o. It seems that, inside the royal court, the Naeguk often produced many bottles of hongsoju to use as medicine and as gifts.[72]

We should note that similar kinds of soju named *hongju* or red soju developed in China, too.[73] The red soju in China is called *hongqujiu* 紅麴酒 because they used a red yeast fermentation starter called *hongqu* 紅麴 (Korean *hongguk*, Japanese *akakōji*), which differs from Korea's hongju which uses chach'o. A great scholarly debate relates to the originality of hongju that has used chach'o: whether Korean kinds of hongju also used hongqu or chach'o in their early stages of use, or whether its makers first used hongqu and later replaced it with chach'o. Chang Chi-Hyun examined the scholarly arguments surrounding this debate and concluded, agreeing with Ch'oe Namsŏn, that while they received the method of making red soju from China, it is highly likely that people in the Chosŏn period easily developed their own hongju using chach'o. Because the use of hongqu was not popular in China before the late Yuan dynasty, it was difficult for people living in the late Koryŏ dynasty to import it. Moreover, Koreans had used chach'o as coloring material and exported it to other countries.[74] As official chronicles mention that hongju was very often produced and served alongside white soju, it is possible that they developed special methods to make it efficiently using chach'o.

This hongju that developed into a new type of soju with more advanced and complicated techniques is called *kamhongno* 甘紅露 (sweetened hongnoju), a famous Korean soju created after two to three rounds of distillation. Yi Kyugyŏng argues that this soju developed during the Koryŏ

---

[70] Chang Chihyŏn, *Hanguk oeraeju yuipsa yŏngu*, 91.   [71] Ibid., 91–92.
[72] A record dated 1626 in *Sŭngjŏngwŏn ilgi* says that they produced 270 bottles of hongsoju every month. *Sŭngjŏngwŏn ilgi*, Injo 4 (1626): 04–03-03[14].
[73] Chang Chihyŏn, *Hanguk oeraeju yuipsa yŏngu*, 95.   [74] Ibid., 97.

dynasty and continued into the Chosŏn.[75] Since the name *noju* 露酒, a nickname for soju, was coined in a Yi Saek poem of the Koryŏ period and cannot be found in any Chinese work, the drink likely originated in Korea. This drink is also introduced as *kamnoju* 甘露酒 (sweetened noju) in other works, including Sŏ Yugu's *Imwŏn Kyŏngchechi*, Yu Tŭkkong's 柳得恭 (1748–1807) *Kyŏngdo chapchi* 京都雜志 (Capital Gazette), and Ch'oe Namsŏn's *Chosŏn sangsik: p'ungsok* 朝鮮常識風俗 (Common Knowledge of Chosŏn: Customs) (1946).[76] Among these books, Sŏ Yugu's *Imwŏn Kyŏngchechi*, written in the early nineteenth century, introduces kamnoju as a well-known thrice-distilled liquor called *kamhongno* that originated in P'yŏngyang (an important city in the northern part of the Korean peninsula), and offers a detailed recipe for preparing it: if you spread honey and add chach'o at the bottom of the receiving bowl during distillation, the result is a very sweet- and strong-tasting liquor with a crimson color – in other words, a high-quality hongnoju.[77]

This kamnoju recipe in *Imwŏn Kyŏngchechi* is among the many more varied kinds of liquor introduced in the book, which includes the most detailed information about soju recorded during the Chosŏn period.[78] The book introduces the recipes of 170 kinds of alcoholic beverage, including numerous ways of making nuruk, brewing, and distilling alcohol, such as single and multiple distillation and the addition of medicinal herbs. Scholars have pointed out that *Imwŏn Kyŏngchechi* and other contemporaneous books written during the late Chosŏn period also copied much of their content from encyclopedic books imported from China at that time, including *Jujia biycng*.[79] Despite the heavy influence of this practice, many primary sources of information about Korean soju, including kamnoju, show that these books include many details about some of the original developments in Korea's soju up to that time, which developed to engender greater sophistication and diversity in its makeup.[80]

Some evidence of identity transformation in alcoholic drinks appears in these recipes, too. For example, *samhaeju* 三亥酒 is introduced as fermented alcohol in earlier records, such as Yi Kyubo's list of liquors in his early thirteenth-century work, as well as in *San'ga yorok* and *Suun chappang*, two fifteenth- and sixteenth-century books. The term originally referred to ch'ŏngju, which is made by fermenting glutinous rice and

---

[75] Ibid., 3.   [76] Ibid., 93.

[77] Ibid., 93. Kamhongno 甘紅露 is a kind of multi-distilled soju, in which a high-concentrate, high-grade distillate is both sweetened and colored red. It got the name *kamhongno* simply because it is a sweetened hongnoju.

[78] Sŏ Yugu, *Imwŏn Kyŏngchechi: Chŏngjochi*, vol. 4, 130.   [79] For example, ibid., 149.

[80] Sŏ Yugu's *Imwŏn sibyukchi*. Yi Sŏngu, *Han'guk sikp'um sahoesa*, 260–261.

twice adding alcohol during a period of time spanning the first three days of the Year of the Pig that fall on the first three months of the Korean calendar. In time, from the nineteenth century, the spirit earned a reputation as a major kind of soju. A court record clearly explains the change, stating that at a certain point in time, people began to use samhaeju as a base wine to distill soju or a soju brand rather than as fermented wine.[81] We can assume, therefore, that as soju became popular, some regular fermented alcoholic beverages were used as base wines for the distilling phase and gradually transmuted into a variety of soju.

We have seen that the people of the Chosŏn dynasty developed different kinds of distilled liquor like soju, all of them made using different ingredients and distilling methods. Intriguingly, none of the relevant historical sources provide any evidence or clues that suggest a similar developmental trajectory for arak, another distilled liquor that made its appearance on the peninsula during the late Koryŏ dynasty. It would appear, then, that arak won popularity among Koreans at the same time, as seen in the poem by the fourteenth-century literatus Yi Saek, though it is not popular today. However, we may also be looking at a case of mistaken identity. A group of Ming dynasty Chinese and subsequent Chinese-influenced sources in Korean say that soju is the same as arak. The meaning of these references therefore requires clarification; to do this, we must trace the evolution of arak in the late Mongol and Koryŏ eras to the end of the Chosŏn. This is difficult to do, because explicit references to arak are few. However, one record provides a clue. This is *Simyang ilgi* 瀋陽日記 (Chinese *Shenyang riji*; The Diary at Simyang/Shenyang) (1636), the daily record of Chosŏn's Crown Prince Sohyŏn (Sohyŏn *seja* 昭顯世子, 1612–1645), during his captivity as a hostage of the court of the Qing dynasty in Shenyang after the Chosŏn court surrendered to the Qing following its invasion of Chosŏn in 1636–1637. The diary states that a Qing official asked the Crown prince the following: they were considering sending to the court of Chosŏn some large fish and fermented-milk shaojiu (i.e., soju – *lao shaojiu* 酪燒酒, pronounced *naksoju* in Korean), but he wished to know whether people would value them. The Crown prince replied that everything Chinese should be valued in Chosŏn; hearing this, the Qing court sent to Chosŏn two large fish and two pots of *arangju* 阿郞酒 or arang liquor – namely, arak – with a collection of gifts. So here is one example in which a Chinese transcription of the word *arak* – in this case,

[81] Yi Sŏngu, *Han'guk sikp'um sahoesa*, 244, 264. *Ilsŏngnok* 日省錄 (Diary of Self-Examination), 1838.

arang – refers to soju.[82] It can also be assumed that this form of arak prevailed in post-Mongol China, most likely in northern regions, more often as distilled liquor made of fermented milk, and that written records identify it using various transliterations. Another thing to note is that, because we cannot find this particular transliteration of the word *arak* as *arang* in other sources of Chinese literature, it is possibly a Korean transliteration of *arak*. Sŏng Haeŭng 成海應 (1760–1839), who summarized *Simyang ilgi* in his writings, equates *arangju* with soju.[83] In fact, the word *arangju* simply signifies one Korean transliteration of *arak* among many, including *arakju* and *araegiju*. These variations were detected among the peninsula's various Korean dialects in the early twentieth century by Japanese linguist Ogura Shinpei 小倉進平 (1882–1944).[84] Chapter 1 demonstrated the transliteration of the word *arak* into different Chinese characters in Yuan- and Ming-period sources; here, at the same time, phonetic differentiation also appears. We can assumed that while *soju* became the major term for distilled liquor during the Chosŏn period, it was often called *arak* in different regions, including P'yŏngyang and some other regions, including Cheju Island, which probably denotes soju made of fermented milk based on the Mongol legacy, as well as other varieties made from fermented grains. Its pronunciation differs according to the dialects of the different regions.

In sum, distilled liquors that were transferred to Korea from China during the Mongol period expanded into different regions of the Korean peninsula, where it diversified into a variety of reputable local beverages. No such thing as a "national" alcohol existed yet. The people of the peninsula valued their regional brews, and the court regularly added soju and the clear strained wine ch'ŏngju to the gifts they distributed nationally and internationally to maintain their political and diplomatic relationships. As Koreans began to produce and consume more soju under the Chosŏn dynasty, they began to utilize soju for an even greater variety of purposes. Since a large amount of fermented alcoholic beverage must be distilled to produce a small amount of distilled liquor, soju remained a precious beverage at the end of the Chosŏn era. However, it seems that people continued to make it to consume for various purposes. Soju proved to be useful for

[82] See Sohyŏn seja (1612–1645), *(Yŏkchu) Sohyŏn Simyang ilgi* ((Annotated) Sohyŏn's Diary in Simyang), trans. Kim Chongsu (Seoul: Minsogwŏn, 2008), vol. 2, 274 for Korean translation, 456 for original Chinese text.

[83] Sŏng Haeŭng (1760–1839), *Yŏn'gyŏngjae chŏnjip: Oejip* (Complete Works of Yŏn'gyŏngjae: Supplementary Volume), *kwŏn* 60, P'ilgiryu (Han'guk kojŏn pŏnyŏgwŏn).

[84] Chang Chihyŏn, *Hanguk oeraeju yuipsa yŏngu*, 80.

preservation and for medicine, as sources from the Chosŏn show. Let us examine in more detail soju's functions during the Chosŏn period, and the unique characteristics that evolved.

Usages, Benefits, and Adverse Effects of Soju in Chosŏn Society: Rituals, Hosting, Medicine, and Bans

As noted briefly in the section above, the Chosŏn dynasty adopted Neo-Confucianism as its state ideology, which played a crucial role in its popularization in Korean society. As a consequence, this elevated the importance of ancestor worship and other Confucian-oriented family rituals in Korea. In some of these ancestral rites, alcoholic drinks played a key role, along with food. As ritual books of the Chosŏn period show, practitioners basically relied on fermented alcoholic beverages like a clear strained liquor called ch'ŏngju to conduct regular ancestral rituals, because these rituals had been created based on *Zhuzi jiali* 朱子家禮 (Zhu Xi's Family Rituals), a collection of ritual prescriptions system- atized and recorded by the Chinese philosopher Zhu Xi 朱熹 (1130– 1200) in the twelfth century.[85] Distilled liquors at this time were rare; there is no evidence of their existence even during the life of Confucius (551–479 BCE) centuries before. Intriguingly, however, we can see that some sources of the late Chosŏn era discuss the use of soju as a function of ancestral rites.[86] For example, *T'aektang chip* 澤堂集 (Collection of T'aektang) (1674) clearly states that people had to use soju to conduct rituals during a special subdivision of the season corresponding to June and July called Yudu il 流頭日 (Day of Yudu).[87] This part of the hot summer is precisely the time when a strained wine like ch'ŏngju easily

---

[85] This book about the manners and rituals of noble families was brought to the Korean peninsula at the same time as Neo-Confucianism's introduction at the end of Koryŏ period. After the Chosŏn was established, other people used it as an important reference not only for noble families generally but also for the royal family rites. Especially in the late seventeenth century, as interest in rituals increased, people began to publish com- mentaries about the book and Neo-Confucian rites became popular among ordinary people. See Yi Sugin, *"Chuja karye* wa Chosŏn chunggi ŭi cherye munhwa- kyŏlsok kwa paeje ŭi chŏngch'ihak"* (Jujagarye and Sacrificial Rites in the Middle Chosun: Politics of Unity and Exclusion), *Chŏngsin munhwa yŏn'gu* 29, no.2 (2006): 35–65.

[86] See Yu Changwŏn (1724–1796), *Sangbyŏn t'onggo* (General Examination of General Changes), *kwŏn* 23, *cherye* (ancestral rituals); and Kim Changsaeng (1548–1631), *Sagye chŏnsŏ* (Complete Book of Sagye), *kwŏn* 41, Ŭirye munhae (Theories about Manners), *cherye* (ancestral rituals), 時祭 (seasonal rites), Han'guk kojŏn pŏnyŏgwŏn, http://db .itkc.or.kr (accessed January 5, 2019).

[87] Yi Sik (1584–1647), *T'aektang chip* (Collection of T'aektang), Pyŏlchip (Separate col- lection), *kwŏn* 16, *chapchŏ* (various writings), Han'guk kojŏn pŏnyŏgwŏn, http://db .itkc.or.kr (accessed January 5, 2019).

spoils.[88] Soju, in contrast, simply will not spoil, even during an extended and intense heat. Obviously, people were willing to modify the original rules of Neo-Confucian ritual to a certain degree for practicality's sake. In any case, in order to utilize soju in ritual, each house had to brew and distill alcohol for the rituals of ancestor worship and entertain guests who gathered for the rituals.

On Cheju Island, people definitely used soju, not ch'ŏngju, in all family rituals, because the island's high humidity was not only bad for fermented liquors but also unsuitable for rice farming. It was difficult for the islanders to brew ch'ŏngju because they did not have enough rice to make it; instead, they used locally available foxtail millet. The island's humid weather made it difficult for inhabitants to preserve ch'ŏngju for very long. As seen in Chapter 3, the island's Mongol influence was profound, since the Khan's forces occupied it for over a century after defeating the Sambyŏlch'o in 1273. It is highly likely that, in the century that followed, Mongols transported soju, arak, and their distillation techniques to Cheju. Once there, distilled liquor became popular almost as soon as it had arrived, and locals subsequently sought to learn how to produce it. According to the section about Cheju in *Sinjŭng Tongguk yŏji sŭngnam* 新增東國輿地勝覽 (Newly Expanded Overall Survey of the Geography of the Eastern Country) (1530), Cheju inhabitants referred to soju as *tayong soju* 多用燒酒 ("soju for many purposes"), meaning that they used the spirit for many purposes, including family rituals and medicine.[89] Here is a clear example of local environments influencing the localization of the ingredients and methods used to produce distilled liquors.

As seen in other examples, people of the Korean peninsula also consumed soju for medical reasons. Soju's utilization as medicinal was typical in all premodern East Asian societies and differed from the later and narrower use of alcohol by Western doctors for disinfection. Many historical medical accounts mention soju, define it, and explain its characteristics. Evidence of soju's medicinal use appears early. The royal physician Hŏ Chun compiled the earliest Chosŏn-era medical account that introduces soju's functions along with its positive and negative effects, based on most

---

[88] A record in 1725 in *Sŭngjŏngwŏn ilgi* says that, in the summer, the consumption of soju is ten times higher than that of ch'ŏngju, so people go around with soju everywhere.

[89] Yi Haeng, *Sinjŭng Tongguk yŏji sŭngnam, kwŏn* 38 (1530), Chejumok p'ungsokcho 濟州牧風俗條 (Section about the Customs of Cheju). Kim Chŏng's 金淨 (1486–1521) *Cheju p'ungt'orok* 濟州風土錄 (A Record of Cheju, Early Sixteenth Century) also states that Cheju people used soju much for many purposes because ch'ŏngju was precious on Cheju Island due to the shortage of rice. O Yŏngju, "Tong Asia sok ŭi Cheju parhyo ŭmsik munhwa" (Jeju Traditional Food Fermentation Culture in the Environment of East Asia), *Cheju-do yŏn'gu* (Journal of Cheju Studies) 32 (2009): 164–165.

of the medical works available to him in his day.[90] Completed in 1613, *Tongŭi pogam* introduces soju and its medicinal effects by discussing a variety called *Xianluo jiu* 暹羅酒, "Thai wine," a double-distilled liquor imported from Southeast Asia, well known for its alleged ability to break down indigestion and kill intestinal parasites.[91]

A passage in *Tongŭi pogam* explicates soju's origin and characteristics, including the dangers of drinking it to excess: "*Soju* began to appear beginning in the Yuan period. Its taste is extremely intense. Immoderate drinking will ruin your health."[92] As Chapter 2 notes, Hŏ claims in this text that soju originated in the Yuan period, which agrees with claims stated in earlier medical books written in China, such as *Bencao gangmu*, an indication of the influence some Chinese medical works had on their counterparts in Korea (though Hŏ never cited *Bencao gangmu*). Many academic books written during the Yuan and Ming periods in China were transported to Korea at about the same time; indeed, medical books created before Li Shizhen produced *Bencao gangmu* likely crossed cultures through Yuan–Koryŏ diplomatic channels. In any case, Li's work and many other Chinese texts discuss soju's poisonous effects, warning against overconsumption because of its high alcohol content.

While such information about soju in books like *Tongŭi pogam* drew their influence from medical books,[93] many of them from China, Koreans also learned about the effects of soju from their own experiences. For example, in addition to information from Chinese sources, *Tongŭi pogam* also cites Chosŏn's folk medicine to explain how to eliminate toxin from soju.[94] This is also evident in many Chosŏn court chronicles and literary works written before Hŏ Chun compiled *Tongŭi pogam*. For example, Kwŏn Kŭn 權近 (1352–1409), a key figure in the Chosŏn dynasty's creation, introduces in his writings an episode that illustrates soju's efficacy in treating indigestion. He recalls the story of Official Kim (Kim *pansa* 金判事), who once received a guest who appeared to be dying from acute indigestion and suffered intense pain. The official advised his guest drink two cups of soju, saying that it would bring a quick cure. As soon as his guest drank the soju, he became drunk, which caused him to cough

---

[90] In particular, soju is introduced in *Tongŭi pogam*, regarded as one of the most important medical compilations in East Asian history, based on collected medical works and knowledge, and experience accumulated in China and Korea, whose compilation began under King Sejong's reign and was completed under King Sŏnjo in the fifteenth and sixteenth centuries.

[91] Chang Chihyŏn, *Hanguk oeraeju yuipsa yŏngu*, 110.   [92] Ibid., 41.

[93] These books were read and quoted by intellectuals of the Chosŏn dynasty, although there are debates about when these books found their way to the Korean peninsula.

[94] Hŏ Chun, *Tongŭi pogam* (Precious Mirror of Eastern Medicine), trans. Tongŭi munhŏn yŏn'gushil (Tongŭi Literature Lab) (Seoul: Pŏbin munhwasa, 2012), 589.

hard and in turn to vomit lumps of meat filled with live insects. Suddenly, the guest's symptom's disappeared.[95] Chosŏn chronicles contain many such references to soju's effectiveness as a medicine, as well as the dangers of overdrinking. For example, an episode introduced in the dynasty's veritable records as having occurred during the reign of Sŏngjong 成宗 (r. 1469–1494) illustrates both effects. According to this documentation, a legal official requested the court to interrogate Hong Yunsŏng 洪允成 for forcing Yi Sunam 李壽男 to drink himself to death. Then Hong Yunsŏng appealed to the king, saying,

Because I suffer from diarrhea caused by heat, I therefore am always taking soju. One day, Yi Sunam came [to me] drunk. While I was chatting with him, I offered him a couple of cups of soju, and I did not expect that it would result in this [his death], so I await your Majesty's punishment.

The king refrained from punishing him, saying that even if alcohol is forbidden, drinking it for medication is not harmful. Then Hong Yunsŏng gave thanks to the king by bowing.[96]

Other official documents provide concrete examples of court doctors using soju as a medicine. For example, a record dated 1647, during the reign of Injo 仁祖 (r. 1623–1649), states that the royal drug department made a request to prepare soju as medicine during the hot season, but the king would not allow it, because the court had already reduced the number and size of dishes served to him due to the heat.[97] Another item from the drug bureau's official records states that doctors invited the king to drink the seed pit of a peach (an herb used in traditional Chinese medicine) mixed with one cup of soju, and he complied.[98] Another item describes making medical soju (yaksoju 藥燒酒) for Chinese envoys, adding details about medical medicaments to mix with the soju to boil and drink.[99]

The cases mentioned above all represent the medical use of soju by royal families, nobles, and other elites. Still other sources demonstrate soju's medical use among ordinary people. The diary of Yi Munkŭn 李文楗, a Chosŏn nobleman of the sixteenth century, shows that servants used soju to treat any kind of disease, probably because they had less access to other forms of medicine. Shin Dongwon offers a story in Yi's diary that illustrates this. Yi's servant, called Mansu, was suffering from excreta

---

[95] It is found in vol. 98 of *Tongmunsŏn* 東文選 (A Collection of Eastern Writings), an anthology of works by Korean poets and writers from the late second to the late-nineteenth centuries CE.

[96] *Chosŏn wangjo sillok*, Sŏngjong 2 (1471): 02–06-05[05].

[97] *Chosŏn wangjo sillok*, Injo 25 (1647): 25–05-18[02].

[98] *Sŭngjŏngwŏn ilgi*, Injo 22 (1644): 22–07-18[06].

[99] *Sŭngjŏngwŏn ilgi*, Injo 25 (1647): 25–02-16[04], Injo 25 (1647): 25–03-03[10], Injo 26 (1648): 26–02-15[06]

poison (*pyŏndok* 便毒), a headache due to heat, malaria, and some other kind of illness, and Yi Munkŭn offered Mansu medicine to recover. Whenever Mansu suffered from some summer ailment, he tried to recover by drinking soju. However, this time it did not work; he grew constipated as his headache intensified. Seeing his servant's worsening distress, Yi Munkŭn convinced him to seek a doctor's prescription for a different medicine. Here, Shin Dongwon argues that people of the lower classes probably consumed soju as a cure-all, unlike nobles, who could obtain sophisticated remedies from reputable medical doctors.[100] Another diary record reports that some blind fortunetellers visited Yi Munkŭn's house to offer their services, for which they received a fee that included soju and medicinal pills, as well as rice, red beans, paper, fans, and liquor bowls.[101] We can assume that soju was available at noble houses for use by family members and servants as well as others as gifts. Another nobleman's diary describes a servant who once laughed scornfully at him, saying that noblemen suffered from the measles and smallpox, and used all variety of medicine, yet died from the side effects, yet simple soju and dried meat could have saved them.[102] Intriguingly, this view of soju as a universal remedy for all varieties of illness also appears in official court records.[103]

As soju grew in popularity at all levels of society in mid-Chosŏn, some literati began to interpret this as an important social phenomenon. For example, an early seventeenth-century writer says,

Soju is a liquor that arose in the time of the Yuan dynasty. As it was taken only as a medicine, it was not used haphazardly. Due to this, it became custom that small cups were called soju cups. In the present day, however, those of upper status drink great quantities to their hearts' content; in the summer, they drink much soju from large cups. Drinking their fill and becoming drunk like this has caused many a person to suddenly die.[104]

This passage details the characteristics as well as the dangers associated with soju overconsumption, a situation that developed in society during

---

[100] Sin Tongwŏn (Shin Dongwon), *Chosŏn ŭiyak saenghwalsa* (A History of Medical Life in Chosŏn) (P'aju: Tŭllyŏk, 2014), 341–343, 386.

[101] Ibid., 479.

[102] Ibid., 626–627; Hwang Yunsŏk 黄胤錫 (1729–1791), *Ijae nan'go* 頤齋亂藁 (Disordered Drafts by *Ijae*) (1786).

[103] For example, a record dated 1786 in *Ilsŏngnok* cites a medical doctor who stated that all the various diseases of that year were cured by using only soju. *Ilsŏngnok*, Chŏngjo 10 (1786): 10–05–29[07].

[104] Yi Sugwang (1563–1628), *Chibong yusŏl* (Topical Discourses by Chibong) (1614), a reproduction of the original text (Seoul: Kyŏngin munhwasa, 1970), 19; English translation from Michael J. Pettid, *Korean Cuisine: An Illustrated History* (London: Reaktion Books, 2008), 119.

the Chosŏn as elites began to consume a spirit originally intended for use as a medicine.

Many documents recorded during the Chosŏn dynasty testify to cases in which people suffered injury to their body or even died because they drank alcohol to excess. Many passages of the Chosŏn chronicles introduce the story of generals who had been overdrinking soju and died or made others drink soju to fatal excess.[105] Despite the continuous warnings of doctors and literati, it seems that many people continued to drink soju without restraint. The situation would have been much the same for ordinary people as well. This led in 1758 to the enactment of a strong Liquor Prohibition Law (*Kŭmjuryŏng* 禁酒令), also called the alcohol ban (*chugŭm* 酒禁), which successive courts enacted frequently throughout the history of the Chosŏn dynasty.[106]

The Liquor Prohibition Law appears often throughout the Chosŏn era.[107] The prohibition applied not only to soju but also to other kinds of fermented wine, such as ch'ŏngju. Some scholars argue that the reason why people began to refer to ch'ŏngju as *yakchu* 藥酒 ("medicinal wine") is related to this issue. While the Chosŏn dynasty enacted the Liquor Prohibition Law several times, they excluded cases involving medicinal-herb wines such as ch'ŏngju. As a result, the privileged class drank a lot of ch'ŏngju under the pretext of that it was yakchu – "medicinal wine" – to avoid the ban. After that, *yakchu* replaced ch'ŏngju, and it is still a common collective term today. There are other theories, too. For example, according to *Imwŏn Kyŏngchechi* (Encyclopedia of Rural Life), a nobleman brewed fine ch'ŏngju, and because his house was located in a place called Yakhyŏn 藥峴, people began to call the spirit *yaksanch'un* 藥山春. As an alcoholic drink named *ch'un* 春 was considered a good liquor with higher alcohol content, *yaksanch'un* became *yakchu*. This is a story about a place associated with a fine wine, rather than a story about miraculous curative properties.[108]

Yet health was not the primary issue for the Liquor Prohibition Law. Primarily it sought to save grain. In this way, soju became the main cause of the promotion of this Liquor Prohibition Law because it consumes more grain, not to mention that it also brought more health issues. As we have seen earlier, Chŏng Yagyong of the late Chosŏn dynasty proposed

---

[105] For example, a record dated 1417 recalls an episode in which the king dismissed two officials from their posts because they let another official drink to excess and die. *Chosŏn wangjo sillok*, T'aejong 17 (1417): 17-축 05-04[01].

[106] *Sŭngjŏngwŏn ilgi, Yŏngjo* 4 (1728).

[107] The first official record about this is found in *Chosŏn wangjo sillok*, T'aejo 2 (1393): 02–12-05[01].

[108] Yi Sŏngu, *Han'guk sikp'um sahoesa*, 238–239; Sŏ Yugu, *Imwŏn Kyŏngchechi: Chŏngjochi*, vol. 4, 66.

that the court prevent the consumption of grains in order to find a way to stabilize people's lives during times of disaster or famine.[109] Here he says that soju is the most crucial thing to be banned. Other scholars, like Yi Ik (1681–1763), made similar claims.[110]

Similar alcohol bans also appeared in China and Japan for similar reasons.[111] Yet they executed these bans more systematically because they could control production and consumption by imposing taxes on the trade in alcoholic drinks in many commercial cities, wherever restaurants, lodging, and shops developed.[112] In Korea, in contrast, where cities and their commerce were less developed, the Chosŏn government simply tried to control production and consumption of soju by official bans. It seems that such alcohol bans continued even when the economy and production of grains improved in the eighteenth and nineteenth centuries. As the commercial activities of buying and selling goods such as rice and cloth grew frequent, many taverns, like restaurants and inns that also sold soju, flourished commercially. Yet such a weak foundation of the commercial system ultimately made it less prepared for encountering new outside challenges from foreign cultures, including the colonial approaches of imperial Japan in the late Chosŏn period. While soju developed much through the localization process inside Korea during the Chosŏn dynasty, it also received some stimulations and influences sometimes through contact and exchange with other countries when many other societies in the world were experiencing major change in all spheres, including politics, the economy, science, and technology. Let us examine that aspect.

Soju's Chosŏn-Era Interactions with the Expanding Modern World

A careful examination of available sources suggests that, since soju was introduced to Korea at the end of the Koryŏ period, people in the Chosŏn dynasty exchanged soju and books and information related to soju and distillation through contacts with other countries in East Asia throughout the long dynastic period. We can guess, therefore, that people in this

---

[109] In one of his poems, Chŏng Yagyong says he is trying to be friendly with takju and growing distant from soju. Chŏng Yagyong, *Tasan simunjip* (Literary Collection by Tasan) (Seoul: Minjok munhwa ch'ujinhoe, 1982), vol. 5.

[110] See Yi Ik, *Sŏngho sasŏl* 星湖僿說 (Rough Discussions of Sŏngo) (1790).

[111] *Bu nongshu* 補農書 (A Supplemental Agricultural Handbook), which reflects the specific situation of the agricultural economy and level of technological development in China during the sixteenth and seventeenth centuries, not only shows the profits gained by the soju industry but also the problems that arose from grain shortages, given the importance of fermentation to the production process. Ch'oe Tŏkkyŏng, 218, 339.

[112] Chu Yŏngha, *Ŭmsik inmunhak*, 206–212.

region received new ideas that developed their understanding of soju production through these interactions. This engagement reached its peak at the end of the Chosŏn period; however, evidence of its existence dates back to the beginning of the dynasty in 1392.

The largest documented means of soju's international circulation and exchange appear to have been in the form of diplomatic gifts. Official records show that the court sent many bottles of soju to the courts of the Ming and Qing dynasties in China, the Tokugawa shogunate in Japan, and the Kingdom of Ryukyu 琉球 (1429–1879) on Okinawa as diplomatic gifts. When the Chosŏn dynasty replaced the Koryŏ, it became a tributary vassal of the Ming dynasty, which had replaced the Yuan just decades before, and so it sent many tributary goods to the Ming court regularly.[113] Thereafter, whenever the Korean kings dispatched their envoys or welcomed China's embassies, they submitted many gifts in tribute to the Chinese emperor. Many of the luxury items that traveled west included bottles of soju, which were often divided into white and red varieties. The Chinese offered shaojiu in return. Some of the Chosŏn envoys who visited the Chinese imperial capital recorded their comparisons of Chinese shaojiu and Korean soju, usually highlighting soju's strong taste. Trade relations most likely developed on the heels of diplomatic relations; however, to date it is difficult even to estimate the actual scale, much less detect it.

At the same time, turning east, the Chosŏn court sent many bottles of soju to the Tokugawa government in Japan and to the Ryukyus. The Chosŏn dynasty's official chronicle reports that they sent ten to 100 bottles of soju, along with ch'ŏngju and other gift items, to various rulers in Japan, including Tsushima and the Japanese main islands.[114] As Chapter 6 shows in more detail, a relatively large influx of soju from Korea perhaps exerted some form of influence or stimulus that accelerated the development of soju making in Japan. The Chosŏn court also sent many bottles of soju to the Ryukyu Kingdom, an independent realm in the Okinawa islands that sent regular tribute to both the Ming and Qing courts of China and the Tokugawa shogunate Japan. This dual tribute reflects the Ryukyuan court's strategy of dealing with its two large neighbors, who maintained closed-door policies on foreign trade, by engaging actively in transit trade in order to import goods from Southeast Asian countries and resell them to China, Korea, and Japan.[115] It was during this time that the Ryukyuans developed a local spirit they called awamori.

---

[113] Lee Ki-baik, *A New History of Korea*, 189.

[114] For example, see *Chosŏn wangjo sillok*, Sejong 6 (1424): 06–02–07[04].

[115] As a result, Shuri 首里 – then Ryukyu's capital – showed off its status as an international city where merchants from East and Southeast Asia gathered to trade. Ha Ubong,

While the influence behind this drink may have come from somewhere in Southeast Asia, it could just as likely have been inspired by Korean soju or Chinese shaojiu as a consequence of direct diplomatic interactions with its immediate neighbors.

Just as the Chosŏn court could influence its neighbors, it also subjected itself to foreign influences that affected the evolution of Korean liquor. These include a Southeast Asian liquor called *Ch'ŏnch'ukchu* 天竺酒 (Indian liquor, Chinese *Tianzhu jiu*), which first traveled to the Ryukyus, along with other tributary gifts, through a diplomatic envoy.[116] A Chinese term, *Tianzhu* means "India," which East Asians generally considered a country at the distant end of the southern sea-trade routes. It has been described as "a liquor made from palm tree juice, whose flavor is fragrant and intense; drinking only two cups would get one drunk all day long."[117] This most likely was a distilled variety of liquor that became popular in South and Southeast Asia during early modern times and that the Manila galleons transported regularly from the Philippines to Mexico.[118]

Ryukyu's connection with Southeast Asia was close, certainly close enough to deserve careful attention. Many recent studies have revealed the history of early modern Ryukyu, in particular its active role in diplomatic and trade relations with countries in East and Southeast Asia. As previous chapters have shown, the distillation of alcoholic beverages probably first developed among Indian Ocean societies and later transferred to China via Southeast Asia. Chinese first called the Southeast Asian spirit in question "Thai wine" (*Xanluo jiu*) during the Song dynasty; most likely they were referring to arak, as some sources suggest.[119] Another Chinese distilled liquor, Nanfan shaojiu, also known as arak, probably found its way to South, Southeast and West Asia from China thanks to Chinese émigrés who had fled the turmoil created by the violent transition between the Yuan and Ming dynasties in the mid-1300s. Interestingly, Nanfan shaojiu became popular among Korean literati who read the many Chinese encyclopedic works that found their way to Korea as imports. Whenever they discussed soju's origins, literati like Yi Tŏkmu and Yi Kyugyŏng identified them as South or Southeast Asian, though it must be remembered that in discussions of spirits like awamori, native to Ryukyu and Kyushu in the southern part of Japan, they refer to

---

Chosŏn sidae pada rŭl t'onghan kyoryu (Exchange via the Sea during the Chosŏn Dynasty) (P'aju: Kyŏngin munhwasa, 2016), 168.

[116] Chang Chihyŏn, *Hanguk oeraeju yuipsa yŏngu*, 103–105; *Chosŏn wangjo sillok*, Sejo 7 (1461).

[117] Chang Chihyŏn, *Hanguk oeraeju yuipsa yŏngu*, 104; *Chosŏn wangjo sillok*, Sejo 8 (1462).

[118] For more details, see Chapter 6 of this book. [119] See Chapter 1.

the eighteenth and nineteenth centuries in which they lived rather than to the Yuan era. Contemporaneous developments in distilled-liquor production in eighteenth- and nineteenth-century East Asia probably inspired Yi Tŏkmu and Yi Kyugyŏng to investigate soju's origins further. Whatever their motivation, Yi Tŏkmu discovered a Japanese source that claimed a novel source for the now-popular spirit: the Netherlands.[120] This suggests another factor that could have influenced the development of East Asian distilled liquors and must be considered.[121]

During most of the Chosŏn dynasty, Korea's main contacts were China and Japan. However, at the dawn of the sixteenth century, in the middle of the Chosŏn period, Europeans bent on global expansion entered Asian waters and found their way to East Asia.[122] East Asian countries that had opened their doors to the broader Eurasian world under the Mongols now enforced comparatively isolationist policies, allowing only limited, major interactions to occur under state supervision. Chosŏn had no direct contact with these Europeans. However, as political conflicts with China and Japan erupted – ranging from dynastic upheavals to invasions and war – a few new European objects and ideas entered Korea, including science and religion. Although Koreans managed to prevail in maintaining traditional culture, some important and transformative innovations regarding foodways did occur, such as chili peppers that scholars assume were transferred to Korea by Portuguese traders from the Americas through Japan beginning in the late sixteenth century, which transformed the taste of kimchi (fermented pickle) and other Korean national foods.[123] While it is not easy to find immediate material evidence of the stimulus that transformed Korean alcohol-drinking culture, we can assume that it existed, at least once Europeans began to make bold moves to open Asian markets. Yielding to the unrelenting pressure of Europeans to open their country to trade, the Chosŏn court signed a treaty in 1876 that opened two ports to international trade. In 1883, Inch'ŏn opened. Soon, foreign objects and ideas began to enter the Korean peninsula, including alcohol.[124] Foreigners or Koreans who contracted with them were shipping European whiskeys to Korea near the end of the Chosŏn era. Perhaps this was a response to Westerners who were recording their observations of Chosŏn Korea and noting how many

[120] Yi misspelled "Netherlands."

[121] For more discussion on this topic, see Chapter 6 of this book.

[122] The world was changing rapidly at that time; Europeans were coming to East Asia in search of spices and silks, and to establish new contacts for direct trade.

[123] Stanley Marianski and Adam Marianski, *Sauerkraut, Kimchi, Pickles & Relishes* (Seminole, FL: Bookmagic, 2012), 45; Brian R. Dott, *The Chile Pepper in China: A Cultural Biography* (New York: Columbia University Press, 2020), 23–24.

[124] Chu Yŏngha, *Ŭmsik inmunhak*, 206.

people there loved strong booze. For example, the German Ernst Jakob Oppert (1832–1903), who is infamous for having caused the Namyungun body-snatching incident in 1868, recollected in his 1886 memoir *Ein verschlossenes Land: Reisen nach Korea* (A Forbidden Land: Voyages to Korea) that when he landed at Chosŏn and met officials, they were delighted when he gave them a bottle of whiskey.[125] As Joo Young-ha argues, Chosŏn officials were well accustomed to soju and thus predisposed to like the taste of whiskey and gin.[126]

At the same time, many scholars in Korea broadened their interests to include a variety of themes, including everyday lives at home and cultures abroad. Chŏng Min argues that these changes are related to many other changes that occurred worldwide during the eighteenth century, as many books communicated new ideas influenced by modern Western ideologies and cultures. Imports to Chosŏn had a considerable influence on a group of scholars often identified as adherents of pragmatic types of scholarship in the eighteenth and nineteenth centuries.[127] Among them were literati who left behind some record of soju, like Yi Tŏkmu and Yi Kyugyŏng, who read new books from China and Japan that offered new information about the widening world.

However, while Chosŏn intellectuals spontaneously accepted external stimuli, their ruling court was already in decline, overrun by corruption and vulnerable to invasion by foreign powers like China, Japan, Russia, the United States, Great Britain, and France. Caught in a vortex, the Chosŏn court also suppressed new ideas from outside, even Christianity. In the end, it became a colony of Japan, winner of the competition to colonize among the great powers of the day, and with that, the long Chosŏn dynasty came to an end. Since then, the Korean peninsula has faced many new challenges – political, economic, and cultural – even in the area of foods. In this context, the development of soju fostered a new opportunity to transform areas of technology, commercial distribution, and law in Korea, a topic for further exploration in the following chapter.

---

[125] Oppert tried to secretly excavate the bones of Namyungun, the father of Regent Yi Haeung (Taewŏngun) who held de facto political power at that time, and to hold them for the purpose of blackmailing the regent into removing Korean trade barriers. Chu Yŏngha, *Ŭmsik inmunhak*, 180–181.

[126] Ibid., 181.

[127] Chŏng Min, *18 segi Chosŏn chisigin ŭi palgyŏn* (Discovery of the 18th-Century Korean Intellectuals) (Seoul: Hyumŏnisŭt'ŭ, 2007), 52, 57–58, 64. These scholars contributed to a sequence of reform-minded Chosŏn Confucian thought from the late-seventeenth to the early nineteenth centuries. Some historians coined this late Chosŏn-period scholarship "practical learning"; however, recently some have criticized this characterization, which they say is at odds with the scholars' fundamental concerns, which are based on Neo-Confucian themes.

Conclusion

The present chapter has argued that soju arrived in Korea from China in its distilled form during the Koryŏ period and developed in scale and technological sophistication, adapting to local conditions throughout the subsequent Chosŏn period. It even became the most important alcoholic drink in locations like Cheju Island, where fermented wines typically turned bad quickly in the humid environment. On the peninsula, soju was neither as popular nor as widely consumed as fermented wine; nonetheless it developed into an important part of Korean culture. While a variety of sources, including culinary books, stipulate the many different kinds of liquor – most of all ch'ŏngju, whose varieties were used in ancestral rites – that developed during the Chosŏn era, soju varieties were always among them. Indeed, their share of Korean alcohol consumption gradually increased over time, so much so that people increasingly began to use it as a substitute during summer when clear strained wines like ch'ŏngju easily turned bad. Its production spread quickly thanks to a simple Mongolian-style distillation technique. In time, the tools to make soju evolved, until the original simple form morphed into the relatively complicated stills called *kori*. By the late eighteenth century, as Korea's economy was expanding, soju had grown popular nationwide. It gradually diversified over time as producers applied new ingredients and distillation techniques, some of which were documented in culinary books. Distilled liquors were identified as soju, although, intermittently, soju was also referred to as arak in various Korean transcriptions. In any case, the Korean term *soju* predominated in official documents and literary works.

Upon its arrival in Korean society, the new spirit began to exert a range of effects. According to official annals, the royal court began to use it for both gift giving and medical use. To meet growing demand, it had its own bureau to distill soju, ostensibly for internal use. At the same time, it required ordinary households to submit soju as tribute to the court, a fact that did much to promote soju's spread throughout the peninsula and into ordinary households. This inevitably raised the importance of soju in a home-brew culture in each household, in contrast to China, where a different set of policies under most of its dynasties promoted large-scale industrial distilleries. While most of this production supplied elites called *yangban*, who obtained soju to support their hosting culture and for applications in classical medicine, soju was also available to ordinary people, who began to use it as folk medicine. The government did not try to control these developments by means of systematic tax collection efforts; instead, it simply banned the consumption of soju and other distillates. It did so out of concern for the deleterious effects of alcoholism that it perceived in Korean society, as well as the large quantities of grain bled from the food supply to fuel alcohol production. In the end, this measure apparently did not work.

While documentary evidence demonstrates an ongoing process of soju localization within Korea during the Chosŏn dynasty's five centuries, it likewise shows that through soju Koreans also interacted with other societies to varying degrees. Soju was regularly transported as official gifts or trade goods, just as new forms of distilled liquor traveled to Korea from China, Japan, and the Ryukyus. As Europeans began to expand into Asian waters, Korea also gradually opened to the world at large. Europeans traveling to Korea observed heavy drinking, especially strong alcohol, which probably meant soju. At that time, still, soju was the second or third most popular drink in Korea. Strained wine was the first variety to appear on the Korean peninsula, according to the evidence, except for Cheju Island, where soju has long been the first drink of choice. Soju's Chosŏn stage of history ended, but it had prepared the way for a major transformation in the twentieth century in which it transformed into Korea's first national drink.

5 Challenges of Modernity
The Rise of Modern Industrial Soju and the Revival of Traditional Soju

We have observed many centuries of the evolution of soju on the Korean peninsula. From its rapid rise in the fourteenth century under the Koryŏ dynasty to its integration into the fabric of Korean life under the Chosŏn, soju developed, diversified, and localized as the spirit gradually earned its reputation as a national drink. Here we examine another important transformation in the evolution of soju, this time during the twentieth century, which laid the foundation for the flowering of the current soju culture in the twenty-first century. This transformation was both directly and indirectly affected by enormous changes that are the hallmarks of the modern period in global history, which include the Scientific Revolution beginning in the eighteenth century, industrialization, modernization, colonialism, and globalization.

The first transformation to consider is the modernization of the soju distillation industry, which involved the introduction of new technologies and methods, during the colonial rule of Korea by imperial Japan from 1910 to 1945. During this period, "traditional" soju, according to distillation methods and recipes made conventional during Chosŏn times (1392–1910), declined gradually as Japanese companies introduced and popularized a modern industrial form to Korean consumers. Instead of using a traditional distillation apparatus like a soju kori, the new industrial form of soju introduced by the Japanese has used a continuous still (also known as a patent still or column still) for mass production at low cost in large factories that innovators in European societies developed during the eighteenth century and later introduced to Japan in the late nineteenth century (see Figure 5.1A).[1]

The second transformation occurred amidst the new environment of rapid economic development and state building that defined Korea (South Korea in particular) between its independence in 1945 and its

---

[1] This modern still has continuously developed with modern technology from a variety of stills consisting of two or more columns, which are called different names, including column still, continuous still, patent still, and Coffey still. For more information about modern distillation technology that has developed in Europe, see Forbes, *Short History of Distillation*, 99–362.

Figure 5.1 A: ethanol distilling columns in Korea in the twenty-first century. Photograph courtesy of the Korea Alcohol & Liquor Industry Association. B: industrial soju factory in Cheju in 1943. Photograph from Sajin ŭro yŏngnŭn 20 segi Cheju si (2000). Photo courtesy of Jeju city.

realization of developed-nation status in the late twentieth century. During this half-century, Koreans faced many chaotic and difficult situations that presented them with major challenges, including the Korean War (1950–1953), rapid modernization, and industrialization led by a dictatorial government, followed by a fiercely determined democratization movement, a trend not unlike other developing countries of the late twentieth century. Early in this period, as modern industrial soju continued to develop and its companies continued to grow, the newly independent government of the Republic of Korea mobilized a campaign to promote a Korean national alcoholic beverage dubbed *minsokchu* (literally, "national folk liquor"), which encouraged a revival of traditional soju, whose consumption had declined during the era of Japanese rule. In the third and most recent transformation between the late twentieth and the early twenty-first centuries, a globalizing economy created unprecedented connections between soju producers and virtually every part of the world, representing a radical form of multiple cross-cultural transfer beyond Korean culture. Along with the rapid industrial development of major companies during the recent Korean Wave (*Hallyu*) in the last two decades, soju also began to expand to other countries and compete with other kinds of distilled liquor both at home and in global markets. Korean companies successfully appropriated foreign spirits, brews, and mixing culture, and popularized new ways of drinking soju mixed with beer or in a cocktail, while foreign makers boldly appropriated Korean spirits, leading to innovations in soju that could compete in Korea as well as abroad. This recent transformation during the past century catalyzed the development of today's soju as a new global brand.

This story of soju's recent transformations raises fundamental questions about the identity of soju and the meaning of such identifiers as "tradition" and "modernity" that have been regularly applied to this ever-changing spirit. Different approaches to interpreting the consequences of soju's transformation, from its traditional form using traditional stills to current forms like industrial soju, have led to debates about what constitutes "traditional" soju. Simply put, can we call modern forms of soju – or even self-consciously traditional ones – *authentic*?

The answer in this book is yes. Once industrial soju developed, Koreans never stopped consuming it. And now the world appears to have followed suit. Distilled soju is largely responsible for soju's successful global debut in the 2000s, a transfer of culture every bit as shaped by world events as the drink's original transfer to Korea under the auspices of the far-reaching Mongol Empire centuries before. In fact, the same pattern of cross-cultural transfer observed in the Koryŏ era also appears in the colonial era of the early 1900s, when Japanese companies first introduced the technology and techniques necessary to make the beverage, and Koreans began to both consume and develop it, laying the basis for a new distilled-soju market that would one day go global. Indeed, even the Japanese technique relied on contributions from outside its cultural sphere, having borrowed the distillation technologies that could make new soju varieties from Europeans, whose ancestors had assimilated them centuries earlier from societies in eastern Eurasia. This search for soju's origins can extend even further: into the past, through the Middle Ages to antiquity and beyond; and into the future, as global soju companies continue to study people's cultural attitudes and consumption patterns and respond with experiments in ingredients and technologies that try to adapt soju to the diverse and changing tastes of people all over the world. We cannot exclude these new manifestations of soju from any discussion of the history of the spirit, not if we intend to identify soju's place in global history.

The Rise of Modern Industrial Soju during the Period of Japanese Colonial Rule

In a way, the thirty-five years of Japanese occupation of Korea from 1910 to 1945 seem to resemble the Koryŏ era, when Korea was also subjected to domination by a foreign power. While it is difficult to compare these two periods, given their different global historical contexts, there seem to have been fundamental differences between them. The long Mongol conquest of Koryŏ in the thirteenth and fourteenth centuries did not aim at the incorporation of Koryŏ into the Mongol Empire, and instead sought only the kingdom's compliance as a vassal state. As a consequence, the Koryŏ dynasty enjoyed de facto autonomy during its last century of existence, thanks to its kings, who successfully advanced the status of their royal family by marrying Mongol royal princesses and aligning the country with its giant neighbor.

In the case of Korea's twentieth-century occupation, in contrast, imperial Japan imposed a colonial form of rule that systematically enforced total dominion over all sectors of Korean society, including its economy and culture. Colonialism, an institution spawned by European societies in the wake of their global exploration beginning in the sixteenth century, reached the peak of its global power by the early twentieth century. The Japanese are notable for assimilating colonial and other European institutions in an effort to maintain their autonomy as other Asian countries were losing theirs. Having ended its official isolation after observing China's humiliating defeat by Great Britain in the Opium War in 1842, and after Commodore Matthew Perry's "black ship" forced Japan to enter into trade with the United States in 1853, Japan swiftly initiated sweeping westernizing and modernizing reforms. In time, it also adopted the imperialist policies that had become the dominant trend among world powers, and so it was inevitable that it would begin to colonize other countries. The closest country that promised fulfillment of Japanese interests was Korea. After securing from its colonial competitors a dominant position over the country, imperial Japan seized and then later annexed Korea in 1910. Once in control, Tokyo established in Seoul the Office of the Governor-General of Korea (Chosŏn Ch'ongdokbu 朝鮮総督府, 1910–1945), which exercised full political authority over Korea for the next thirty-five years and, during that time, transformed many sectors of Korean society with modernizing reforms directed at its economy and culture.[2] Korea's complete loss of sovereignty to its neighbor happened a mere fourteen years after the Chosŏn court renounced its longtime client relationship with China, proclaimed the Great Korean Empire (Taehan jeguk 大韓帝國, 1897–1910), and attempted partial modernization and westernization initiatives in various sectors of society, influenced by modernized Western countries such as the United States.[3]

The issues of Japan's responsibilities vis-à-vis colonization, which Koreans often regard as a harsh and humiliating experience, have lingered

[2] See Lee Ki-baik, *A New History of Korea*, 313–327.   [3] Ibid., 300–313.

in Korea to the present due to the continuous turmoil to which Koreans have been subjected since immediately after Japan's surrender and Korea's independence in 1945 and the Korean War that began in 1950. It ended in 1953 with the division of the country into North and South.[4] There are fierce debates over how to clear the Japanese colonial legacy, yet allow the modernity achieved under colonialism to continue progressing.[5] In such a charged climate, scholars studying the history of science, technology, and other forms of culture in colonial Korea began to turn their attention to colonial-era changes and developments that had considerable influence on Korean society yet could be examined from more objective points of view. This has produced a new body of studies that examines various aspects of "mutuality" in the colonial relationship rather than viewing it as a singly directed form of oppression.[6]

We should also recall that many aspects of the effort to modernize before the colonial era happened under the initiative of Koreans who sought out and cultivated contacts and relationships with foreigners during the late nineteenth century, including initiatives by the Chosŏn court, a trend that did not cease with the Japanese conquest. Many societies in the wider world in this colonial period interacted with each other in various fields and through various experiences, which led to a rush of new globally developed materials and concepts into Korea, where they were further modified by Koreans locally.[7] Koreans had been assimilating new things and ideas from imperial Japan since the early twentieth century. We can find similar cases at the same time in

[4] For example, the 1965 agreement with Japan that South Korea's third president, Park Chung-hee, made, entitled Treaty on the Basic Relations between Japan and the Republic of Korea, tried to settle lingering problems regarding property and claims once and for all in order to promote economic co-operation between the two countries. Glenn D. Hook, *Japan's International Relations: Politics, Economics, and Security* (London: Routledge, 2001), 491–492. However, many Koreans have objected to this Act, which the president made without seeking national consensus, and this has contributed to persistent anti-Japan sentiments.

[5] For example, see Yi Yongch'ang, "Ilche singmin chanjae wa ch'inil munje" (Issues around the Vestiges of Japanese Imperialism and Pro-Japanese Activities), *Kukhak yŏn'gu* 7 (December 2005): 297–328.

[6] For example, Jung Lee's studies of botanizing in colonial Korea elucidate both coerced and voluntary changes made in the Korean science education system in the course of assimilating Japanese (or "Western") science and technology into Korea's colonial environment. Jung Lee, "Mutual Transformation of Colonial and Imperial Botanizing? The Intimate yet Remote Collaboration in Colonial Korea," *Science in Context* 29, no. 2 (2016): 179–211.

[7] At that time, Korea truly came out of isolation; became exposed to the world at large, which led to increased contacts and more intensive interactions with foreigners; and fully assimilated and modified new things and ideas, though it lost the opportunity to do so of its own will and often did so with limited choices given the specific conditions of the colonial setting.

China, such as the import of many Western books and ideas from Japan by Chinese intellectuals who studied in Japan;[8] however, Koreans were exposed to more fundamental change in all aspects of their tradition because of the peculiar circumstances of Japanese rule. Such aspects include traditional food culture. Joo Young-ha provides convincing examples of Japanese foods, such as Kikkoman brand soy sauce, *ajinomoto* (a chemical seasoning), or instant ramen, which led to the disappearance or distortion of longtime culinary tastes among the occupied cultures of the Japanese Empire. These foods developed first in Japan during the nineteenth century; the Japanese took them to Korea and Taiwan and ate them there; from there, they were exported abroad, where they quickly developed a presence.[9] In another prominent example, distilled liquor experienced great change when producers began to forsake traditional Korean distillation methods for the multi-column rectifying system, which had first developed in Europe before it was adopted in Japan and transferred to Korea, where it became the conventional system for soju production until the present day.

The development of a modern alcoholic-beverage industry in Japan was among the radical changes made in areas like the economy and science as part of the Japanese government's effort to achieve rapid modernization and full integration into the global economy. As Chapter 6 describes, Japan developed its own alcohol production and drinking culture during premodern times thanks to the assimilation of key influences from China and Korea. In the early seventeenth century, a Dutch factory set on a small island in Nagasaki's harbor became Japan's communication window to the world at large; through this portal it began to receive new commodities, ideas, and technologies from Europe. By the twentieth century, Japanese producers had begun making major changes in the manufacture of alcoholic beverages by importing modern distillation machines from Great Britain.[10] By this time, it made sense also to make these changes in Korea, which, as Japan's new colony, was assigned to support its colonizer's economy. Soon, starting in 1919, new factories using continuous-type stills were producing distilled liquors on a large scale (see Figure 5.1B).[11]

[8] Joanna Waley-Cohen, *The Sextants of Beijing: Global Currents in Chinese History* (New York: W. W. Norton & Company, 1999), 166–205.

[9] Young-ha Joo, "Imperialism and Colonialism in the Food Industry in East Asia: Focusing on Instant Ramen," *The Newsletter* (IIAS), no. 75 (Autumn 2016), 19–20.

[10] The Portuguese, Spanish, English, and Dutch came to Japan from the sixteenth century, yet only the Dutch were eventually allowed to enter Japan and live on Dejima island, located in modern-day Nagasaki prefecture. See Marius B. Jansen, *The Making of Modern Japan* (Cambridge, MA: The Belknap Press of Harvard University Press, 2000), 63–85.

[11] After the first continuous-type still was installed in P'yŏngyang in 1919, six factories came into operation in 1925 and soju production was increased. Ch'oe Hansŏk, "Saengmyŏng

In addition to these material changes, the Japanese also initiated several major social reforms that together transformed the industry, including the Liquor Tax Act and a modern hygiene system. These changes affected Korea and Japan differently. In Japan, earlier regimes like the Tokugawa shogunate had imposed taxes systematically on breweries and distilleries, and so the new tax system in the early twentieth century caused fewer shocks to the industry. Despite some blows, many small sake breweries and distilleries in Japan attempted to overcome the new challenges as soon as possible by devising innovations to their systems. As a consequence, traditional distilled liquors developed (mainly in the southern part of Japan) hand in hand with their factory-produced counterparts. The traditional breweries and distilleries in Korea, however, received these changes with great shock, which led to sudden decline in some areas of traditionally distilled manufacturing.

The earliest academic monograph to discuss the history of soju in Korea, *Chōsen shuzōshi* (1935), is also the most useful source to help us gauge the many fundamental changes in traditional liquors and their production that were caused by these new conditions in Korea. It was created to describe the general situation of alcohol production in Korea before 1935, highlighting major changes made by Japanese policies. This book was written by the Committee of Alcoholic-Beverage Making in Korea (Chōsen shuzō kyōkai 朝鮮酒造協会), based on large-scale systematic field research by Japanese scholars who came to Korea to conduct academic examinations and practical experiments in various fields, including science, as well as anthropology and folklore, under the supervision of the Office of the Governor General of Korea.[12] This report helps us to understand the overall situation of alcoholic beverages, breweries, and distilleries in Korea in the early 1900s, thanks to its analyses of traditional distillation practices, its introductions to distilled liquors from other countries, and its discussions of modernizing the industry they were helping to introduce.

First of all, *Chōsen shuzōshi* describes the distilled liquors that producers developed during the early Chosŏn period and sustained to the end of the dynasty. For example, it describes soju as follows: "also called 'fire liquor.' Its alcohol content varies. Its taste is clean and fresh, and it is much consumed in the northwestern area of Chosŏn."[13] The report also

---

ŭi mul, chŭngnyuju! Urinara ŭi chŏnt'ong soju wa kŭ ch'in'gudŭl" (The Water of Life, Spirit! Korean Traditional Soju and Their Friends). *RDA Interrobang* 168 (2016): 3.

[12] For an example of the reports published on folk customs and culture at that time, see Murakami Tadayoshi, *Chōsenjin no ishokujū* (Clothing, Food, and Housing in Chosŏn) (Kyŏngsŏng 京城: Chōsen sōtoku fu, 1916).

[13] *Chōsen shuzōshi*, 59.

shows that, while more people consumed soju in northern Korea, they consumed it more generally across the entire peninsula during the summer. The book introduces varieties of soju consumed in Korea, such as *pakch'wi* 粕取 soju, made by distilling the lees of clear strained wine; soju made by distilling the mash of seeds or its clear strained wine; soju made of nuruk (traditional Korean fermentation starter made of wheat, rice, and barley) as the most traditional Chosŏn-period soju; *koji* soju made from Japanese koji (a yeast prepared in Japan from rice inoculated with the spores of a mold) instead of traditional nuruk in order to maximize quantity and reduce production cost; and new-style soju made by a column distillation apparatus using sweet potatoes as raw material. Along with this diversification came foreign liquors, such as European whiskey and Chinese *gaoliang jiu* (a local sorghum-based liquor).[14] According to the Japanese report, both koji soju and modern soju began to dominate the soju market by the 1930s, overtaking traditional soju made with nuruk. Intriguingly, *Chōsen shuzōshi* does not cite local nicknames for soju, like *arangju* and *araegiju*, clearly transliterations of *arak*, though they can be found in earlier Chosŏn-era sources like *Simyang ilgi* (The Diary at Simyang) (1636) and in more contemporary anthropological research conducted by foreign scholars like G. J. Ramstedt.[15] Given its expansive scope in covering all varieties of Korean beverages, the report does not present the kind of detail about consumption that one might find in an academic study.

The first thing to note about *Chōsen shuzōshi* is its evaluation of the soju industry in Korea as economically and socially depressed and inferior to Japanese standards of modernization and technology. Its argument can be summarized as follows.[16] Because it was customary for the Chosŏn dynasty to despise artisanship and industrial work, the Korean people, particularly women, brewed alcohol at home or ran alcohol production as a side business. Therefore the industry's scale of production during the Chosŏn period was small, its methods unprofessional, and its technology inferior. Those most engaged in the industry lacked progressive ideas to improve the situation.[17] Because each restaurant that made soju used only a small still called a kori, the quality of the product was poor in general, so much so that people in the northern border regions imported gaoliang from China (valued at 2 million won every year at that time). In such a situation, people imported more and more industrially produced spirit (the main ingredient to make soju) from foreign countries through port cities like Inch'ŏn and Wŏnsan which had been open to foreign trade

---

[14] Ibid., 59–61.   [15] Chang Chihyŏn, *Hanguk oeraeju yuipsa yŏngu*, 56–58.
[16] *Chōsen shuzōshi*, 217.   [17] Ibid., 38.

since the late nineteenth century, and mixed it with traditional soju to sell at the market. While this report tries to highlight outmoded methods of production and supply in Chosŏn-era soju making in an effort to justify the import of new systems and technology from Japan, we can still benefit from its content if we read between the lines. The report provides a rich body of statistics and systematic field reports that together reflect the general situation of soju production and consumption in Korea, a situation characterized by small-scale handcraft workshops that could not meet the growing demand for high-quality distilled liquors at that time.

Through these research reports on traditional Korean methods, analysts for imperial Japan also concluded that Korean soju was made without any regulation of hygiene, which led to its critical views of the country's home-brew culture. Compared to Japan, which traditionally registered, taxed, and regulated private breweries and distilleries, Korea had many more brewers, given its home-brew culture; this included private homes, which were free of tax obligations, because each household could brew alcoholic beverages for entertaining guests and conducting ancestral worship.[18] Soon after its annexation of Korea, Japan began to impose regulations on soju making in its new colony, aiming particularly at promoting sanitation and imposing a tax and license system. Starting in 1910, imperial Japan initiated and enhanced the Liquor Tax Act, which considerably increased the tax rate on all alcoholic beverages and banned home brewing, an important Korean tradition for centuries.[19] While the 1935 Japanese report argues that the Liquor Tax Act advanced Korea's alcohol industry in terms of both quality and quantity, many Koreans regarded it as a great blow to their alcohol traditions.[20] Based on an examination of relevant statistics, current

---

[18] Ibid., 49. While traditional Japanese breweries and distilleries suffered blows from state policies promoting modernization and large-scale industrialization, their counterparts in Korea felt it even more acutely, owing to their home-brew culture. Yi Hwasŏn and Ku Sahoe, "Ilche kangjŏmgi chuseryŏng ŭi shilch'e wa munhwa chŏk hamŭi" (The Actual and Cultural Implications of the Liquor Tax Act during the Period of the Japanese Occupation), in Han'guk ŭi sul 100 nyŏn ŭi kwaje wa chŏnmang (Challenges and Prospects for 100 Years of Korean Alcoholic Beverages), ed. Chŏng Taeyŏng, Ku Sahoe, Chŏng T'aehŏn, Chŏng Sŏkt'ae, Kwŏn Sŏngan, Chŏng Ch'ŏl, Yi Sŏkchun, and Yi Hwasŏn (Seoul: Hyangŭm, 2017), 26, 31–32, 37.

[19] Chong Dae Song, Chosen no sake, 178–93.

[20] The Rural Development Administration (Nongch'on chinhŭngch'ŏng) argues in a report entitled "Traditional Korean Soju and Its Friends," written in 2016, that the causes of traditionally distilled soju's disappearance from the market were the Liquor Tax Act, the use of the Japanese black koji, and the production of ethanol. The report says that in the year 1909, the number of traditional soju distilleries reached 28,000, yet the Japanese government issued the Liquor Tax Act five times between 1909 and 1934, restricting the production of traditionally distilled soju (which had been freely made in the private

scholars argue that the Liquor Tax Act led to a great increase in tax income for Japan's Office of the Governor General of Korea.[21] However, it also undeniably undermined the popularity of many once-famous spirits like traditionally distilled soju.[22]

The 1919 introduction of modern industrial distilleries, mandated by Japanese policy, dealt another great blow to Korea's traditional home-brew culture. Japan had already been operating modern industrial distilleries at home for more than two decades.[23] As seen above, the Japanese book *Chōsen shuzōshi* introduces this as new-style soju, along with other kinds of soju that were consumed in Korea in the 1930s.[24] Using new industrial column stills for multiple distillations at large scales, these distilleries first made high-percentage alcohols (up to 95 percent) like ethanol spirit from cheaper materials like potatoes or tapioca instead of grains like rice, and then added water and sweeteners such as sugar, glucose, citric acid, saccharin, amino acids, and sorbitol to enhance flavor.[25] This event marks the rise of the modern form of industrial soju in Korea, which has remained the country's dominant variety. Many large-scale factories that produce new-style soju on a mass scale were first built in Korea in places like Cheju Island off the south coast of the peninsula, near Japan.[26]

As Chapter 6 shows, the Japanese also created a new (though similar) rendition of the Chinese term *shaojiu*, in which they replaced the trad-itional character for alcohol, 酒 (Chinese *jiu*), with another Chinese character 酎 (Chinese *shu*), which refers to high-proof alcohol. Korean

sector) only to factories at standard scales. On this, see Ch'oe Hansŏk, "Saengmyŏng ŭi mul, chŭngnyuju!", 1–20.

[21] Yi Hwasŏn and Ku Sahoe, "Ilche kangjŏmgi chuseryŏng," 30–31.

[22] In 2016, the Korea Suul Institute (Uri sul munhwawŏn), which researches Korean alcoholic beverages, convened a conference on the centenary of the Liquor Tax Act's promulgation in Korea and published the presented papers as an edited volume. See Chŏng Taeyŏng, Ku Sahoe, Chŏng T'aehŏn, Chŏng Sŏkt'ae, Kwŏn Sŏngan, Chŏng Ch'ŏl, Yi Sŏkchun, and Yi Hwasŏn, eds., *Han'guk ŭi sul 100 nyŏn ŭi kwaje wa chŏnmang* (Challenges and Prospects for 100 Years of Korean Alcoholic Beverages) (Seoul: Hyangŭm, 2017). To read about the conference, see ibid., 14.

[23] Takahashi Kōjirō, "Nihon no jōryūshu, shōchū" 日本の蒸溜酒、焼酎 (The Japanese Distilled Alcohol "Shochu"), in *Shōchū higashi mawari nishi mawari* (Shochu around the World), ed. Tamamura Toyo'o (Tokyo: TaKaRa Alcohol Beverage and Life Research Institute, 1999), 238.

[24] *Chōsen shuzōshi*, 60–61.

[25] For a succinct survey of modern industrial soju in Korea, see Chu Yŏngha, *Sikt'ak wi ŭi han'guksa: menyu ro pon 20 segi han'guk ŭmsik munhwasa* (A Korean History of the Dinner Table: The 20th-Century Cultural History of Food Viewed through Menus) (Seoul: Hyumŏnisŭt'ŭ, 2013), 483–492.

[26] For details of these soju distilleries and the factories that developed during the period of Japanese colonial rule, see *Chōsen shuzōshi*, Chapter 11; Pae Sangmyŏn, trans., *Chosŏn chujosa* (Korean translation), 431–512.

companies adopted this latter character, too, pronounced *ju* (*chu*) in their language. Because of this, *ju* 酎 in the word *soju* 燒酎 continues to be used to this day. At the same time, the compulsory introduction of a Japanese yeast called "black *kōji* 麴" (*Aspergillus oryzae*) to replace the Korean fermentation starter, nuruk, traditionally used to distill liquors in Korea, caused great change in the process of making traditional Korean soju. Using a cheap and stable form of Japanese yeast made from black beans instead of Korean's expensive and unstable traditional yeast only further undermined the Korean tradition.[27] The usage rate of traditional yeast in soju, which had reached 99 percent in 1923, fell to 5 percent in 1932. During that time, the taste of soju began to change. The Japanese report *Chōsen shuzōshi* viewed this change positively, as a sign that Korea's soju industry was beginning to systematize and grow more commercially competitive, though some later scholars argue that this simply triggered traditional soju's decline.[28]

How did Koreans actually react to these changes in their traditional drink? Unlike many traditional Japanese distilleries that worked to overcome the effects of their government's intervention in product development, and continued to develop their traditional shochu by simply improving its quality, Korean distilleries, due to their colonial status, found it difficult to deal with Japan's sweeping changes.[29] Therefore, while traditional distilleries in Japan survived, in Korea they rapidly declined. Imperial Japan also compelled a number of Japanese to move to Korea and develop advantages there through industrial investment.[30] Many freely moved or traveled to Korea to seek business opportunities that the new colony presented to Japanese. This included the alcohol industry, in which opportunities had emerged thanks to the vacuum created by the decline of traditional manufacturing caused by the new liquor laws. While it is true that many Koreans resisted Japanese rule – some even organized independence campaigns like the 3.1 Movement in 1919 in which many Koreans participated, and still others organized military resistance – it is also true that many Koreans co-operated with the Japanese. Indeed, co-operation appears to have occurred quite regularly in the case of the alcohol industry. According to research by Chŏng T'aehŏn, the percentage of Korean industry specialists and investors who succeeded in the alcohol industry was higher, as statistics confirm, than in

[27] Ch'oe Hansŏk, "Saengmyŏng ŭi mul, chŭngnyuju!", 3–4.
[28] Indeed, available sources suggest that the soju distillery industry boomed in scale. Chong Dae Song, *Chosen no sake*, 187–188.
[29] Yi Hwasŏn and Ku Sahoe, "Ilche kangjŏmgi chuseryŏng," 25–66.
[30] Michael J. Seth, *A History of Korea: From Antiquity to the Present* (Lanham: Rowman & Littlefield Publishers, 2011), 265–267.

many business sectors in which Koreans gradually lost their positions to the growing tide of Japanese. This happened because big capitalists, though new to the brewing business, could easily grow with little competition given that the governor general had already forced out many small breweries.[31] Large companies like Jinro (Chillo 眞露) dominated the liquor business during the period of Japanese colonial rule, and because of this they continued to succeed in the soju market by developing solid cartels, a trend that has continued to the present.[32]

Korea had the chance to develop this new distillation industry even further after it won independence with imperial Japan's surrender to the Allies following the atomic bombing of Hiroshima and Nagasaki in 1945. After the Japanese left the Korean peninsula, the new Korean leadership enacted various social reforms throughout the country. However, during thirty-five years of Japanese imperial rule, newly created forms of material culture and ideas from Europe that had reached the peninsula through Japan had grown deeply rooted in local society and retained their influence after Japan's departure, whether people there liked them or not. This phenomenon looks similar to the continuing legacy of the Mongol–Koryŏ period, but Korea's more recent experience with Japan's imperialist rule caused much greater antagonism among the generations of Koreans who lived during and after it. However, it is worth noting that various elements of science and culture, which had entered the Korean peninsula through colonial rule at that time and have continued to influence Korean society, had much earlier origins in world history than the Japanese era. The same goes for soju – the distillation technology was a collective product of development through contact and exchange in Afro-Eurasia since ancient times. After being introduced and established in Korea during the Japanese colonial period, new factory-made industrial soju developed without much interruption on the Korean peninsula.

Large Industrial Soju Companies and the Revival of Traditional Soju

After independence in 1945, Korea continuously endured challenging times. Once the USSR and the US occupied the Korean peninsula, they

---

[31] Chŏng T'aehŏn, "Ilche kangjŏmgi chujoŏp kwa chuse chŏngch'aek" (The Alcohol-Producing Industry and the State Tax Policy during the Japanese Occupation), in Chŏng Taeyŏng, Ku Sahoe, Chŏng T'aehŏn, Chŏng Sŏkt'ae, Kwŏn Sŏngan, Chŏng Ch'ŏl, Yi Sŏkchun, and Yi Hwasŏn, eds., Han'guk ŭi sul 100 nyŏn ŭi kwaje wa chŏnmang (Challenges and Prospects for 100 Years of Korean Alcoholic Beverages) (Seoul: Hyangŭm, 2017), 78–79.

[32] Yi Hwasŏn and Ku Sahoe, "Ilche kangjŏmgi chuseryŏng," 53.

divided it into two zones governed by separate states, the North (Democratic Republic of Korea) and the South (Republic of Korea). Right away, efforts were undertaken to wipe out vestiges of Japanese colonial rule. Focus on this recovery did not last long, however. Five years after division, in 1950, war broke out, which plunged the Korean population into miserably poor conditions. Under its first president, Rhee Syngman (1875–1965), the Republic of Korea in the South had to continue imposing limits on the development of traditional soju production based on fermented grains because war and consequent poverty produced endemic crop shortages.

The war ended in 1953, and recovery could finally be pursued, at least in the South. By the 1960s, under the presidency of Park Chung-hee, South Korea began to experience a rapid economic advance that its people call the "miracle on the Han River," an event that marks the beginning of the country's ascent to developed-nation status.[33] Despite these gains, the economy had a long way to go to develop, and when grain shortages persisted, the government was compelled to enact the Act Banning the Making of Alcoholic Beverages Using Grains, passed in 1965, which led to the complete suppression of traditional distillation methods. Koreans would have to wait a few more decades before their traditional soju returned to the public sphere.

In fact, in the 1970s, the firm foundation for a full-scale development of modern industrial soju was established thanks to active government intervention. By the 1970s, there were about 300 domestic soju companies in twenty-four provinces nationwide. In 1973, however, each province was forced to foster a major soju company and brand that represented its region (see Figure 5.2, left), and all other companies had to be incorporated into major companies or shut down.

The government justified it as necessary for quality control, but in fact it did so in order to advance the efficient control of tax revenue. Because the portion of alcohol taxation was very large in the national taxation system at that time, the Office of National Tax Administration led to the increase in size of these companies in order to manage tax collecting easily.[34] For example, there were thirty-three soju companies in South Chungcheong province before 1973, yet after this act they were unified into one big soju company called Sŏnyang soju (now Mackiss Company).

In addition, the government started administering ethanol spirit, the main ingredient in soju, and each company was given a designated

---

[33] This consists of many new economic policies, such as the New Village Movement (Saemaul Movement). See Seth, *A History of Korea*, 381–394.

[34] Chu Yŏngha, *Sikt'ak wi ŭi han'guksa*, 483–485; Yi Chihyŏng, *Soju iyagi: Isŭl kwa pul kwa ttam ŭi sul* (A Story of Soju: Alcohol of Dew, Fire, and Sweat) (P'aju: Sallim, 2015), 52–53.

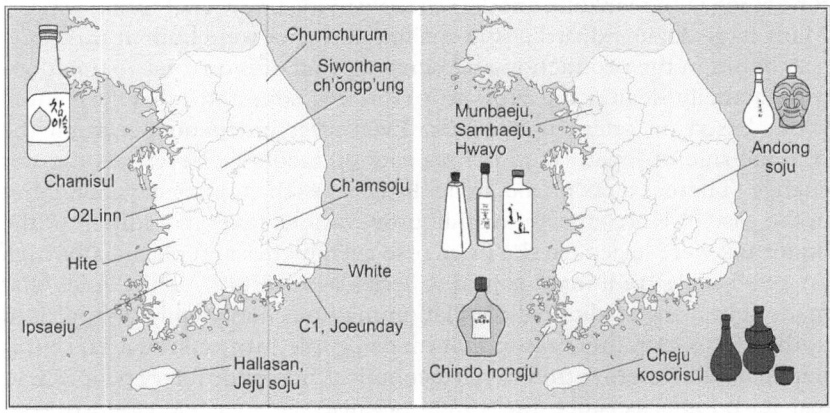

Figure 5.2 Major modern industrial soju companies and brands representing their regions (left), and major revived traditional soju companies and brands representing their regions (right). Drawing by Matilde Grimaldi.

amount. From this point on, the state's monopoly of ethanol spirit for soju production began. The raw materials of the alcoholic beverages used to make the ethanol spirit were tapioca, potato, and sweet potato, which were more productive than cereals such as rice, barley, wheat, or corn.[35] In addition to the proliferation of the ethanol spirit distribution system, another tactic strengthened the state monopoly of soju; that is, the mandatory requirement that each provincial alcohol wholesaler purchase more than 50 percent of their soju from within their own province. Later, it became a means to protect the monopoly system of the soju industry until it was judged unconstitutional in 1995 and banned.[36]

We should also look at the gloomy social background for the increased consumption of modern industrial soju since the 1960s: difficult lives of laborers in the rapid economic development and despair under the regime's dictatorship during the reign of President Park Chung Hee. In order to overcome hardship, laborers who worked hard at factory jobs drank cheap industrial soju that dominated the alcohol market at that

---

[35] Yi Chihyŏng, *Soju iyagi*, 18–24; Joo Young-ha (Chu Yŏngha) argues that industrial soju produced earlier in the 1960s used ethanol spirit that included fusel oil that is not good for health, yet thanks to the development of technology for making pure distilled spirit in the early twentieth century, the concern about fusel oil decreased. Chu Yŏngha, *Sikt'ak wi ŭi han'guksa*, 489–491.

[36] Yi Chihyŏng, *Soju iyagi*, 52–55.

time along with makkŏlli, another cheaply made traditional turbid wine.[37] Many large-scale industrial soju-making factories were built at this time.

A boom in the production and consumption of industrial soju nation-wide actually dealt a big blow to traditional soju and other fermented wines based on grains and nuruk. Even after independence from the colonial rule of Japan, the high tax rate on alcoholic beverages did not change. There is a tax on alcoholic drinks in each country in general, but in the case of Korean traditional liquors, various taxes in addition to the liquor tax were levied, so the price rose even higher and placed a burden on small-scale traditional soju distilleries as a result.[38] Distilleries that made soju using traditional distillation methods completely disappeared in the 1970s. Another reason that made people turn to industrial soju is that makkŏlli, another traditional alcoholic drink for ordinary people, had lost its popularity since the 1960s because its price increased due to government regulation banning rice materials and the imposition of a quota on the amount produced by each region. Therefore groggeries came to sell cheap industrial soju to workers from rural areas who moved to Seoul.[39] Such a situation changed people's tastes, too.[40] These circumstances eventually led to the result that modern industrial soju spread all over the Korean peninsula, not only as the dominant soju brand but also as the single most popular alcoholic beverage in the country.

However, a chance to revive traditional soju appeared in the 1980s. The government began to recognize the grief that many Koreans felt about the disappearance of traditional spirits like soju due to the excessive involvement and control of the government's national alcohol production policy. Before hosting unprecedented major international sports competitions, including the Asian Games of 1986 and the Olympics of 1988 in Seoul, the government worked to advance the development of traditional alcohol varieties in each province, calling them *minsokchu* ("national folk liquor"). As the government promoted traditional alcoholic beverages, the traditional form of distilled soju received a new opportunity to revive again after several decades in obscurity.[41] The government gave special licenses to those trained in proper distillation techniques, and they began to innovate traditionally distilled soju using fermented grains, which could be sold for higher prices at local markets and used as gifts and for other local rituals. Since then, as interest in traditional forms of liquor increased, the number of distilleries used to make them in many regions increased dramatically.

[37] Ibid., 27–35.   [38] Chong Dae Song, *Chosen no sake*, 187–188.
[39] Chu Yŏngha, *Sikt'ak wi ŭi han'guksa*, 311–312.
[40] Young people also began to prefer beer to makkŏlli. Ibid., 335.
[41] Chong Dae Song, *Chosen no sake*, 15.

It is difficult to trust that all the revived national folk liquors were authentically based on regional specialties. As Joo Young-ha convincingly points out, while many breweries, seeking to secure their positions, presented the state with cookbooks that they claimed to have received from ancestors who had developed special recipes during the Chosŏn dynasty, these recipes are often similar to each other as they are copies of earlier circulating recipes; thus it is difficult to give credit to their authenticity. Against such a backdrop, some of the places famous for making soju made efforts to connect to the traditions of the past in their regions. This includes Andong soju, which began to revive in the Andong area and came to represent traditional Korean soju.

Lee Sang-Hoon points out that some scholars, like Yi Seong-wu, claim that Andong soju originated during the Koryŏ period, when the region was the site of one of the Mongol army camps stationed in Korea at that time, along with Kaesŏng, Chindo, and Cheju. Many major national folk liquor companies located in these regions during a period of minsokchu revival in the late 1980s also emphasized their connections to ancient Koryŏ as a means of marketing – brands like Munbaeju (originating in P'yŏngyang), Chindo hongju, and Cheju kosorisul (see Figure 5.2, right).[42] As we do not have empirical support for their Koryŏ origins, some scholars, like Lee Sang-hoon and the author of this book, began to question the authenticity of their claims to them. Yet, in the market, they could claim Koryŏ origins with the help of suggestive trademarks.

Despite their emphasis on connections to the past, it is strange that instead of using the characters *soju* 燒酒 (Chinese *shaojiu*) for the traditional soju and shaojiu of Korea and China, the word *soju* in "Andong soju" is written as *soju* 燒酎 (Chinese *shaozhou*), replacing the 酒 character (Korean *chu*, Chinese *jiu*) with 酎 (Korean *chu*, Chinese *zhou*), a character that means a stronger liquor that has been used by the Japanese for their shochu and that was introduced to Korea along with modern industrial soju in the early twentieth century. Not many Koreans probably recognize this because the Chinese characters *jiu* 酒 and *zhou* 酎 are both pronounced the same way in Korean – *chu*. This "misspelling" was probably caused by an error made in the course of the rapid revival of Andong soju as a form of minsokchu promoted by the government in the 1980s. For example, the government designated Ms. Cho Okhwa, a famed expert in making *tongdongju* (also called "floating alcohol"), an intangible cultural asset as a specialist in Andong soju (the region she represented, North Kyŏngsang province, was once famous for Andong

[42] Yi Sanghun, "Urinara kanghwa parhyoju," 77–82.

soju).[43] Many other intangible cultural assets and national folk liquors arose as a result of similar political drives, without considering deeply the actual historical background.

Despite these weaknesses in its claims to Koryŏ-period origin, Andong soju had several advantages that predisposed it to become famous as a nationally representative variety. It held high appeal as the most authentic form of soju because it had developed in Andong, a place where many *yangban* or noble families maintained ancestral rites and host culture using foods and alcoholic drinks based on Neo-Confucian forms from the Chosŏn dynasty. The fact that the earliest extant cookbooks including soju recipes were written in the Andong region lent it some special advantages for connecting Koreans to an earlier tradition.[44]

Scholars like Yi Seong-wu point to somewhere else which they argue was a major place in soju's development during the late Koryŏ period: Kaesŏng, once the capital city of the old kingdom and now located in North Korea. As a Koryŏ-era center of political and commercial activities, we can only assume that soju developed there for the purpose of consumption by court nobles, soldiers, and other rich people. The 1935 Japanese report *Chōsen shuzōshi* also recognized Kaesŏng as a famous site of soju production. Kaesŏng soju has continued to develop, and has become a representative soju brand in North Korea. This makes soju the only alcoholic beverage that continues to be developed in both South and North Korea, legitimizing any claim for the title of authentic national liquor.[45]

All sojus branded national folk liquor, which were titled traditional soju, have high alcohol content, around 45 percent, in contrast to the low alcohol content of around 20 percent for modern industrial soju. While old-style distilled soju distinguished itself from modern industrial soju, we should not neglect some modern technologies that began to be applied to the traditional soju industry that developed from the late twentieth century. During the period of Japanese colonial rule, new methods and alternative yeasts were introduced from Japan to traditional

---

[43] Being designated as an intangible cultural asset, she gained the exclusive right to manufacture and sell Andong soju using traditional distillation methods from the Chosŏn period. Still other regional distilleries making soju began to claim their rights to make Andong soju as well. See Chu Yŏngha, *Ŭmsik chŏnjaeng munhwa chŏnjaeng* (Food War and Culture War) (Seoul: Sagyejŏl, 2000), 163–166.

[44] Nowadays, they try to promote some cookbooks to UNESCO World Heritage. For example, see the cases of *Suun chappang* (Various Methods of High-Class Food Culture) and *Ŭmshik timibang* (Recipes for Tasty Food) discussed in Chapter 4 of this book.

[45] Chong Dae Song, *Chosen no sake*, 13, 110. For a situation of alcoholic-beverage consumption in North Korea, see ibid., 18–20.

soju distilleries in the hope of making the taste of soju more uniform and smoother. Even after independence, another innovation would come to Korea from Japan. At the time of Korea's revival of national folk liquors (minsokchu), Japan had advanced distillation technologies further, unlike Korea, where traditional distillation was discontinued. By continuing to develop shochu using both traditional and modern technologies, the Japanese developed a greater variety of modern distillation machines over time, reaching a peak in development during the 1980s, at just the moment when Japan experienced a new boom in demand for shochu. Many of the technological aspects of various types of modern distillation machines created in Japan by that time are analyzed in an academic volume written by Japanese specialists, entitled *Honkaku shōchū seizō gijutsu* 本格燒酒製造技術 (Manufacturing Technology of Distilled Shochu) (2001), which was published by the Japanese Brewery Association (Nihon Jōzō Kyōkai 日本釀造協會).[46] This book was also introduced to Korea in a translation produced by the Korean Liquor Research Center under Pae Sangmyŏn, who admits in his introduction that they had to translate this Japanese work because Korea had no book giving such detail about soju and, moreover, Japan indeed possessed more advanced distillation technology.[47] These modern machines for pot-still distillation developed by Japanese companies (different from the multiple distillation apparatus such as the column still that creates industrial soju) grabbed the attention of small-scale distilleries in Korea, which began to produce distilled soju true to the methods of the Chosŏn period.[48] Unlike the old home brewers, small-scale restaurants, and breweries of early twentieth-century Korea, these new small distilleries aimed at a wider market, so they adopted modern technology in order to realize expected sanitary levels and maintain an even flavor and level of quality. They must also continuously buy new machines with updated technology to help them improve production methods, fine-tune flavors, and develop new model varieties of soju that are modified from more authentic traditional ways. For example, the Cheju Kosorisul Company has recently replaced its old distillation machines with new ones that help to make good-quality traditional soju in a more efficient way. Once accomplished, they

---

[46] This is acknowledged by the Korea Liquor Research Institute. Their book about the Japanese shochu industry, translated by this Korean institute, records the development process of Japanese spirits and the results of alcohol research at that time.

[47] *Chosŏn chujosa*, 8–9.

[48] Ilbon yangjo hyŏphoe (Nihon Jōzō Kyōkai), *Pon'gyŏk soju chejo kisul* (*Chŭngnyusik soju chejo kisul*) (Manufacturing Technology of Distilled Shochu (Soju)), trans. and ed. Pae Sangmyŏn (Bae Sang-myun) (Seoul: Pae Sangmyŏn churyu yŏn'guso, 2003), 3–6.

Figure 5.3 Traditional soju brands by the Cheju Kosorisul Company, including a bottle that resembles a *kosori*, a unique still used in Cheju (left), and outdated machines exhibited in the garden at the company's front entrance (right). Photographs by the author.

exhibited their old machines in the garden at the company's front entrance (see Figure 5.3).

Although some of these national folk liquors became quite popular in Korea thanks to government promotion, they would never replace the cheap industrial soju that many Koreans grew accustomed to in terms of price and taste. After all, cheaper soju varieties brought down prices, and that made it even more popular. Instead of ch'ŏngju, the clear strained wine that was more widely used at home during the Chosŏn period, soju became the most representative alcoholic beverage consumed in restaurants and bars in modern Korea. Once the new type of popular liquor changed people's taste, it became more difficult to change again. As a consequence, industrial soju became the bona fide national drink of Korea.

The developments of the early twentieth century have shaped soju's evolution to the present day. For example, many of the firms that were founded during the period of Japanese colonial rule, like Jinro (now HiteJinro), still flourish as major companies in Korea and still develop new varieties. In fact, Korean soju companies brought a new innovation to soju production that was not achieved in Japan, namely massively increased production. For example, most soju companies produce soju at 600 bottles per minute (BPM) (see Figure 5.4). Some big soju companies like HiteJinro run lines that produce even 1000 BPM.

Figure 5.4 The Mackiss Company soju factory that produces 600 bottles of soju per minute. Photograph by the author.

They have also departed from the specialization system that divides the production of ethanol spirit from the soju-bottling process to combine ethanol spirit, water, and flavours, which has produced one of the cheapest distilled liquor in the world. (A bottle of soju, for example, can sell for one to two dollars at supermarkets in Korea.) Thus price has become a major factor in the competitive market for Korean industrial soju. In short, soju has become a modern commodity, responding to the modern global economy in ways similar to other commodities.

The biggest weak point of these mass-produced forms of industrial soju lay in their nearly identical flavors, despite the fact that different soju companies produced them. These soju companies do not distill alcohol using a fermented wine base; they simply purchase ethanol spirit from the Office of National Tax Administration and dilute it with water before adding flavors to distinguish their brand. Because of the government's one-province, one-soju policy, consumers tended to decide on brand loyalty less on the merits of particular varieties or their distinguishing characteristics and more on the association between a particular brand and region. As a consequence, the soju companies of each province were able to enjoy monopolies in production and marketing, at least for a while. Some brands won national fame with their special characteristics. For example, Hallasan soju, one of the two most representative soju brands on Cheju Island, became famous for its 22 percent alcohol content, which is higher than other industrial soju brands.[49] However, most soju brands were known for their province rather than for their special characteristics.

The spread of cheap, common, and similar-tasting forms of soju created a new kind of soju drinking culture. Because distinction in flavor was no longer an important factor compared to traditional soju, people began to mix it with other things in order to improve its simple taste. Many Koreans drink soju by mixing it with beer in certain proportions depending on their preferences. In modern times, as beer found its way into Korea, many young people began to pass up traditional Korean beverages in favor of newer Western ones at restaurants and social gatherings. However, beer tasted bad in the early stages of its production in Korea, as did industrial soju, so many began to mix the two. The easiest way to do this was to make a *p'okt'anju*, which translates as "bomb shot" or "bomb cocktail."[50] The blend does more than increase flavor. The mixture of soju and beer increases the alcohol strength, so that people can get drunk quickly – hence the name "bomb shot." A bomb shot (also "depth charge") is an international name – its Western equivalent is a boilermaker – for a classical mixed drink; it implies either putting a shot glass (typically filled with spirit) into another, bigger glass full of beer or stout, or simply mixing beer and strong alcohol like whiskey. Some experts have assumed that the origin of this term dates back to the culture of command and discipline that was particularly popular among prosecutors and police during the military dictatorship of the 1980s.[51] Whatever the case, the bomb shot became a popular item at drinking parties in Korea. Because of this, it has been rumored, beer companies cared less about making delicious beer. However, thanks to globalization, imported foreign beer of higher quality stimulated further interest among Koreans in drinking only beer. Expectations have increased as a result. In response, Korean beer companies stepped up efforts to produce better-tasting beers, while a number of new beer brands emerged in the twenty-first century, leading to better reputations and a growing taste for beer without

---

[49] Beginning in the Japanese colonial era, soju companies established bases on Cheju Island. Two brands, Cheju soju and Hallasan, continued to develop as the most representative soju labels based on the island.
[50] See Yi Chihyŏng, *Soju iyagi*, 69–80.   [51] Ibid., 70–72.

a soju mix. This change had no effect on bomb cocktails; the drink parties continue. This particular drinking culture that has evolved since the 1980s has become a hallmark of soju consumption, which has often intrigued foreigners interested in Korean culture.[52]

In fact, this popularization of cheap industrial soju in Korea created an advantage for the companies that would create a new kind of soju brand for the global market in the late twentieth century. Local companies had grown large because they had enjoyed monopolies in their home regions for decades. Thus they could take advantage of their large-scale industry and research capacities to develop new brands of soju with better quality and lower price.[53] As a consequence, once the global push began, some major companies like HiteJinro and Lotte grew to gigantic proportions. While modern industrial soju was transmitted from Japan, the companies that produce it there have never been so great in size, though there are more of them than in Korea, and they are more widespread regionally. The big soju companies gradually developed sufficient know-how to make liquor smooth in as quickly as a few days. The smooth taste was achieved in the past only by maturing liquor for a period of ten to twenty years, during which impurities, fusel oil, and bad odors were removed (nowadays smaller companies that produce national folk liquor have a three-month to one-year time frame for this process). The knowledge to obtain smoothness without the wait of a long maturing process has been one of the rapid-production techniques that the Koreans have developed since the country's economic surge began in the 1960s. This increased production capacity has been met with a rise in consumption. During the second half of the twentieth century, in the midst of profound economic and cultural change, nationwide consumption of modern industrial soju dramatically increased. In the early years of the twenty-first century, soju has become the number-one alcoholic drink consumed by Koreans. Indeed, its popularity is no longer confined to Korea. At the same time as its domestic surge, soju has begun to be exported overseas to become a global brand.

---

[52] For a description of how to mix soju and beer (Korean *maekchu*) to make *somaek* (a soju-beer bomb cocktail), see Norman Miller, "Soju: the Most Popular Booze in the World," *The Guardian*, December 2, 2013, www.theguardian.com/lifeandstyle/wordofmouth/2013/dec/02/soju-popular-booze-world-south-korea (accessed December 2, 2018), cited in Yi Chihyŏng, *Soju iyagi*, 79–80.

[53] For example, HiteJinro developed into a gigantic company that was able to lower the cost of soju production. See Joshua Hall, "Soju Makers Aim to Turn Fire Water into Liquid Gold," *Wall Street Journal*, October 17, 2014, www.wsj.com/articles/BL-KRTB-6764 (accessed December 2, 2018).

Korean Wave (*Hallyu*) and the Global Promotion of Soju in the Twenty-First Century

Exports of soju to other countries have increased sharply since the beginning of the twenty-first century. Not surprisingly, the spirit spread to its neighbors in East Asia and to those countries with whom it has had long-term, continuous intercultural contact, such as in North America and Europe.[54] However, bottles of soju have even begun to be exported to countries with whom Korea has little cross-cultural exchange historically, such as in Africa.[55] HiteJinro is not the only large soju company to export; others have tried to expand their markets overseas as well. Three factors in particular have prompted this globalizing trend: (1) the gradual spread of Korean cuisine thanks to the world's globalization trends, the popularization of Asian and other ethnic foods, marketing and promotion campaigns, the establishment of Korean restaurants run by Korean immigrants, and Korean government promotion; (2) the rapid spread of Korean culture thanks to the Korean Wave boom of the early twenty-first century, which includes cultural content such as dramas, television programs, movies, and K-pop; and (3) soju's cheap price and its continuous supply in global markets made possible by unprecedented mass production. In this context, the early twenty-first century turned into the heyday of soju's internationalization.

It has only been in recent decades that Korean foods have become more familiar to people worldwide. As Korean immigrant communities in the United States and other countries have continued to grow, the Korean restaurants they have established have played an important role in making Korean foods popular in their host societies. Kimchi, a traditional vegetable side dish flavored like spicy and salty pickle, has become the most popular and representative Korean food worldwide. In addition, some popular menu items like *pulgogi* (also *bulgogi* – "fire meat") and *pibimbap* (also *bibimbap* – "mixed rice") have helped to increase awareness, and some dishes have even come to be included in English-language dictionaries.[56] Some more recently popular meat dishes such as *samgyŏp-sal* (also *samgyeop-sal*, a grilled barbecue of sliced pork belly) are consumed both at restaurants and at home, often accompanied by glasses of soju, a scene often reflected in Korean dramas. The relationship

[54] For example, see Miller, "Soju: The Most Popular Booze in the World."
[55] Pak Miyŏng, "Ap'ŭrik'a esŏdo chillo soju masinda (People in Africa Are Also Drinking Jinro Soju)," *DigitalTimes*, April 18, 2016, www.dt.co.kr/contents.html?article_no=2016041802109976798002 (accessed December 2, 2018).
[56] For example, *bulgogi* is found in Merriam-Webster's dictionary (www.merriam-webster.com/dictionary/bulgogi) and in Oxford's *Lexico* dictionary (www.lexico.com/en/definition/bulgogi).

between soju and Korean meat dishes can be compared to that between Japanese sake and sushi as a good match in terms of a drink and food combination. As soju appeared with meat dishes in Korean restaurants in other countries, it began to grow popular worldwide. A Korean restaurant in New York City put a note on its menu that, while its selection of wines is poor, people should instead try its variety of sojus.[57] Korean government and public institutes also exerted efforts to advertise Korean foods by publishing Korean-food cookbooks in English, Korean, and other languages.[58]

The dramatic rise of soju in the world owes a great deal to the massively popular "Korean Wave" or *Hallyu*. Beginning with the export of drama and pop music that first gained great popularity in China and Japan in the 2000s, Korean Wave has since spread to Southeast Asia, the Middle East, Europe, and the United States in various forms – television and movie drama, pop music, and even gaming.[59] Dramas and advertising often convey scenes of social gatherings in Korean society, where people look natural drinking soju and eating classic grilled meat dishes. The placement of soju in such dramas has dramatically increased overseas demand and export, and this trend will continue as long as the Korean Wave persists. Even in countries like Mongolia, where Mongolian and Russian vodka and traditional arkhi and airag still dominate the supermarket shelves, a small section of Korean industrial soju can be found next to them, a situation that developed when people began to drink it with popular Korean grilled-meat dishes like samgyeop-sal.[60] It seems only natural that soju companies hired Korean Wave stars such as Psy or popular actresses and girl-group members to promote soju on posters advertising various soju brands all over the world. While a bottle of industrial soju from Korea may sell cheaply at three to four US dollars in Korean markets, the same bottles sold in luxurious Korean restaurants in the United States go for much higher prices.

---

[57] For example, see a menu at a Korean restaurant in New York City called "Madangsui: Korean BBQ Restaurant" that was introduced by Sam Sifton at the *New York Times* on December 9, 2009.

[58] For example, see Mun Pyŏnghun, *Han'guk ŭmsik, segye nŭl hyangan tojŏn* (Korean Foods' Challenge to the World) (P'aju: Idambooks, 2009).

[59] Chu Yŏngha, *Ch'ap'on·chanp'on·tchamppong: Tongasia ŭmsik munhwa ŭi yŏksa wa hyŏnjae* (Ch'ap'on, Chanp'on, and Tchamppong: Past and Present of East Asian Food Culture) (P'aju: Sagyejŏl, 2013), 275.

[60] I saw this Korean section during my trip to Mongolia in July 2018 and learned more about it later from professors Batdorj Batjargal and Dashdondog Bayarsaikhan at Mongolian National University. Also, huge numbers of Mongolians went to work in Korea in the grey economy, and many of them brought their acquired dietary habits back with them. There are also many Korean restaurants and cafes in Ulaanbaatar.

Indeed, during the 2010s, Korean soju entered the global market with low visibility compared to whiskey and vodka but soon began to attract fame. Some newspaper articles testify to the fact that the consumption of soju began to increase in the US market with many advantages of its own. For example, one NPR (National Public Radio) online newspaper article compared soju with vodka in the US market: while vodka can only be sold by alcohol sellers who possess a special license to sell distilled spirits, soju can be sold by those who have only a regular alcohol license thanks to its low alcohol content; in other words, it can be treated like wine.[61]

In fact, Korean industrial soju did not always have such a low level of alcohol. When soju companies decided the alcohol percentages of their brands, it was generally taboo to go as low as 20 percent. However, the domestic alcohol market, especially the soju market, has been trending toward lowering alcohol strength since the year 2000, reducing it from an average of around 30 percent in the 1970s, to 20 percent in the 2000s, and finally to 16.5 percent in 2009.[62] This change happened as the result of efforts by industrial soju companies to develop a wide variety of soju that could satisfy the needs of a greater variety of people to increase both national and international demand. By lowering alcohol levels, they also hoped specifically to target women and young people.[53] Nowadays, almost all soju brands have alcohol levels of around 18 percent, except for Hallasan of Cheju province, which has 22 percent. The generally low alcohol content in Korean soju today appears to have facilitated its growing sales in countries like the United States as a result. This is a good example of how it mass-produced industrial soju, rather than its traditional predecessors, began to do well in the global market. No longer confused with Chinese shaojiu or Japanese shochu, soju became known as soju and took its place among the distinctive alcohol brands of the world.

The globalization trend did more than simply help soju and other Korean foods break into global markets. It also opened the way for imports of foreign liquor to Korea, where they began to attract the attention of Koreans. Western distilled spirits such as whiskey remained confined to high-end restaurants, but cheap and tasty beer became widely consumed nationwide. In response, the companies that manufacture soju also began to make their own beer, some of which would become famous

---

[61] See a newspaper article by Tom Dreisbach, "Move Over Vodka." The low-alcoholic features of industrial soju, as in Jinro's case, make it more competitive than vodka in stores that sell alcoholic beverages. In the United States, Korean industrial soju belongs to the same category as wine.

[62] Cho Sŏnggi and Yu Kimok, *Soju ŭi tosu chŏngch'esŏng hwangnip pangan yŏn'gu* (A Study of the Establishment of the Identity of Soju's Alcohol Content) (Seoul: han'guk churyu yŏn'guwŏn (Korea Alcohol Research Center, 2010), 9.

[63] Some soju alcohol levels dropped to 17 percent.

as Korean beer brands. For example, Jinro Company, which created its famed Jinro soju, began to produce its own beer brand called Hite; they even changed the company's name to HiteJinro.

However, Koreans who remember the deep taste of traditional liquor were not satisfied with the flavor of these factory-made industrial sojus and beers, and they often mixed the two beverages to make bomb shots in order to remedy their shallow taste. Yet, as more varieties of liquor, including beer, wine, and vodka, came to be imported to Korea, the drinking ways and kinds of liquor in Korea began to influence a new socializing culture. For example, while traditionally minded people were drinking excessive quantities of soju or bomb shots whenever they ate grilled meat, younger people began to enjoy imported wine and beer at Western restaurants, as well as vodka at clubs. People in Korea, like those in other countries, are now concerned with excessive drinking, which causes health problems, and drunken driving, which leads to car accidents.[64] Such a change in drinking culture is motivating the production of new kinds of liquor, including more varieties of soju.[65]

New ideas to innovate novel varieties of industrial soju to sell overseas came from Japan, too. In one example, many industrial soju companies tried to sell a cocktail soju and *chūhai* (lemon flavored with a shochu base), which had grown popular first in Japan. Although sake remains the mainstream alcoholic drink, shochu got a boost in the 1980s. Both modern and traditionally distilled forms of shochu benefited.[66] By the late twentieth century, affected soju companies in Korea began to produce fruit soju. Liquor restaurants and bars, responding to the change, created new kinds of soju in the form of soju mixed with citrus fruits, such as grapefruit and lemon. These soju cocktails immediately won popularity among people, including women, who like mild flavors. Soju companies also launched new varieties of fruit cocktail soju that are ready to drink.[67]

Looking at developments among a variety of distilled liquors in modern Japan and Korea, we should remember that, as Chapter 4 shows, Koreans

[64] In Western societies, alcohol drinking provides people with a good opportunity for social gatherings. This drinking culture in which people drink alcoholic beverages slowly over a long conversation strikes a great contrast to the drinking culture in Korea, in which people try to solve stress by drinking a lot and to forget the difficult routines in their lives. Western drinking culture presented an opportunity to change the Korean culture of drinking to excess. Experts argue that, in order to compete with other famous distilled liquors, soju should meet this new drinking culture. Yi Chihyŏng, *Soju iyagi*, 78.

[65] Alcoholism has become a serious social problem in Korean society. Chu Yŏngha, *Ŭmsik chŏnjaeng munhwa chŏnjaeng*, 141.

[66] The Japanese call industrial shochu "no. 1 type soju," compared to "no. 2 type soju," which means traditionally distilled soju. Chu Yŏngha, *Ch'ap'on·chanp'on·tchamppong*, 155.

[67] Cho Hanbyŏl, "Talk'omhan kwail soju" (Sweet Fruit Soju), *JoongAng Ilbo*, September 20, 2015, https://news.joins.com/article/18702462 (accessed December 9, 2018).

in the Chosŏn period had already developed many varieties of soju using ingredients and flavorings of various kinds, like medicinal herbs, as well as various distilling methods, including one-time distillation and multiple distillation. Traditionally distilled soju has not received enough attention, despite its rich history since its rise in the fourteenth century, because the modern industrial forms became so rapidly popular after the twentieth century, and because its tradition was cut off for many decades of the twentieth century. While Japanese makers successfully bridged modern industrial shochu to traditional shochu, the Koreans still have much potential to do so with soju, too, by taking advantage of its long tradition.

As we have seen, modern distillation has its origin in the traditional form that developed in Afro-Eurasia. If we look deep and wide, we can similarly connect Korea's modern industrial soju to its legacy of traditional distillation. Now, in the twenty-first century, as industrial soju continues to flourish, traditional soju makers have tried to spark a revival with some success. However, they have encountered many obstacles, as the revival of interest in traditional soju has tended to position the spirit's two types – traditional and modern – on opposite sides in a relationship of competition as well as coexistence.

This dilemma between tradition and modernity can be found in many other cultural items. It can be difficult to know when something should be seen as a tradition, and when it should be seen as the product of a transformation affected by modern elements. In other words, the controversy over traditionalism has begun. What is soju? What makes one soju traditional and the other modern? Experts in the field have begun to grapple with this identity issue.[68]

Studying the methods of establishing the identity of soju's alcohol strength in their 2010 book, Cho Sŏnggi and Yu Kimok conclude that it is necessary to attempt to define soju's identity based on its alcohol percentage in order to resolve the current confusion about Korea's most representative liquor. Cho and Yu point out that, since the early 1900s, producers of liquors with international reputations such as wine, whiskey, and vodka have been striving to maintain the essential characteristics of their products by strictly restricting alcohol content. Little effort was made in the case of soju during this time, resulting in confusion about its identity.[69] The authors conclude that defining soju's alcohol content within a range between 16 and 37 percent would make it a distinctive and competitive liquor compared to other alcoholic drinks such as beer, wine, and sake, which contain 1 to 15 percent alcohol, and rum, gin, whiskey, and vodka, which contain 37 to 50 percent, though they are concerned that

[68] Cho and Yu, *Soju ŭi tosu chŏngch'esŏng*, ix.   [69] Ibid., 2.

this could limit people's understanding of soju's varieties.[70] They also point out that, compared to other distilled liquors such as whiskey and vodka whose alcohol content people can control by adding tonic water or other drinks based on their tastes and preferences, soju, with its lower alcohol percentage, could be drunk straight, without such mixing.[71] That is to say, soju as a drink is distinctive on its own. Cho and Yu's argument makes sense when we turn our focus to modern industrial soju. This definition would not work with traditionally distilled soju because its alcohol content often exceeds 40 percent (for example, Andong soju contains 45 percent alcohol). It also conflicts with the government's single category for soju, including both modern and traditional, in the definition created by the Office of National Tax Administration in 2013.[72]

Against such theoretical efforts to define soju, companies have continued to experiment with the production of new varieties, tinkering with alcohol percentages and flavorings and responding to consumer choice. Soju companies have also been trying to loosen the boundaries between modern and traditional by mixing a small amount of traditionally distilled soju – as little as 1 percent – into their industrial soju. Some success has been realized with a new brand called Mackiss of Sŏnyang, which has boosted the traditional portion to 5 percent. Moreover, regional soju companies have been trying to differentiate their soju brands by introducing new technologies to the production process. Since the state abolished the one-province–one-soju policy in 1989, local soju companies have gone to great lengths to make their specialty soju do well in nationwide competitions and to develop and obtain patents for new technologies such as far infrared rays, sonic waves, and oxygen injections that are helping them to add some distinctive flavors to their local brands. For example, Mackiss Company, a soju producer in Taejŏn (also Daejeon), developed the technique of homogenizing alcohol by spraying oxygen into a mix of water and ethanol spirit in order to soften the final product's flavor. Mackiss soju even published its research results in the academic journal *Alcoholism* after doing clinical tests.[73] However, Mr. Kim Choong Hyeon, a director of the company's Technical Research Center, has pointed out that it is difficult to scientifically prove that spraying in oxygen to change a liquor's molecules changes the quality of the taste, because it

---

[70] Ibid., viii.   [71] Ibid., 7.

[72] See the categories of liquor in Korea defined in the Liquor Tax Law, revised April 5, 2013, www.law.go.kr/%EB%B2%95%EB%A0%B9/%EC%A3%BC%EC%84%B8% EB%B2%95 (accessed September 2, 2018).

[73] In-hwan Baek, Byung-yo Lee, and Kwang-il Kwon, "Influence of Oxygen on Pharmacokinetics of Alcohol," *Alcoholism: Clinical & Experimental Research* 34, no. 5 (2010): 834–839.

is difficult to examine the influence of such technology on the taste of soju on a molecular level. He argues, therefore, that this signals a need to approach soju's taste emotionally. Efforts to improve technology and techniques in order to improve taste are also evident in the production of traditionally distilled soju by small enterprises. For example, the Cheju Kosorisul Company produced a new soju brand by modifying their original brand of kosorisul, adding ginseng and other medicinals. They also experimented with newly obtained oak barrels normally used to make European whiskeys in order to test possible ways in which they could apply new methods to soju production. Given the long history of experimentation with soju since its introduction to the Korean peninsula during the late Koryŏ period, it seems reasonable to maintain a broad and flexible approach to the issue of identity.

To this environment in flux came a new challenge to soju's identity from outside. An experienced brewer and distiller named Brandon Hill, based in Brooklyn, NY, in the USA, created a new soju brand that he called Tokki soju (see Figure 5.5).

Interested in the Korean industrial soju that was becoming popular in his country and all over the world, Hill visited Korea to learn about it directly. Soon he fell for traditionally distilled soju. Hill then returned to Brooklyn, where he established a company that would use modern distillation machines similar to those used for whiskeys to make soju according to the same recipes used in the Chosŏn period, including rice and nuruk. Then he

Figure 5.5 A new soju brand, Tokki soju, created in the United States by Brandon Hill (left), and modern machines used to make it in the factory in Brooklyn, NY, USA (right). Photograph courtesy of Brandon Hill.

named his new brand Tokki, which means "rabbit" in Korean. Hill's soju became popular in the US market and has found its way onto the menus of Korean restaurants all over the United States. In fact, it eventually was smuggled into Korea and sold illegally there thanks to its fame.[74] As its popularity grew, an American freelancer writer named Joshua Schenkkan sparked a controversy over traditional soju in a newspaper article he wrote, which begins with a question: can we call this *traditional* soju, if it is made by an American using a modernized machine (similar to that made for whiskey) in Brooklyn, USA? Schenkkan interviewed different kinds of people for his article, and found that some agree and others disagree.[75] A national crafts-man of samhaeju declares that Tokki soju cannot be recognized as tradition-ally Korean simply because it is not made in a traditional Korean way in Korea; a member of the Korean parliament argues otherwise, saying that it certainly falls within tradition.[76] Just as interesting, Schenkkan asked several Korean restaurant owners and soju-related businessmen to explain the definition and origin of soju, but no one was able to answer him in any detail.

Perhaps it is best to stop and review this controversy over authenticity based on our understanding of the history of soju discussed so far. As Chapter 1 explains, the distilled liquors that developed as soju (Chinese *shaojiu*, Japanese *shōchū*) in East Asia and as arak in Afro-Eurasia at large have evolved since the ancient period on a world-historical scale. In other words, distilled liquors developed in various forms like soju and arak as a consequence of constant interaction between societies that led to trans-fers of goods and technologies that in turn enabled the development of an alcohol culture generally across this Afro-Eurasian exchange network and specifically, by the fourteenth century, on the Korean peninsula. Since their introduction at the end of the Koryŏ dynasty thanks to Mongol rule and a close relationship with China, distilled liquors became part of Korea's national drink culture during the Chosŏn dynasty that followed. We do not know exactly how people at the end of the Koryŏ dynasty actually made soju; however, sufficient grounds for educated speculation exist based on Chosŏn-dynasty records that indicate the likely original use of portable stills popularized by the Mongols over the course of their presence on the Korean peninsula. The distillation techniques they passed on to local Koreans continued to develop after the Koryŏ and Mongols faded from Korean history and the five-century history of the

---

[74] The success of Tokki soju in the United States led to the rise of other American brands of soju, such as West 32 Soju (www.west32soju.com, accessed April 2, 2020).

[75] Joshua Schenkkan, "What Is 'Traditional' Soju? A Spirited Debate," *Serious Eats*, October 3, 2017, www.seriouseats.com/2017/10/what-is-traditional-soju-korea-tokki-brandon-hill.html (accessed October 2, 2018).

[76] Ibid.

Chosŏn dynasty began. Soju makers experimented with different kinds of still made with different materials, including metal pots originally used for cooking other foods and earthenware stills that they had created as part of the rise of earthenware tools in the eighteenth and nineteenth centuries.

Considering this pattern of evolution, it may also seem natural to regard new distillation apparatus as representative of a new stage of modernization and globalization, which also raises questions about the identity and authenticity of many soju brands in Korea today. Many of the spirits restored in modern times as Korean national folk liquors are made using modernized machines; few rely on the soju kori used in the Chosŏn dynasty. A similar logic can be argued in the case of traditional rice cake: as most traditional rice cakes in Korea are made with modern machines instead of the traditional *maettol* millstone because labor and other costs prevent it, few would ever say they are not traditional rice cakes. As for the Tokki soju controversy, it would be unreasonable to say that it is not a traditional liquor just because it was not made by Koreans in Korea using a traditional soju kori. Yet such an authority as the national craftsman of samhaeju denies Tokki soju as a traditional soju. What should we think? To respond, we should keep in mind that samhaeju during the early Chosŏn era was a fermented wine. As soju became popular again, soju makers began to use samhaeju as a base wine for distillation. This affected samhaeju's evolution so that gradually it would become best-known as a distilled spirit. In short, few foods, if any, remain authentic. Just as European-style wines and beers are no longer made only in Europe, soju is no longer made only in Korea; indeed, both can be made in any other country and still retain their identity. The controversy over traditionalism surrounding this soju will continue, but it should be balanced.

The flexibility of soju's development is suggested not only from its still technology, but also in the various distillation methods involved in its making, such as the single-distillation method, the multiple-distillation method, and using different ingredients such as herbs to alter flavor, a technique developed during the Chosŏn period. This tradition originated in the broader Afro-Eurasian distillation development scheme that soju shared with arak and shaojiu. As this global development of distillation continued, distillation methods and raw materials in Korea continued to change, too. There are too many different ways of making soju to adequately define what a real soju is. Western distilleries of spirits such as whiskeys have innovated their distillation technology, adapting modern technological innovations whenever they have appeared; this has produced a standard of whiskeys that is world-renowned. This form of adaptation did not take place in Korea during the Chosŏn period. In

Japan, the government of Ryukyu sponsored Awamori shochu in order to make them standardized and commercialized. Meanwhile, the Chosŏn government, having emphasized the harmful effects of alcohol and grain shortage, concentrated on raising bans on alcohol; in such a political environment, no efforts could have been made to systematically develop the alcohol industry. When the Japanese introduced industrial soju to Korea, it became immediately popular, and soju made with traditional distillation methods fell far behind. Intriguingly, it was the modern form of industrial soju, and not a traditional form, that came to be the national liquor in Korea as well as a global brand today.

The rapid changes that affected soju brought many challenges, too. Soju gourmands who like the deep flavors and textures of distilled liquors are not satisfied with cheaply made industrial soju. Some Korean soju distillers have been trying to revive traditional soju. The market remains considerably small; however, it has been growing at the remarkable rate of 100 percent over the last two to three years.[77] The traditional alcohols that have come to represent Asia recently have been famed spirits from China and Japan like Maotai jiu and Awamori shochu. Soju shares with them a similarly deep history, so it has the potential to develop its stature in a similar way. Perhaps this will occur. Along with efforts to globalize Korean food in the twenty-first century, the government and businesses have made efforts to export traditional alcoholic drinks and further develop them for the global market. This no longer happens in the form of national campaigns like those mobilized under the military dictatorship; still, many liquor institutes and cultural centers feel the same imperative and in that spirit are actively promoting select brands. Many brands are surviving in the global arena, including several famous Korean folk-liquor brands like Andong soju, Munbaeju, Chindo hongju, and Cheju kosorisul. Additionally, some of the new traditionally distilled soju brands, such as Hwayo, are on the rise thanks to soju-company investment, which is seeing results from their high quality and global marketing. The recent situation in the rice market, in which rice is producing surpluses due to changing eating habits among Koreans who are turning increasingly to bread, favors makers of traditionally distilled soju, who rely on the grain.

Experts say that, despite such efforts in both technology and commerce, the biggest problem lies in the fact that, for decades, the drinking culture surrounding distilled spirits has declined, and therefore there is no popular culture of consuming them to stimulate demand. As foreign liquors like whiskey or vodka have become popular in places all over

---

[77] Kim Choong Hyeon, based on personal conversation, August 8, 2018.

world, including Korea, people's tastes have changed and with it their drinking cultures. However, Korean spirits may have a chance to create another culture that departs from the culture of the Chosŏn dynasty. Some experts claim that it is possible to apply Korean distilled spirits to cocktail culture through the Japanese *wari* ("to mix," "to blend") method.[78] Another way to modify soju mixes soju with fruits or other types of juice, similar to Japanese *chuhai* but with Korean characteristics. This may appeal to consumers of ready-to-drink beverages, as Cho and Yu suggest.[79] Developing a new product by orchestrating new demand while adding product value in response to variable consumer preferences may prove an important strategy for Korean soju to follow as it endeavors to appeal to the whims of the global market. Some of these new marketing efforts may even prove effective in rekindling interest in traditional soju of the past.

It is intriguing to see that, while many foodways have changed in response to new technologies, such as the refrigerator, some premodern purposes such as preservability continue to support the viability of making foods in traditional ways. For example, Mackiss soju also produces in its factory traditionally distilled soju with fifty percent alcohol; it is not to be drunk by individuals but rather used by food companies that add it to specific food items like hot pepper paste as a preservative.[80] Even today, some East Asian medicine adds soju in order to enhance efficacy, a testimony to soju's continuing function as a medicine.[81]

Conclusion

The period from the twentieth to early twenty-first century witnessed the most dynamic and transformative changes in the development of soju to date. Characteristic of many unprecedented world-historical processes, such as rapid globalization, soju not only assimilated important influences from the world at large but also reciprocated by exerting cross-cultural influence on a grand scale, thanks to the global appeal of Korean movies,

---

[78] Ibid.
[79] Cho and Yu, *Soju ŭi tosu chŏngch'esŏng*, 7. Ready-to-drink (often known as RTD) packaged beverages are those sold in a prepared form, ready for consumption. For other kinds of RTD cocktails, see "Ready-To-Drink (RTD) Cocktails Fact Sheet," *Cocktail Times*, www.cocktailtimes.com/indepth/rtd/rtd.shtml (accessed December 1, 2018).
[80] Kim Choong Hyeon, personal conversation.
[81] For example, see a list of liquors for medicinal use in O Yŏngju, "Tong Asia sok ŭi Cheju parhyo ŭmsik munhwa" (Cheju Traditional Food-Fermentation Culture in the Environment of East Asia), *Cheju-do yŏn'gu* (Journal of Cheju Studies) 32 (2009): 169–170.

music, and, in this case, alcoholic beverages. New factors like modern industrial soju, with its radically different methods and technology, and the Liquor Tax Act imposed by the Japanese colonial government, dealt a near-fatal blow to the traditionally distilled soju that its makers had developed over centuries, through the late Koryŏ and the Chosŏn eras. The governmental policies issued after Korea's independence in 1945 perpetuated colonial-era liquor laws to earn much-needed tax revenue and to promote industrial soju as a part of its broad policy to push rapid economic development, making it difficult for traditionally distilled soju makers to maintain a foothold in the industry, much less encourage a revival of interest. This left it marginalized for decades thereafter. As a consequence, a new type of industrial soju, mass-produced using modern industrial technology, became a mainstream drink in Korea, as it continues to be today. Meanwhile, traditionally distilled soju continued its struggle to survive and rebound. In the present situation, it is not unreasonable to divide soju into two categories: the current traditional soju made using a single-distillation method and an industrial soju of modern provenance made by means of modern continuous-distillation methods to produce a pure ethyl alcohol (or neutral spirit). Remember, however, that industrial soju made using column stills of the highest technological level may be a modern phenomenon, but it also has its roots in a global pattern of distillation development that dates back to ancient times. That is, from a long-term historical perspective, the technology of industrial soju proved essential to the recent trend in which soju has generally diversified. However, massive quantities of cheaply produced industrial soju, mass-produced by first mixing a form of ethanol spirit made from potatoes, water, and selected flavors at gigantic soju factories in Korea, led to a simplification of Korean tastes. In any case, the modern trend in soju production has created an end product that radically differs in basic features, like taste, from traditionally distilled soju.

It was only at the end of the twentieth century that the government, in line with international trends, began to shift its policies away from promoting modern industrial production toward rescuing traditional liquors with the promotion of national folk liquors (minsokchu). The government's efforts at revival have targeted potential consumers beyond Korea and Korean culture, as part of its appeal to the globalizing world. Along with small-scale traditional distilleries, relevant governmental departments and academic institutes have presented new ideas for defining and promoting a sustainable Korean soju market. That said, a recent trend shows that such changes are made slowly and organically, as people change their tastes and accept new varieties in response to changing preferences. A bomb shot of soju, a soju cocktail, a low-alcohol soju, and a growing array of minsokchu brands manifest naturally as they go with the historical flow. By tracing variations over a defined period of time, therefore, soju's evolution continues, thanks to people's ability to assimilate outside cultural objects and ideas into local environments and develop a synthesis of the two over time. Technological innovations continue to matter, such as changes to column stills, so that even makers of traditionally distilled soju now employ modernized technological machines, leading to the further development and diversification of soju varieties. Such changes are not new to the history of soju, of course; indeed, they have been part of a long-term pattern of historical development that dates back to premodern times in both Korean and global history.

An important recent change occurred when Korean society bestowed its stamp of approval on soju with the title "National Liquor of Korea." Its profile increased as it was sold in Korean restaurants, drunk conspicuously in Korean television or movie dramas becoming popular overseas, and advertised by Korean Wave stars like Psy. Soju then became identified as the most representative of Korean liquor both nationally and globally. Of course, soju is not the only liquor to have played a role in Korea's history. The turbid wine makkŏlli more recently has been experiencing a popular revival and now has begun to export to other countries, like Japan. Nonetheless, soju best exemplifies the rich history of cultural exchange that connects past to present and Korea to the wider world. Additionally, soju today holds the distinction of being the most popular drink on the entire Korean peninsula, South and North. It has attracted so much interest overseas that even lesser-known traditional soju attracted an American brand. Of course, the culture of soju continues to change. As its interest in soju has grown, Korean society has also responded to other kinds of alcoholic drink, including imports from other countries. With them have come new drinking cultures into Korean society, as well as new information about the functions of alcoholic drink that may affect both the production environment and the drinking culture of soju. In short, the dynamic development of soju in the twenty-first century continues to epitomize the process of change affecting a cultural item under conditions of globalization. If analyzed properly, it can provide a framework for understanding other, similar, cultural items. The story of soju is not over yet. It will continue to develop, along with other spirits, into the future.

As for its significance to global history, soju's story also begs comparison to other globally popular spirits that have experienced similar developmental patterns. To this end, the final chapter examines two cases of possible distillation-technology transfer that appear comparable to the story of soju. However unique its many qualities may be, perhaps the patterns underlying its evolution reflect conditions common not only to many spirits but also to many cultures facing similar forces that have driven the development of a global world.

6   Alcohol Globalism
Distillation Technology in Afro-Eurasia and Other
Areas of the World: The Cases of Japan and Mexico

As distilled liquors like soju and arak spread to the Korean peninsula from China and Afro-Eurasia at large during the Mongol period, distilled liquors of a similar kind may also have spread to other world regions and influenced alcoholic drinking cultures there. It is difficult to find well-documented cases like those examined in this book, in which one can draw clear connections between events, from the introduction of distillation technology into Korean society via China and the Mongol imperium in the thirteenth century to the reverse transfer from Korea to the global economy seven centuries later. However, many sources written after the Mongol period hint that distilled liquors continued to develop in different areas of Afro-Eurasia, including Central Asia, Southeast Asia, and the Middle East. For example, the name *arak* appears in sources written in the Timurid empire immediately after the Mongol demise in 1368 and in Turkey during the seventeenth century, although we need further study to determine their provenance.[1] Today, the beverage defines a cultural zone that stretched from Central and Southeast Asia to the Mediterranean Sea, where arak maintains its position as a common spirit, along with *rakı* in Turkey and Greece and arrack in South and Southeast Asia.[2] Arkhi, a distillate of fermented mare's or cow's milk discussed in previous chapters, also continued to develop in Mongolia and Central Asia; this is no doubt a legacy of the beverage's Mongol-era development.[3] Another topic that needs further study is the possible set of connections between the rise and popularity of vodka and the trajectory of the

---

[1] Sharaf al-Dīn 'Alī Yazdī (d. 1454), *Zafarnama* (Book of Victory) (1424–28) (Tihrān: Kitābkhānah, Mūzih va Markaz-i Asnād-i Majlis-i Shūrā-yi Islāmī, 1387 (2008)), vol. 2, p. 1266.
[2] As Braudel has shown, several Europeans traveling outside Europe noted the presence of arak (*arrequi* in Algiers and *arac* in Gujarat) in the seventeenth century. Braudel, *Civilization and Capitalism*, 247. For the most comprehensive study of the history of arak, see El-Asmar, *The Milk of Lions*.
[3] Luo Feng, "Liquor Still and Milk-Wine Distilling Technology," 501–518.

Mongol's two-century rule of Russia.[4] These are all subjects that lie beyond the scope of this book, yet suggest the benefits that further investigation would bring to advancing an understanding of the development of distilled liquors in global history. Instead of stopping at the case of Korea, however, this book examines two cases that stimulate discussions of the transfer of distilled liquors in a global context: the case of shochu in Japan and of tequila in Mexico.

As an island country located east of Korea and China, Japan developed a unique culture that nonetheless has shared strong similarities in the East Asian cultural zone, sharing various aspects such as written Chinese characters, Buddhism, and food influenced by Chinese culture from the ancient period. Japan did this not only by means of its contacts with China but also by means of its neighbors on the Korean peninsula, which sits between China and Japan.[5] A limited number of Japanese sources written after soju began to develop in Chosŏn Korea talk about *shōchū*, the Japanese pronunciation of the written characters that correspond to Korean *soju* and Chinese *shaojiu*. Shochu has developed into a similar but distinct distilled liquor of Japan, which has grown in popularity up to the present.[6] As seen in Chapter 4, the official historical records of Chosŏn explicitly mention many times that the Chosŏn court sent many bottles of soju as gifts to different Japanese regimes, including those of Tsushima and Ryukyu, and therefore we can assume some influence from Chosŏn on the development of Japanese shochu culture through official exchange. This leaves little doubt that many bottles of soju/shaojiu and

---

[4] A hypothesis that Alexandrhas Gorokhovskiy proposed in his paper presented at the Fifteenth International Conference on the History of Science in East Asia, Chonbuk National University, Jeonju, Korea, July 19–23, 2019. He argues that, contrary to the common belief in its original transfer from Western Europe, distilled-alcohol technology in Russia could have disseminated there as a result of Batu's invasion of Russia in the thirteenth century. Confirmation of this hypothesis requires further examination of a greater variety of sources, including Polish sources, in order to verify and, if true, qualify Mongol influence on the development of vodka. The argument is convincing, given that the earliest records that demonstrate the development and popularization of distilled liquors in Europe date back to only the fifteenth and sixteenth centuries. Moreover, as the contemporary soju-making process – which first makes high-percentage distillates with up to 95 percent alcohol created from cheaper materials like potatoes or tapioca instead of grains and then adds water and sweeteners to create flavors – becomes virtually identical to modern Russian vodka production, it would be useful to compare similar patterns of historical development in both Korea and Russia. Based on personal conversation with Alexandrhas Gorokhovskiy.

[5] There have been numerous works on the East Asian relations. For example, see Jiang Wu and Lucille Chia, eds., *Spreading Buddha's Word in East Asia: The Formation and Transformation of the Chinese Buddhist Canon* (New York: Columbia University Press, 2016).

[6] Hishinuma Hayato, *Sake: The History, Stories, and Craft of Japan's Artisanal Breweries* (Singapore: Gatehouse Publishing, 2015), 323.

the technology of distilled liquor making transferred from Korea and China to Japan at the same time by way of private trade. Some sources hint that Japanese shochu was also sometimes called *araki* (a Japanese transliteration of a Chinese word for arak). Accordingly, the three East Asian countries, China, Korea, and Japan, all share the use of two terms, *shaojiu/soju/shōchū* and *arak*, to refer to their distilled liquor, though the Chinese term, *shaojiu/soju/shōchū*, became much more dominant in the East Asian cultural milieu than did the foreign word *arak*, which has been transliterated into a different set of Chinese characters. The rise of shochu in Japan certainly suggests the possibility of transfer and expansion of the distilled liquor from China and Korea to Japan soon after the Mongol period. Yet the period witnessed a growing global connection at an unprecedented scale because of the coming of Europeans to Asia by way of maritime routes. Therefore we can assume another route for the transfer of distilled liquors from other regions, such as Southeast Asia, and from Europe in the early modern period. This chapter discusses these various possibilities.

Another intriguing case in which we can observe connections and comparison with soju is mescal (also spelled mezcal) and tequila, the most widely representative of Mexico's distilled liquors. This case also suggests some possibilities of the transfers of distillation technology from Asia to the wider world, though the times and methods are still the subject of scholarly debate. Some scholars, like the ethnologist Henry J. Bruman, suggest that Asian-style stills were first transferred from the Philippines to Mexico through the "Manila galleon" trade (1565–1815), operated by the Spanish during the early modern period, which linked Manila (and China, including a then large Chinese population in the Philippines) to Colima province and then the city of Acapulco, two places on Mexico's Pacific coast.[7] More recent research has identified a wider variety of transfer routes that ideas about distillation took to Mexico, carried there by Asian immigrants to Mexico who arrived as *Indios Chinos* in the thousands, not only from the Philippines but also from other places in Asia, because Manila was a central market for the regional slave trade.[8] Yet other scholars even propose a possibility of pre-Columbian origins. While we

---

[7] Henry J. Bruman, "The Asiatic Origin of the Huichol Still," *Geographical Review* 34, no. 3 (July 1944): 418–427. On the Manila galleons, see William L. Schurz, *The Manila Galleon* (New York: Dutton, 1959).

[8] I appreciate Ana Valenzuela's insights on this. See Tatiana Seijas, "Indios Chinos in Eighteenth-Century Mexico," in *To Be Indio in Colonial Spanish America*, ed. Mónica Díaz (Albuquerque: University of New Mexico Press, 2017), 123–131; Tatiana Seijas, *Asian Slaves in Colonial Mexico: From Chinos to Indians* (New York: Cambridge University Press, 2014); and Edward R. Slack Jr., "The Chinos in New Spain: A Corrective Lens for a Distorted Image," *Journal of World History* 20, no. 1 (2009): 35–44.

cannot deny that the idea of distillation first developed independently in Mexico (as in the cases of China and India), discoveries of many traditional stills similar to those that developed in Asia, not to mention the growing connections between Afro-Eurasia and the Americas after 1500, suggest strong possibilities for direct transfer of some important elements in distillation technology thanks to early modern European connections. The second part of this chapter provides a review of previous scholarship that suggests different theories about the development of distilled liquors in Mexico, and in response proposes a new scheme based on the findings of the present book: as with the case of transfers from China to Korea, these East Asian stills (both Mongolian and Chinese types) spread to Southeast Asia after the Mongol period and then found their way to Mexico through the Columbian exchange. Unlike the case of Japan, there is no way that Korean stills had a direct influence on Mexican stills; however, the case of Korea suggests ways in which East Asian stills, which had influenced Korean stills, may have influenced distillation in other parts of the world, like Mexico, in ways that have been less well documented than in Korea.[9] This discussion, therefore, provides a way to discuss the history of distilled liquors from both connective and comparative perspectives.

By examining the cases of Japan and Mexico, this chapter explores how the idea of distillation continued to develop and transfer through broader channels of political, commercial, and cultural contacts from various parts of Afro-Eurasia and other parts of the globe. The cases of Japan and Mexico show that both countries were open to the influence – directly, indirectly, and to varying degrees – of other distillation traditions on the Afro-Eurasian continent. Through an examination of two of these cases, we can consider possible directions of future research into the historical development of distillation in world regions as a consequence of the Mongol Empire.

Japan's Shochu: Possible Transfers from the Afro-Eurasian Continent, including Korea, to Japan

The evolution of shochu in Japan began with the introduction of fermented starter from Korea during the ancient period, and then developed

---

[9] Ana Valenzuela and her colleagues attempted to compare types of still being used for traditional Japanese shochu and Mexican distilled liquors in order to explore a link between the development of distillation in Asia and on the Mexican Pacific coast after the thirteenth centuries. Having discovered several similarities, they called for further studies. See Ana G. Valenzuela-Zapata, A. Regalado, and M. Mizoguchi, "Influencia asiática en la producción de mezcal en la costa de Jalisco: El caso de la raicilla," *México y la Cuenca del Pacífico* 11, no. 32 (2008): 81–116.

with a brew, drunk in a fashion reminiscent of sake, similar to Korean clear strained wine of the medieval period. This was followed by the development of shochu as a distilled liquor imported from outside, from places like China and Korea, during the early modern period. This section presents the general arc of this narrative, paying attention to the historical contexts surrounding it.

Few records detailing ordinary cultural life have survived since antiquity; however, documentary and archaeological evidence of Japanese cross-cultural contact with neighboring countries like China and Korea suggests significant influence on Japanese culinary culture in many areas, including alcoholic beverages. Experts on Korean and Japanese alcoholic beverages argue that the Japanese first assimilated systematic techniques of alcoholic-beverage production from Koreans, particularly those living in the kingdom of Paekche (18 BCE–660 CE). An ancient Japanese record claims that a person from Paekche named Susukori 須須許里 (Susubori in Korean) brought to Japan *nuruk* – a form of malted rice used as yeast in Korea – and taught the people there how to use it to brew alcoholic drinks. That document even declares that the legendary Japanese emperor Ōjin 応神 (201?–311? CE) grew drunk and sang after consuming Susubori's drink. From that moment, according to legend, Susubori became the god of alcohol in Japan.[10] The assumed influence of Paekche on alcohol brewing in Japan may constitute one in a variety of cultural influences that Paekche exerted on Japan, a fact to which the many remains of Paekche culture lingering today in Japan still testify.[11]

Having assimilated these new techniques in alcohol production, the Japanese began to develop their unique alcohol culture, affecting both production and consumption. This led to the development of sake, a new, Japanese form of brewed alcohol. Striving to imagine sake's origins, scholars depict a humble scenario that began with the first alcoholic beverage, possibly *kuchikami no sake*, the production of which would have required a producer to masticate rice and other forms of starch in order to release the enzymes necessary to convert starch into sugar and begin fermentation.[12] While such early forms of alcohol making are not specifically documented, signs of significant change in alcohol making in

---

[10] Sakaguchi Kin'ichiro, *Nihon no sake* (Tokyo: Iwanami shoten, 2007), 115. This is the earliest Japanese record to contain the earliest Chinese character for alcoholic drink, *shu* 酒 (pronounced *shu* in modern Japanese, *jiu* in modern Chinese). See Ilbon yangjo hyŏphoe (Nihon Jōzō Kyōkai), *Pon'gyŏk soju chejo kisul* (*Chŏngnyusik soju chejo kisul*) (The Manufacturing Technology of Distilled Shochu [soju]), trans. and ed. Pae Sangmyŏn (Bae Sang-myun) (Seoul: Pae Sangmyŏn churyu yŏn'guso, 2003), 18.

[11] Bailey G. Sansom, *Japan: A Short Cultural History* (New York: D. Appleton, 1943), 150–159.

[12] Hishinuma, *Sake*, 17; Sakaguchi, *Nihon no sake*, 112.

some documents testify to major political changes in medieval Japan. After the major political changes of the Taika Reform in 645, the Japanese royal court began to promote all industries and cultural activities under its control; this included alcoholic-beverage production, of course.[13] Since then, the Japanese have researched the brewing techniques they possessed and advanced their research in hopes of developing even further. The Japanese court assigned some families to alcohol production and others to cultivating the brewing base as part of its policy of labor specialization. It is highly likely that the government also assigned experts and artisans from Paekche to develop the brewing industry in Japan.

Japanese sources only write the name of their beloved brew as *sake*, whose Sinitic character gives us its written word – *jiu* 酒 (Japanese *shu*) – which means nothing more than "alcoholic beverage," though it is colloquially translated as *rice wine*. As no distillation is traced in ancient Korea, the alcoholic beverage that has been developing in Japan since traditional times was most likely similar to the Korean strained wine called ch'ŏngju (written 清酒 but pronounced *seishu* in Japanese).[14] That is, sake was not a distilled alcohol but rather a brewed fermented drink. Over time, sake developed to become Japan's most representative alcoholic beverage – and ultimately a national symbol, earning the name "Japanese sake" (*Nihon shu* 日本酒) among Japanese. In recent times, Japan's national drink has also become a global one.[15] Given its similarity to Korean ch'ŏngju in terms of both taste and production methods, we can safely assume that the original methods of sake production hailed from Korea. In fact, signs of cross-cultural influence persist to the present: Korea's most representative form of ch'ŏngju, called *chŏngjong*, was influenced by a form of Japanese sake named Masamune; indeed, the Korean word *chŏngjong*, the Japanese *masamune*, and the Chinese *zhengzong* represent simply three different pronunciations of the same Chinese characters – 正宗, meaning "authentic," a name that has held since the late nineteenth century.[16] The history of Japanese sake bears some similarities to that of Korean alcohols like chŏngjong in its evolution since medieval and early

---

[13] Sakaguchi, *Nihon no sake*, 115.
[14] For more details on the Korean strained wine called *ch'ŏngju*, see Chapter 2 of this book.
[15] Hishinuma, *Sake*, 17.
[16] The term *masamune* 正宗 (*zhengzong* in Chinese) has been a common term to describe strained wine in Japan ever since the Japanese government established its modern trademark regulations in 1884. After that, alcohol-brewing companies in Korea, influenced by this Japanese form, began to make the same kind of beverage. It became popular nationwide under the brand name Chŏngjong, a Korean transliteration of Masamune and synonymous with *ch'ŏngju*. See Kim Soyŏng and Kim Hyeju, *Sak'e, Ryu* (Sake for Beginners) (Seoul: Altent'e buksŭ, 2009), 22–23.

modern times, which should be reviewed briefly in order to better contextualize the rise of distilled liquors in Japan.

The earliest detailed documentary evidence of sake production based
on rice and *kōji* mold (malted rice used as yeast, similar to nuruk in
Korea) appeared during the Nara period (710–794).[17] This text was
written at about the same time as the earliest Japanese chronicles – such
as *Kojiki* 古事記 (Records of Ancient Matters) (completed 712) and
Nihonshoki 日本書紀 (The Chronicles of Japan) (completed 720), and
legal texts such as *Ryōnoshūge* 令集解 (Commentary on Yōryō Legal
Code) (*c.* 868) and *Engishiki* 延喜式 (Procedures of the Engi Era)
(completed 927) – which shows that the state took the lead in building
a systematic brewing system.[18] The Department of Breweries
(*Jōzōkyoku* 醸造局), for example, produced fifteen kinds of sake to use
at important state ceremonies like sacrificial rites. Such activity
extended beyond the state, of course. Shinto temples began to brew
sake in the tenth century. From the end of the Heian period (795–1185)
and into the Kamakura (1185–1333) and Muromachi (1336–1573)
periods, urban development and commercial activity flourished, and
sake was distributed as a commodity with economic value, like rice.
Goshu no nikki 御酒之日記 (Book on How to Brew Sake) (completed
1489), the first private document about brewing sake, provides detailed
instructions on methods of lactic acid fermentation using kōji, rice, and
water, which formed the basis of a unique Japanese system of brewing.[19]
This book also suggests that a brewing industry flourished as the
demand for alcohol in Japan greatly increased. By the sixteenth century
in the Nara region, sake making included a barrel capable of brewing ten
koku 石 (approximately 144 kilograms of husked rice) as part of a mass
production system – indeed, during this time, Japan experienced an
"industrial revolution" in its sake economy.[20] At the same time, the
manufacturing and management of liquor broke free of the monopolies
of temples and shrines.[21] Breweries were created in each region, and
competition in quality intensified. This begat a growing variety of sake as
a consequence. Japanese society also witnessed the making of
a complete prototype of sake making called Moro Haku.[22] To this can
be added the creation of guilds, another important milestone in the

---

[17] This period is important to the history of sake. See Kim Soyŏng, *Sak'e, Ryu*, 38;
Sakaguchi, *Nihon no sake*, 115.
[18] Sakaguchi, *Nihon no sake*, 122.
[19] This book describes a classic multi-stage brewing method using two separate applications
of fermented rice followed by an application of lactic fermentation and the use of
charcoal. Kim Soyŏng, *Sak'e, Ryu*, 38; Sakaguchi, *Nihon no sake*, 130–131.
[20] Kim Soyŏng, *Sak'e, Ryu*, 38.   [21] Sakaguchi, *Nihon no sake*, 127–128.
[22] Kim Soyŏng, *Sak'e, Ryu*, 37.

development of the sake economy.[23] The government encouraged such developments and sought to benefit from them; for example, it provided special authorization to those who professionally made alcoholic beverages and imposed a tax on their work.[24] Later, in the Edo period (1603–1868), sake continued to develop rapidly. People began to brew sake several times during a year – varieties like Shinshu ("newly brewed sake"), Aishu, Kanmae Sake, Kanzake ("cold sake"), and Haru Zake ("spring sake"). They discovered that the taste of Kanzukuri ("cold brew") is best achieved when made in winter and subjected to long fermentation at low temperatures.[25] Brewers of the time advanced pasteurization methods and enhanced alcohol preservation techniques. Apparently, this affected the scale of production; according to records, more than 27,000 breweries dotted the Japanese landscape in 1698. Better distribution of sake and its ingredients helped in this growth as well. In the middle of the Edo period, for example, people distributed the sake of Nara to the Edo (Tokyo) area thanks to the development of sea shipping. In the late Edo period, people became more aware of the importance of water, low iron levels, and rich mineral content.[26] These are just a few examples of the sake industry's inexorable development in Japan over many centuries, which continues today.

In the twentieth century, sake culture expanded beyond Japanese society to become a global commodity; the trend continues today. Sake and *izakayas* (pubs) have become a big deal on both the East and West Coasts of the United States, where soju is slowly also gaining ground. The same is true in Korea. Since 2008, for example, the amount of sake imported by Korea from Japan has increased sharply. The fact that sake suffers less regulation than do other liquors affects the distribution process in ways that contribute to the increase in the spirit's imports. The popularity of Japanese alcohol has grown in tandem with the *izakaya*, an informal form of Japanese pub now found in every neighborhood in Korea's cities, where the epicenter of the sake boom lies. Even wine lovers have begun to turn their eyes toward sake, exploring its varieties.[27] As Japanese foods

[23] The guild for alcohol production was created during the Muromachi period. Sakaguchi, *Nihon no sake*, 127–128.

[24] Ibid., 127–128.

[25] Most breweries still brew in the winter, except for those breweries that have adopted modern facilities and mass production.

[26] Kim Soyŏng, *Sak'e, Ryu*, 40.

[27] Sake has grown popular in Korea along with the boom of Japanese *izakaya*. In Japan, the Liquor Research Institute, an independent administrative agency, conducts new sake competition fairs and administers the dissemination of information about liquor research and technology in an effort to improve the quality of sake and increase its export to other countries. In recent years, sake has been increasing in status as an independent category in overseas food competitions. Kim Soyŏng, *Sak'e, Ryu*, 9–10, 68–69.

like sushi gained popularity in the global food market – with the help of Japanese government promotion – so did sake (a case similar to Korean soju and cuisine).[28] As a consequence, a great variety of sake of various kinds, qualities, and prices is appearing in the Asian markets and Japanese restaurants that can be found today in many of the world's countries and most of its cities.

A good grasp the basic characteristics of sake as a specific type of alcoholic drink is important to understanding its evolution in the history of alcoholic drinks in Japan. The alcohol content of sake generally ranges between fifteen and seventeen percent because, as a strained wine, it is not subject to distillation. As with strained wine in Korea, which people have historically used not only in ancestor-worship ceremonies but also in cooking and cuisine, sake became a popular alcoholic beverage in Japanese cuisine because it compliments foods well. Japanese sake, like Korean strained wine, spoils – goes flat and loses flavor – easily.[29]

Shochu and sake followed different developmental paths. Distilled liquors that were good for longer-period preservation were probably less needed in Japan because different kinds of sake were produced and distributed in society systematically, and there were no long military expeditions during which soldiers needed to bring alcoholic drinks that kept longer. While never widely popular, some distilled spirits nonetheless developed in Japan as an important kind of traditional alcoholic beverage from the sixteenth century.

The earliest evidence that testifies to the presence of shochu in Japan relates an episode that took place in Kyushu, the Japan's southernmost main island, which sits close to Korea. In 1559, a carpenter employed at a shrine in the city of Kagoshima in southern Kyushu complained that he worked hard there, but the shrine's owner did not offer him shochu to drink. An intriguing anecdote; however, without other sources, it can only be interpreted as an isolated episode. Subsequent to this episode, it is difficult to locate a record that shows when and how shochu spread and grew popular on Japan's main islands. Still, this anecdote is concrete enough to suggest that shochu spread in Kyushu and that ordinary people there began to drink it. We also learn people were already making shochu using sweet potatoes, at least in southern Kyushu.[30]

---

[28] Sake's largest global consumer is the United States, and it has also rapidly gained popularity in Korea. See Tara Nurin, "Sake Sales Soar as Brewers around the World Defy Ancient Japanese Traditions," *Forbes*, February 28, 2017, www.forbes.com/sites/taranurin/2017/02/28/u-s-sake-sales-soar-as-brewers-around-the-world-defy-ancient-japanese-traditions/#6fbae005380e (accessed December 30, 2018).

[29] Sakaguchi, *Nihon no sake*, 224.   [30] Takahashi, "Nihon no jōryūshu, shōchū," 236.

Intriguingly, the language of this first document about shochu in Japan presents the alcohol very differently from the terms used in China and Korea. The written word 焼酎, pronounced *shōchū* in Japanese, *shaojiu* in Chinese, and *soju* in Korean, divides into two parts: the first character 焼 – pronounced *sho, shao*, and *so* – means "burning," no doubt a reference to the craft's process. The second part of the name, the character 酎 *chū*, means triple-brewed liquor; this character differs from the one used to describe alcohol in general, 酒, pronounced *shu* (or *sake* in *kun'yomi*, the Japanese reading of Chinese characters), *jiu*, and *chu/ju* in the three languages.[31] Takahashi Kōjirō argues that, according to a 1597 book entitled *Yodarekake* (A Bib), thrice-fermented sake came to be called "triple sake"; at the same time, the Japanese word 酎 *chū* came to signify "triple sake."[32] This means that people heat sake to create vapor and capture its condensing droplets, filtering the alcohol three times. Takahashi argues that this word *chū* most likely referred to distilled liquor. Is it possible that this shochu was transferred from Korea or directly from China? It is important, then, to consider different theories about the origin of shochu in Japan.

Considering that the most famous traditional shochu brands in Japan developed on its southern islands – namely the southern main island of Kyushu and the island of Okinawa in the Ryukyu chain – we can surmise its direct transfer from Korea and China. In other words, distilled liquors and distillation technology were introduced from outside. Some experts on Japanese shochu argue that its distillation technology developed first in the Ryukyu islands and probably later transferred to Kyushu.[33] Such technology existed in the southern island chain: most notably, a special kind of distilled liquor called *awamori* developed there and has since become globally famous.[34] Some Japanese experts claim that it originated in Thailand,[35] although others counter that it moved to Japan from China.[36] The name *shōchū* puts heavier weight on the latter view. Moreover, it appears possible that the Japanese spirit was influenced by Chinese shaojiu and Korean soju after its introduction. Intriguingly, a document influenced by Chinese literature reveals that shochu was

---

[31] Hishinuma, *Sake*, 323. Most characters in Japanese have two readings, one derived from a now archaic Chinese pronunciation, and one indigenous Japanese.

[32] Takahashi, "Nihon no jōryūshu, shōchū," 235.   [33] Ibid., 235.

[34] For more details about the awamori liquor, see Tominaga Asako, *Awamori wa oishi: Okinawa no aji wo sodateru* (Awamori Is Tasty: Raising the Taste of Okinawa) (Tokyo: Iwanami shoten, 2002).

[35] Shurui sōgō kenkyūjo (National Research Institute of Brewing), *Umai sake no kagaku* (The Science of a Wonderful Alcoholic Beverage) (Tokyo: Softbank Creative, 2007), 42, 47–48.

[36] For example, experts contributing chapters to one volume – Nihon Jōzō Kyōkai (Brewing Society of Japan), *Honkaku shōchū seizō gijutsu* (Manufacturing Technology of Distilled Shochu) (Tokyo: Nihon Jōzō Kyōkai, 2001) – espouse different theories.

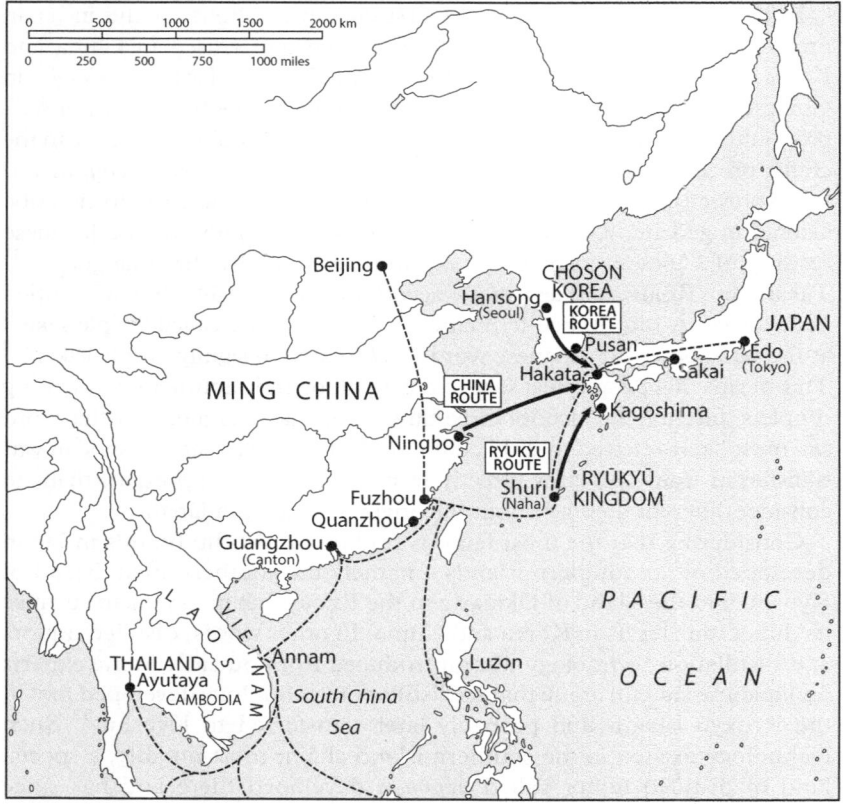

Map 6.1 Possible routes of distilled-liquor transfer to Japan during the early modern period (1500–1750).

also called araki, as it was in China and Korea.[37] Because Japan is an island country, there are various possibilities, all of them involving sea transport, on which people in transit moved – and sometimes fled – to other parts of the world, often bearing official gifts and presents, trade goods, books, and other objects, like technologies, not to mention know-how, or distillation techniques (see Map 6.1).

[37] Terashima Ryōan, *Wakan sansai zue* (Illustrated Sino-Japanese Encyclopedia) (1712) (Tokyo: Heibonsha, 1985–1991), 1786–1787; Ishige Naomichi argues that distilled liquor was once called *araki* liquor using different Chinese characters in Japan: *araki shu* 荒木酒 (pronounced *huanmu jiu* in modern Mandarin Chinese) or *araki shu* 阿剌吉酒 (*alaji* in Mandarin). Here, *shu* refers to liquor (*jiu* in Chinese). Ishige, 122.

Let us consider all the possible ways in which distilled liquor and its technology transferred to Japan, starting with China and Korea. The Mongol period makes a good place to start, especially in the case of Korea. From there, we can expand to larger scales of perspective that would allow, for example, on a regional scale, direct transfer routes from Thailand to Japan via Asia's premodern shipping trade beginning in the fourteenth century and the shipping trade of the Europeans who first arrived in East Asia soon after 1500.

Because Japan had facilitated both official and unofficial trade relations with China and Korea from the seventh century until the early seventeenth century, it is highly likely that distilled liquors consumed there or ideas about distillation were introduced to Japan through commercial relations. During the Mongol period, Japan was free from Mongol domination thanks to the failures of the Mongol expeditions to Japan. Unlike Korea, therefore, Japan offered fewer opportunities for Mongol soldiers and people connected to them to come to Japan and directly introduce new forms of liquor making. Despite the looser political connections between Mongol-ruled China and Japan during this era, seagoing trade vessels sailed to Japanese port cities carrying many trade goods, as attested by the sunken ship excavated at Sinan, which was en route to Hakata in Kyushu. While the excavation site yielded a variety of trade goods, including porcelain pieces such as wine bottles and small dishes, it could not be determined whether the ship's carriers actually brought such things as distilled liquors – which had gradually grown to become widely consumed in China – although we cannot deny the possibility.

Given its close geographical proximity to Japan, it is indeed possible that shochu transferred directly from Korea to Japan. As discussed in Chapters 3 and 4, Cheju Island was under strong Mongol influence from the thirteenth century, and recent studies and archaeological excavations hint that different groups of people in Cheju (and possibly those from the Korean peninsula nearby) left Cheju as political exiles fleeing upheavals in the late fourteenth century, in search of a haven in such places as the Ryukyus.[38] These people could have brought to the islands such things as soju and the distillation technology to make it. Another piece of strong evidence that hints at possible transfers of soju to Japan appears in a record in the official chronicles of Chosŏn, a report that describes many bottles of soju presented to Japanese rulers by Korean royal emissaries as diplomatic gifts. For example, immediately after its defeat of

---

[38] O Yŏngju (Oh Young-Ju), "Cheju chŭngnyuju 'Kosorisul' kwa Ok'inawa' Awamori' ŭi munhwa pigyo siron" (A Comparative Study of the Cultures of Cheju Spirits "Kosorisul" and Okinawa "Awamori"), unpublished paper presented at the symposium of the Cheju–Okinawa Studies Association held at Cheju University, November 4, 2017.

Koryŏ, the Chosŏn court sent many bottles of soju to the ruler of Tsushima island near Cheju, which at the time was a *han* (fief controlled by a *daimyo*, or territorial lord) of Japan's Muromachi shogunate (1338–1573).[39] The chronicles also report official gifts to the rulers of Ryukyu and the main islands of Japan. An episode shows that Shimazu, Satsuma's feudal lord in the seventeenth century, received a distilled drink from Ryukyu and presented it to the shogun of the Muromachi shogunate. The Japanese scholar Edagawa Kōichi argues that, given its high quality, distilled liquors captured the attention of high-ranking elites in Japan.[40] However, we need further evidence to determine whether the liquor spread to Japanese commoners.[41]

Some pieces of evidence that hint at the spread of distilled liquors in some regions in Japan on some scale hail from outside: in this case, two accounts, one European, the other Korean. An account about Japan written in 1548 by a Portuguese missionary named Jorge Alvares reports that Japanese were drinking a liquor called *oraka* made of rice.[42] At that time, the word *oraka* meant simply "distilled liquor" in Portuguese. This document suggests two things. First, the Japanese identified by Alvares probably were the people of Kyushu, because he and all other foreigners were not allowed to travel to Japan's main island, Honshu.[43] This supports the reported episode about the carpenter in Kyushu documented in 1559, and suggests that distilled alcohols based on grains like those in China and Korea spread to Japan in the early sixteenth century.[44] Second, this document suggests the possibility that, before Europeans arrived, Japanese living in Kyushu and Ryukyu were already drinking distilled liquors that probably had come to the islands from either Korea or China.

Besides, could the word *oraka* be a transliteration of *arak*, the universal name for distilled liquors throughout Asia? Eugene Anderson proposed

[39] Japanese records show that Korean envoys of the Chosŏn period brought soju to the feudal lords of Tsushima in Japan in 1404 as a gift, because they considered it a precious beverage. Edagawa Kōichi, "Jōryūshu, higashi mawari nishi mawari" (Distilled Alcohol, Bound for East and West), in *Shōchū higashi mawari nishi mawari* (Shochu around the World), ed. Tamamura Toyo'o (Tokyo: TaKaRa Alcohol Beverage and Life Research Institute, 1999), 57.

[40] Edagawa, "Jōryūshu, higashi mawari nishi mawari," 61.

[41] The earliest evidence of an actual transfer of soju made from Korea to Japan appears in a document that reports Korean envoys of the Chosŏn dynasty bringing it to Japan as a diplomatic gift in 1424. *Chosŏn wangjo sillok*, Sejong 6 (1424): 06–01-09[02]. For more on this source, see Chapter 4.

[42] Takahashi, "Nihon no jōryūshu, shōchū," 235. This book could be *Livro que trata das cousas da India e do Japão* (Book about Things from India and Japan). The first Portuguese arrived in Japan in around 1542. Sansom, *Japan: A Short Cultural History*, 416.

[43] Takahashi, "Nihon no jōryūshu, shōchū," 235.

[44] Edagawa, "Jōryūshu, higashi mawari nishi mawari," 61.

a convincing argument that *oraka* indeed derived from *arak* (Arabic *araq*). It begins in medieval Spain, most of which came under the rule of the Arabic and Berber caliphates between the eighth and eleventh centuries, except in the south, where the Nasrid dynasty continued to rule southeastern Spain until 1492. During this era under Arab rule, Spain assimilated a number of Arabic words, including high-culture words – words for ceramic tiles, irrigation ditches, lemons, cotton, sugar, saffron, and almost anything civilized at the time. Neither *oraka* nor *arak* exists in the Spanish language, other than as a loanword from Southeast Asian languages, which suggests a later origin. However, Arabic words for things like stills and alcohol do, and continue to do so today (just as English speakers continue to use the obviously Arabic word *alcohol* today). Thus the linguistic evidence points to an Arab introduction of distilled liquor to Spain and then to its neighboring country, Portugal.[45] Further investigation will determine the concrete routes by which the transfers of liquor, names, and technologies transferred, but the existence of this transfer is clear.

In addition, a Korean source examined in Chapters 2 and 4, *Ch'ŏngjanggwan Chŏnsŏ* by Yi Tŏkmu (1741–1793), also provides an important clue to the puzzle of soju/shaojiu's arrival in Japan. In a section on soju's origin, Yi discusses the shochu of Ryukyu and Satsuma, referring to it as awamori liquor, which he compared to a Chinese liquor he called alaji jiu 阿剌吉酒, using here the same Chinese transliteration for *arak* found in a fourteenth-century Chinese cookbook, *Yinshan zhengyao*.[46] This suggests two things: first, awamori shochu was popularly consumed in Ryukyu and Satsuma and enjoyed fame among Korean scholars like Yi Tŏkmu;[47] second, these Korean, and by extension East Asian, scholars could categorize local varieties of shaojiu/soju/shochu that had developed in the region in the wake of its introduction.[48]

In the eighteenth and nineteenth centuries, Yi Tŏkmu and other Korean scholars interested in Practical Learning read books from the outside world in order to learn more about various issues in the wider world. The books read by these Korean scholars include Japanese works that contain knowledge that the Japanese possessed at that time. One such is *Wakan sansai zue* 和漢三才図会 (Illustrated Sino-Japanese

[45] Based on personal conversation with Eugene Anderson.   [46] See Chapter 2.
[47] Kim Posŏng, "19 segi Chosŏn chisigin ŭi Ilbon· Yugu e taehan insik koch'al: Oju Yi Kyugyŏng ŭi *Sigajŏmdŭng* ŭl chungsim ŭro" (Tha Chosŏn Literati's Recognition of Japan and Ryukyu in the 19th Century: On the Basis of *Shigajumdeung*), *Hanmunhak nonjip* (Journal of Korean Literature in Chinese) 35 (2012): 191–235.
[48] Few studies of Japanese shochu utilized Korean primary sources to discuss the history of shochu in Japan, yet these include some valuable relevant information.

Encyclopedia), compiled in 1712 and based on a Ming Chinese encyclopedia entitled *Sancai tuhui* 三才圖會 (Illustrations of the Three Powers)
assembled by Wang Qi 王圻 (1565–1614) and completed in 1609.[49]
Other books on making shochu that were published in Japan after the
seventeenth century include *Dōmōshuzōki* 童蒙酒造記 (Maniac Text
Book about How to Brew Sake) (*c.* 1686), *Honchōshokkan* 本朝食鑑
(Encyclopedia on Food) (1697), and *Bankinsangyōbukuro* 万金産業袋
(Reports on Various Industries) (1723).[50] Compilations of these books
suggest that Japanese learned about soju through Chinese books about
foods and medicine imported from China to Japan. Basically, people
made shochu by using a sake fermenting mash called *moromi* 醪 and
a strained wine of lower quality. It seems that this moromi-based shochu
spread throughout the entirety of Japan, at least taught by Chinese books.
All these books influenced by Chinese texts, like *Wakan sansai zue*, use
the same Chinese characters for alcoholic drinks, *jiu* in Chinese and *shu* in
Japanese, because these books directly cite *Bencao gangmu*, which introduces shaojiu as *huojiu* (火酒, "fire liquor") or alaji jiu (阿剌吉酒, "arak
liquor").[51]

We should consider another source of transfer: Southeast Asia. Some
Japanese scholars think that the distillation technology for shochu came
from Thailand, because people who conducted field research there in the
early twentieth century found technological similarities between distillers
in Thailand and Japan. Of course, we have to be careful about suggesting
premodern connections based on recent similarities. However, the signs
of medieval and early modern connections between Southeast Asia and
Japan in terms of distillation technology do exist. We have seen that
distilled liquors called arak first traveled from Southeast Asia to China
when people further developed and transferred the technology for East
Asian stills – a term coined by Needham to describe the suspended catch-
alcohol stills of the Mongolian tradition and a Chinese variation that uses
a tube to conduct alcohol from the still's boiler – to Korea and then
Japan.[52] The early modern era witnessed occasions on which representatives of some Southeast Asian regime made a direct transfer of knowledge
by presenting a distilled liquor to the Chosŏn court as a diplomatic gift.
From this, it seems reasonable to assume that other forms of distilled

[49] Terashima Ryōan, *Wakan sansai zue*; Wang Qi (1565–1614), *Sancai tuhui* (Illustrations
of the Three Powers) (1609) (Shanghai: Shanghai guju, 1988).
[50] Takahashi, "Nihon no jōryūshu, shōchū," 236.   [51] Sakaguchi, *Nihon no sake*, 75.
[52] Many ethnic groups and small producers in Asia and America continue to employ
distillation methods and tools of the Asian tradition. Scholars categorize them as
Mongolian or Chinese styles, although more recent studies suggest a need to revise this
earlier categorization. Needham, Ho, and Lu, *Science and Civilisation in China*, vol. 5,
part 4, 103–121.

liquor and distillation technology transferred not only to Korea but also to the Ryukyu islands. This hypothesis is supported by recent studies that have begun to examine maritime trade between China, Southeast Asia, and the Kingdom of Ryukyu in the fifteenth century and point to a direct transfer from Thailand to Ryukyu.[53] We can also speculate that Japanese, Chinese, and Korean pirates called *wako* active between the thirteenth and sixteenth centuries transferred distilled spirits and distillation technology in the course of their trade, which circulated through ports all over East and Southeast Asia.

However, we must also consider the roles played since the early modern period by Europeans, who after Vasco da Gama's pioneering voyage to India advanced to East Asia via Southeast Asia. They too could have transported concepts and tools of distillation that had developed in Southeast Asia as they passed through the region on their way to Japan. It is also quite possible that European distillation technology transferred to Japan. The fact that the Portuguese word for distilled alcohol at that time, *oraka*, is related to the word *arak* suggests that Europeans first assimilated ideas about distilling, which was already advanced all over Asia, as the result of growing contacts in those regions beginning in late medieval times. After Europeans began to advance into East Asia, they most likely brought the distillation technology to Japan. One of the few premodern sources related to shochu in Japan is a set of documents discovered in the archives of the Murakami family (*c.* 18th century), a Kyushu family, many of whom were medical doctors in the Edo period (1603–1868). These documents discuss pharmacy as well as the method of distilling alcohol using a *rambiki*.[54] A small still-like porcelain object forty centimeters high was discovered alongside the documents.[55] As the Murakami family practiced *Langaku*, so-called Dutch learning, it is highly likely that the Dutch introduced the rambiki to them again. It seems that the Murakami family distilled alcohol primarily for use in surgery. The object looks similar to other East Asian stills, including a Korean soju kori.[56] However, this still-like object is too small to be used for actual distillation; perhaps it was meant to be a miniature. Given this, we do not yet know if Japanese actually made and used larger stills of

---

[53] Shurui sōgō kenkyūjo, *Umai sake no kagaku*, 42, 48.

[54] This distillation technology is not found in earlier Japanese documents. See Wolfgang Michel, Endō Jiro, and Nakamura Teruko, *Murakami ika shiryō kanzō no kusuri-bako oyobi ranbiki ni tsuite* (On the Medicinal Box and the *Ranbiki* Distillation Apparatus Kept by the Murakami Archive, City of Nakatsu), Murakami Archive Series 4 (Nakatsu: Nakatsu-shi kyōiku iinkai, 2007), 70–78.

[55] For a photograph and diagram of this small porcelain object, see Needham, Ho, and Lu, *Science and Civilisation in China*, vol. 5, part 4, 114, Figures 1488a, b.

[56] Ibid., 33.

this shape to produce alcohol. In any case, the fact that the name *rambiki* probably refers to *alembic* suggests the possible transfer of European alembic technology to Japan.

The Dutch arrived in Asia during the mid-sixteenth century. Following the Spanish and Portuguese, they sailed to Southeast Asia. Unlike the English and French, who focused on occupying India, the Dutch ventured into Indonesia, where they based their Dutch East India Company. From there, these Europeans continued on to East Asia. Japan at that time was open to Europeans to a limited extent, which opened the way not only for European merchants but also for European missionaries in the seventeenth century. Jesuit missionaries like Francis Xavier quickly took advantage of the opportunity to travel to Japan and establish missions there. The English followed suit, and soon English and Dutch East India companies were competing in Japan. Eventually, the Dutch won supremacy. However, trouble arrived in 1634, when the Tokugawa government began to introduce its policy of seclusion, which closed the islands to further contact with the West until 1859. Japan allowed only one exception to its ban on Europeans: it allowed Dutch people to live on Dejima island in Nagasaki. During those years, Western artifacts continued to enter Japan through Dutch hands, and as a consequence they circulated throughout Japan through the Tokugawa shogunate. Of course, it is difficult to deny the possibility that the Dutch introduced spirits as well. At that time, distilled liquors such as whiskey and brandies were becoming popular alcoholic drinks in Europe. They became important to European sailors, some of whom endured the long sea voyages to Asia. It would be natural, then, for the Dutch to transfer European spirits and even their distillation technology to Japan.

In short, an examination of all existing sources has yielded no document that demonstrates the existence of native distilled liquor in Japan before 1500, so it is clear that distilled liquors came to the country from outside. The best candidates for such an outside source of influence are China, Korea, and other Southeast Asian countries like Thailand, via maritime trade connections. It also seems likely that, in a second transfer, Europeans brought distilled alcohol as well – either, in and around the Mongol period, alcohol produced in Southeast Asia, or, later, alcohol made in Europe. We have seen that distilled liquors in Japan first flourished in the southern part of Japan, in particular Kyushu and Okinawa, rather than in Honshu, the main island. Once assimilated, they produced shochu using traditional methods of distillation that had evolved in premodern Afro-Eurasia, as well as specific brands of distilled alcohol like Okinawa's awamori.[57] Like many other distilled liquors, the history of shochu in Japan remains complicated.

[57] Hishinuma, *Sake*, 376.

Like Korea, Japan experienced a major transformation in the face of European expansion; from the late nineteenth century, this included distilled liquors. A difference with the Koreans is that the Japanese responded to the consequent changes voluntarily. In 1868, in the wake of the Opium Wars (1839–1842), the Japanese initiated the Meiji Reformation. After that, they actively adopted Western technologies, which included the scientific manufacture of distilled liquors. Even before the Meiji Reformation, the Japanese had undertaken limited study of Western science and technology and even traveled into the wider world in order to actively pursue the knowledge and contacts that set the foundation for their own development of modern sake production. In 1872, the Japanese sent their sake to the Vienna Expo in Austria, exporting to Europe for the first time. The Japanese representatives who went to the fair looked at the prices of sake made in other countries, and at how competitive sake was globally.[58] This contrasts with the situation of Chosŏn, which maintained its isolation policy and resisted constant foreign pressure to open to the world.

Modern innovations occurred in the area of distilled liquors. Besides simply distilling alcoholic beverages, the Japanese soon learned that they could use the European distillation technique to make drinks with a higher alcohol content. In 1822, a doctor called Udagawa Genshin 宇田川 玄真 achieved an alcohol content of 34 percent. He adopted the Chinese characters 亜爾箇児 (pronounced *arukorū*) as a transliteration of the word *alcohol*.[59] By the 1870s, people were using alcohol medically, and they imported much of it from Western countries. Then, in 1899, the Japanese government established tariffs on imported goods, including alcohol, and in 1901 established the Japanese Liquor Tax Act. After this turning point, the Japanese tried to produce more alcoholic beverages domestically rather than rely on imports. Another important movement in the production of alcohol occurred in 1894 when, in the wake of the Sino-Japanese War, Japan began to import from Europe industrial stills, continuous distilling devices that make liquors with high alcohol content. After the Russo-Japanese war ended in 1905, the alcohol industry grew in importance. After that, the Japanese began to undertake serious research on the distillation of high-quality spirits and promoted a modern distillation industry.[60]

In Japan, too, small breweries and distilleries suffered from various regulations, including the Liquor Tax Act, which was enacted in

[58] Japan Sake and Shōchū Makers Association, *Nihonshu no rekishi* 日本酒の歴史 (The History of Japanese Sake) (Tokyo: Japan Sake and Shōchū Makers Association, 2006).
[59] Takahashi, "Nihon no jōryūshu, shōchū," 237.   [60] Ibid., 238.

1901.[61] As in Korea, in Japan the number of small breweries and distilleries rapidly decreased as a consequence.[62] However, unlike Korea's traditional breweries and distilleries, which after 1910 were seriously affected by these sudden blows, Japan's alcohol producers were in a better situation to undertake quick reforms and overcome the limitations established by the new system. Accordingly, for example, sake makers have ever since continually produced new brands. As seen in Chapter 5, this influenced Korean soju. The modernization of Japanese sake brands affected modern Korea's main ch'ŏngju brand, now manufactured on a large scale in factories.

In Japan, the alcohol brewing industry responded to great changes by continuing to develop. Using a continuous distilling apparatus, the Japanese could produce distilled liquors of high quality.[63] This is the beginning of a new style of shochu made with rotten Japanese sake. After that, Kyushu's traditional method of making shochu based on sweet potatoes gradually declined, reinforcing the new technology. At the end of 1900, people added a special germ to make awamori using a double-filtering process, introducing a new kind of shochu to the Kyushu region. In 1912, a Japanese sake-making company imported an industrial still that could use fermented residues from potatoes, barley, and sugar to make odorless alcoholic beverages.[64] Since then, many Japanese liquor companies have started making shochu and developing high-quality distilled liquors that have earned global recognition to this day.[65]

During the period of Japanese occupation in Korea, the elements of a modern shochu industry continued to transfer from Japan to Korea. Shochu continued to be consumed in Japan, but meanwhile, as Chapter 5 illustrates, soju became the dominant form of alcohol in Korea. However, while traditional forms of Japanese shochu, like awamori, became global brands, no Korean minsokchu earned the fame of many diluted soju brands. In Japan, shochu using traditional distillation methods continued to develop with the help of governmental support, as seen from books published by governmental distillation research centers. For a while during the late twentieth century, there was a boom in shochu production and consumption in Japan, though the spirit did not win the national-liquor title. The Japanese also developed *chūhai*, a shochu highball made with a blend of shochu and a soda and lime mixer. These have also influenced soju in Korea. Nonetheless, its fame has been somewhat overshadowed by the global popularity of sake.

[61] Ibid.   [62] Kim Soyŏng, *Sak'e, Ryu*, 41.
[63] Takahashi, "Nihon no jōryūshu, shōchū," 238.   [64] Sakaguchi, *Nihon no sake*, 76.
[65] Ilbon yangjo hyŏphoe, *Pon'gyŏk soju chejo kisul*, 62.

Mexico's East Asian Stills

Another case of distilled liquor that is becoming a popular non-European spirit in the globalized world, and which possibly shares the same tradition of distillation that developed in Afro-Eurasia, is Mexico's tequila. Mexico is geographically distant from the societies discussed earlier in this book. However, since European expansion to the Americas began, people in Mesoamerica have been interacting directly with people in Afro-Eurasia through the Atlantic trade networks and then through the Manila galleons, and through them have received distillation techniques and technology. Indeed, some scholars even assume the possibility that cross-cultural exchange likely occurred before the Columbian exchange ever began. Against this academic backdrop, let us examine the development of spirits in Mexico as an example of remotely spreading distillation. This will help us to expand our understanding of aspects of long-distance cultural propagation, and to compare them with propagation in Afro-Eurasia on a relatively smaller scale.

Like Japanese sake, tequila and mescal are world-famous distilled liquors, and have already established a firm foundation in the global liquor market on par with whiskey or vodka. By the twenty-first century, the popularity of these distilled alcoholic beverages made from a single species of Mexican agave had spread all over the world, probably faster and more extensively than soju. People know tequila, along with mescal, as a representative distilled spirit of Mexico and often confuse the two. In fact, tequila is a kind of mescal, a distilled spirit made from the agave, which is a genus of flowering plant that grows in Mexico.[66] There are actually hundreds of different species of agave, and most of them can be found only in Mexico. Mescal can be made from any one of these species. Not so for tequila, which is a form of mescal made from blue agave particular to the city (formerly a village) named Tequila – hence the name.[67] The use of term *tequila* – "Vino Mezcal Tequila" – is related to a person named Pedro Sánchez de Tagle, who was one of the first producers among the Spanish elite of early eighteenth-century Mexico.[68] The name *tequila* came to be used as the official name for this

---

[66] This is commonly called blue agave and is officially classified as *Agave tequilana*. See www.pastemagazine.com/articles/2017/04/ask-the-expert-is-mezcal-tequila-is-tequila-mezcal.html.

[67] Anthony Blue argues that some historical sources hint at the commercial production of tequila by the Spanish as early as 1600. Anthony Dias Blue, *The Complete Book of Spirits* (New York: HarperCollins), 112.

[68] Eric Zolov, *Iconic Mexico: An Encyclopedia from Acapulco to Zócalo*, vol. 2 (Santa Barbara, CA: ABC-CLIO, 2015), 592.

kind of mescal only in the late nineteenth century.[69] Don Cenobio Sauza, founder of the Sauza Tequila company and municipal president of the village of Tequila from 1884 to 1885, first exported the local drink abroad, starting with the United States.[70] This established the tequila export business, which received a boost from a newly constructed rail line built to connect Mexico to the United States.[71] One brand of tequila was awarded a prize at the Chicago Fair in 1893. Another brand earned an award at the San Antonio Fair in 1910. Since then, the popularity of tequila internationally has soared. In the 1960s, tequila export to the United States surged.[72]

But what was tequila's history before its international development? When and how did its production begin? Scholars began studying the topic in the early twentieth century. In the 1940s, academics like Bruman undertook the first comprehensive and comparative studies of the origins of distilled alcohols in Mexico. According to their research, the tradition of agave distillation in Mexico dates back to the early years of Spanish colonization, while the brewing of fermented beverages stretches back even further. In other words, Mexican society has long seen the agave as a potential source of intoxicating drink. As with societies in Afro-Eurasia, there are many mysteries surrounding the origins of distilling fermented alcohol in Mexico.[73] Sources for their development before the coming of the Europeans are relatively scarce because the Spanish conquistadors destroyed most of the scant written sources that once existed in Mayan and Aztec societies. Also, because it developed in the New World during its premodern isolation from Afro-Eurasia, it is difficult to follow any hints at possible documentary evidence. However, based on comparative examinations of available sources, several scholars argue that the Spaniards first introduced distilled liquors and the idea of distillation to the Americas. Their arguments are supported by the fact that the Europeans were drinking distilled liquors at the time of their arrival and continued to transport them to the Americas and consume them. Let us review the scholarly hypotheses that attempt to explain the appearance of distillation in Mexico, from the post-Hispanic distillation hypothesis to

---

[69] Yoshida Shūji, "Umi wo watatta jōryūki: Mekishiko no jōryūshu" (Distillers Going Abroad: Distilled Alcohol In Mexico), in *Shōchū higashi mawari nishi mawari* (Shochu around the World), ed. Tamamura Toyo'o 玉村豊男 (Tokyo: TaKaRa Alcohol Beverage and Life Research Institute, 1999), 218.

[70] Blue, *The Complete Book of Spirits*, 112.

[71] Joanne Weir, *Tequila: A Guide to Types, Flights, Cocktails, and Bites* (Berkeley: Ten Speed Press, 2009), 5.

[72] For general histories of tequila, see Ana G. Valenzuela-Zapata and Gary P. Nabhan, *Tequila! A Natural and Cultural History* (Tucson: University of Arizona Press, 2003); and Marie Sarita Gaytán, *Tequila! Distilling the Spirit of Mexico* (Stanford, CA: Stanford University Press, 2014).

[73] Ian Williams, *Tequila: A Global History* (London: Reaktion Books, 2015), 56.

the hypotheses that challenge it, from distillation's pre-Columbian spread to its independent rise from antiquity.

Following Henry Bruman, several scholars, such as Yoshida Shūji, have argued that distilled liquors in Mexico emerged about 500 years ago because we can see little possibility that the people of Mexico produced or consumed distilled liquors before the arrival of the Spanish.[74] They did produce and consume alcohol, however. Natives of pre-Columbian Mexico made two different kinds of fermented beverage based on agave plants prior to European contact. The oldest of these, consumed by nobles, is *pulque*, made from fermented sap tapped from the flower stalk of agave, made using a totally different technology than that employed for mescal.[75] Considered a gift from the gods, pulque was used for religious sacrificial rituals, for medicinal applications, and as a libation to celebrate bravery.[76] The second of these is another fermented beverage processed from agave and used for foods, which was known under the general name *mescal* (the word derives from the Nahuátl *metl ixcalli*, literary meaning "cooked agave").[77] While one simply taps the sap from living plants to begin natural fermentation to make pulque, in order to make the cooked agave beverage, one must first roast and crush the agave in order to release the sugars necessary to produce alcohol.[78] This is still commonly done – just as a sweet food.

Pulque and the cooked agave beverage known under the general name *mescal* soon grew popular among the colonizing Spaniards. Surely Spaniards brought to Mexico the alcoholic drinks, including distilled spirits, that they had developed in their native Spain.[79] However, once European supplies ran out,[80] the new colonists began to consider indigenous sources to supply the production of alternative distillates. Somebody had the bright idea of using the sweet sugar syrup extracted from roasted agave stem bases as a feedstock for distilled liquor.[81] To this, the name *vino de mescal* was given, creating the word that eventually came to denote a whole category of strong alcohol. This began the process of development from which mescal and tequila would ultimately emerge.

Naturally, one assumes that the technology used to produce new spirits in Mexico originated in Spain – or, if not there, then somewhere in

[74] Yoshida, "Umi wo watatta jōryūki," 174.   [75] Ibid., 176–195.
[76] Weir, *Tequila*, 3. Bruman, *Alcohol in Ancient Mexico*, 61–64. For more historical and technical details about pulque, see ibid., 64–82.
[77] Regarding fermented mescal and other aboriginal fermented alcoholic beverages in Mexico, see Bruman, *Alcohol in Ancient Mexico*, 83–98; Yoshida, "Umi wo watatta jōryūki," 198–203.
[78] Williams, *Tequila*, 48–49. Bruman also points out that the sizes of agaves used for making pulque and mescal differ. A pulque agave is larger and has fleshier leaves. Bruman, *Alcohol in Ancient Mexico*, 67.
[79] Yoshida, "Umi wo watatta jōryūki," 200.   [80] Ibid., 200.
[81] About distilled mescal, see ibid., 206–209.

Europe. However, two surprising facts derived from continuing archaeological excavations and anthropological field research in Mexico have
raised doubt. First, thanks to this new body of evidence, scholars have
discovered that, in order to distill the cooked agave, people used not only
European but also Asian-style still technology.

The finding that most shook previous assumptions was the discovery of
East Asian stills used by the Huichol tribe, which the anthropologist Carl
Lumholtz (1851–1922) first reported in the nineteenth century.[82]
Following Lumholtz, further archaeological and anthropological
research, including the most recent article by Ana Valenzuela-Zapata,
suggests that further evidence of these Huichol stills may exist.[83] Second,
based on these findings, some scholars, starting with Lumholtz, have
claimed the possibility of the existence of distillation using East Asian
stills in Mexico prior to European contact. Lumholtz even argued the
possibility of pre-Columbian contacts between Chinese and Mexicans.
He failed to present these stills that he used to support his claim to prove
his theory, yet subsequent studies in the past few decades have begun to
reclaim the possibilities he raised based on new findings.

Before critically reviewing the issues Lumholtz raises, we must first
understand that two methods of distillation could have produced distilled
liquor in Mexico at the same time: first, Spanish colonists could have
distilled the cooked agave using a European distilling method, which then
spread to native Mexicans; at the same time, other Spaniards could have
imported an Asian distilling method from the Philippines to western
Mexico via the Galleon trade, which similarly spread among Mexicans
who used the Asian technology to begin distilling mescal privately and
more easily. Through either method, cooked agave-based alcohol transformed from its originally fermented form into a distilled one.[84] This
theory of a European delivery of Asian stills from the Philippines through
the Manila galleons was first argued systematically by Bruman, who
refuted Lumholtz's theory of a pre-Columbian transfer of Mongolian
technology by studying the introduction of distilled "coconut wine" to
Mexico sometime before the seventeenth century (see Map 6.2).[85]

---

[82] An important thing to note is that Asian stills differ from the European alembic in the
cooling part.

[83] Ana G. Valenzuela-Zapata, "East Asian Stills: Distillation Influences in Mezcal
Production in Mexico," in *Tribute, Trade and Smuggling*, ed. Angela Schottenhammer
(Wiesbaden: Harrassowitz Verlag, 2014), 141–151.

[84] Yoshida, "Umi wo watatta jōryūki," 226.

[85] Bruman, "The Asiatic Origin of the Huichol Still"; Henry J. Bruman, "Some Observations
on the Early History of the Coconut in the New World," *Acta Americana* 2 (1944): 200–243;
Henry J. Bruman, "Early Coconut Culture in Western Mexico," *Hispanic American Historical
Review* 25, no. 2 (May 1945): 212–223; Yoshida, "Umi wo watatta jōryūki," 215.

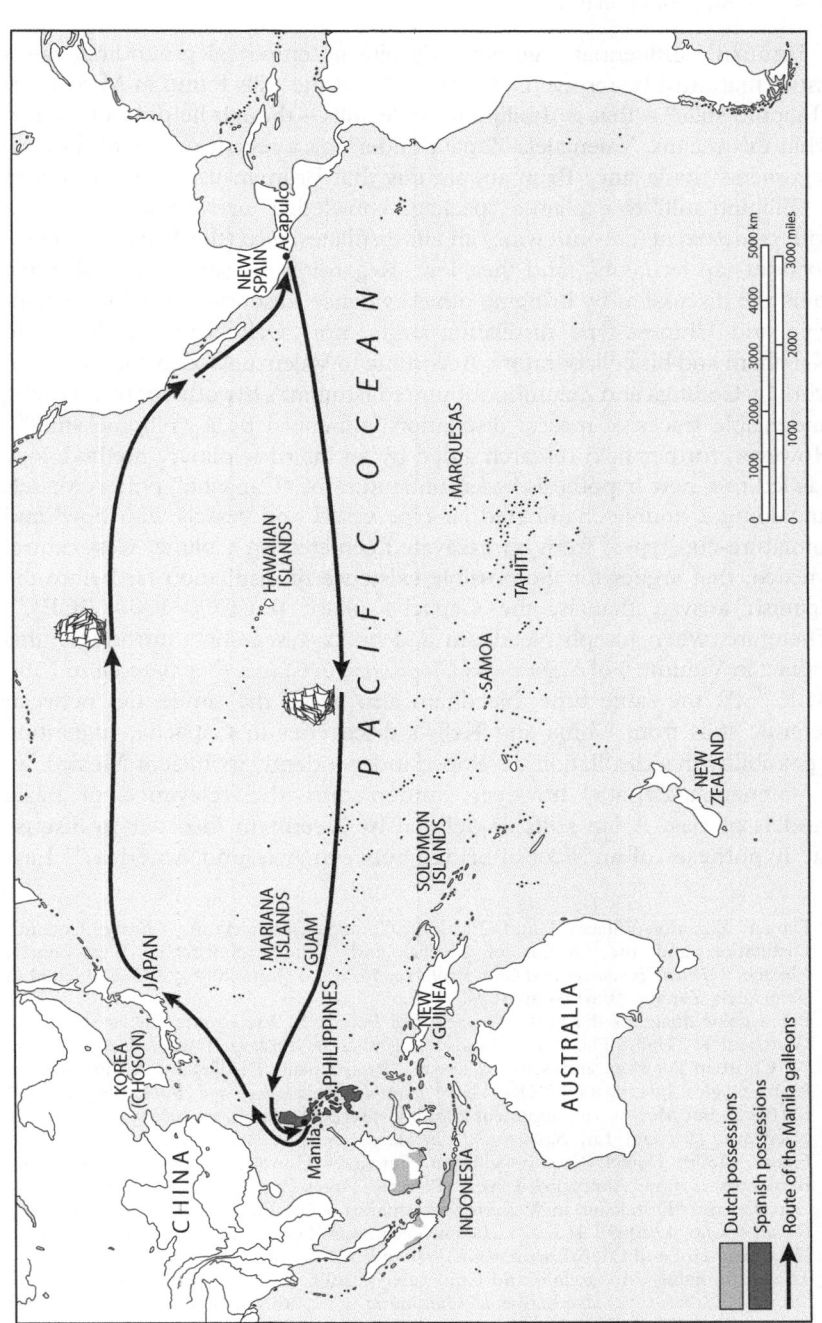

Map 6.2 The Manila galleon route.

Bruman's influential argument, despite its empirical grounding, raises issues that must be resolved. He claimed that the stills found in Mexico are "Filipino stills" – that is, Philippine-style stills – though he did not explain what this means. Valenzuela-Zapata undertook a recent review of the new arguments made since Bruman, arguing that Bruman used the concept of a "Filipino still" to explain a "package knowledge" arriving in connection with *vino de cocos* (coconut wine) and its distillates, *tuba* (the Tagalog term for coconut-sap ferments), and their lore. Regarding the same topic, Bruman joins the discussion by bringing other evidence, discussing his Mongolian-type and Chinese-type distillation units, now well classified thanks to Needham and his collaborators. According to Valenzuela-Zapata, the recent work by Colunga and Zizumbo supported Bruman's hypotheses by revealing undeniable traces of mescal distillation influenced by a "Filipino still."[86] However, further field research aided by an interdisciplinary methodology has led to a new hypothesis based on studies of "Capacha" pottery vessels (including a double-chambered jar-type vessel and vessels with bowl and miniature-cup types) from an excavated cemetery in Colima, west central Mexico, that argues for the possible existence of distillation far before the Spanish arrival; that is, the Capacha phase (*c.* 1500–1000 BCE).[87] Therefore, when Joseph Needham and his co-researchers introduced this artifact in Volume 5 of *Science and Civilisation in China*, they dated it to 1500 BCE.[88] At the same time, Needham also noted the similarities between ceramic stills from China and Kelly's discoveries in Capacha, suggesting a possibility that distillation developed independently in ancient Mexico.[89]

Valenzuela-Zapata, however, underscores the relevance of basic models of East Asian stills as defined by Needham in order to discuss the hypotheses of an alcohol bridge between Asia and America.[90] East

---

[86] Daniel Zizumbo-Villareal, and Patricia Colunga-GarciaMarin, "Early Coconut Distillation and the Origins of Mezcal and Tequila Spirits in West-Central Mexico," *Genetic Resources and Crop Evolution* 55, no. 4 (June 2008), 493–510, cited in Valenzuela-Zapata, "East Asian Stills," 141.

[87] For a color image of the pottery vessels, see Patrick E. McGovern, Fabian H. Toro, Gretchen R. Hall, Theodore Davidson, Katharine Prokop Prigge, George Preti, W. Christian Petersen, and Mike Szelewski, "Pre-Hispanic Distillation? A Biomolecular Archaeological Investigation," *Open Access Journal of Archaeology and Anthropology* 1, no. 2 (2019): 2. For McGovern's argument in his most recent research, see below.

[88] Needham, Ho, and Lu, *Science and Civilisation in China*, vol. 5, part 4, 109–110, Figure 1485b; Daniel Zizumbo-Villareal, Fernando González-Zozaya, Angeles Olay-Barrientos, Laura Almendros-López, Patricia Flores-Pérez, and Patricia Colunga-GarciaMarin, "Distillation in Western Mesoamerica before European Contact," *Economic Botany* 63, no. 4 (2009): 413–426, cited in Valenzuela-Zapata, "East Asian Stills," 141.

[89] Needham, Ho, and Lu, *Science and Civilisation in China*, vol. 5, part 4, 110.

[90] These are mainly Mongolian- and Chinese-style stills, though Valenzuela-Zapata also discusses Indian-type, also known as Gandharan-type, stills. Valenzuela-Zapata, "East Asian Stills," 143–145.

Asian stills, constructed on the principle of vapor condensation, existed in diverse forms throughout the region until Islamic alembic forms supplanted them. Needham also suggested that someone brought East Asian stills from either China or the Philippines. He even advances the possibility of a wrecked junk landing in Mexico.[91] In his book, published in 2000 and based on his dissertation written several decades earlier, Bruman did not cite Needham directly but responded indirectly to Needham's categorization through an illustration in which he compares distillation technology found on both sides of the Pacific for the first time, including Mongolian, Filipino, and Huichol stills.[92]

Against such a backdrop, Valenzuela-Zapata brought new findings from her anthropological fieldwork that would enhance this developing scholarship further. She argues that we should re-examine earlier discussions of East Asian stills used on Mexico's Pacific coast using new interdisciplinary approaches involving ethnobiology, anthropology, and history. Because many people in Mexico's western regions distilled alcohol illegally at the time (now legally as mescal and *raicilla*), Valenzuela-Zapata began investigating "mescal moonshine" practices to compare them with Asian stills. With empirical data from Jalisco in western Mexico acquired from field research and interviews, Valenzuela-Zapata concluded that many Mexicans in local regions have been distilling mescal using Chinese-type stills. While she was not able to see them with her own eyes, she nonetheless has been able to verify the existence of Mongolian-type East Asian stills based on the pictures and descriptions produced by individuals working with native cultures like the Wirraika (Huichol) and Cora.[93] In her most recent article (co-authored with Paul Buell, María de la Paz Solano-Pérez, and myself), Valenzuela-Zapata confirms the influence of Asian stills on distillation development in Mexico with Chinese and Mongolian types.[94] By examining various East Asian stills in Mexico, she discovered that, even among the Huichol tribe, the Mongolian style that Lumholtz identified gradually disappeared, and the Chinese stills with their more complicated features grew to prominence in their stead. She provides three corroborating scenarios to explain ancient distillation in Mexico and its Asian influence, concluding that, while the possibility exists of ancient proto-distillation

---

[91] Yoshida, "Umi wo watatta jōryūki," 215–216. Needham discusses the possibility of the arrival of a wrecked junk from China landing in Mexico. Needham, Ho, and Lu, *Science and Civilisation in China*, vol. 5, part 4, 110.

[92] Valenzuela-Zapata, "East Asian Stills," 149; Bruman, *Alcohol in Ancient Mexico*, 6; Yoshida, "Umi wo watatta jōryūki," 211–212.

[93] Valenzuela-Zapata, "East Asian Stills," 148–151.

[94] Valenzuela-Zapata et al., "'Huichol' Stills."

knowledge in the region as a kind of "Capacha souvenir," we cannot deny that the extant sources suggest that Asiatic stills influenced distillation in western Mexico after the Spanish conquest.

While Valenzuela-Zapata argues, "The alcohol bridge that exists between Asia and America supports hypotheses in a history that still has to be written,"[95] new research by McGovern and his colleagues revisits another hypothesis related to the topic: the possibility of independent development of distillation in Mexico from ancient times, which Needham proposed by putting a possible date for Capacha pottery as far back as 1500 BCE.[96] Using the groundbreaking method of an experimental archaeological approach thanks to modern advanced technologies, McGovern and his colleagues attempted to test the ancient-origin theory by applying biomolecular archaeological investigation to the excavated pottery. By distilling fermented alcohol using replicas of the ancient pottery vessels and comparing molecules (i.e., biomarks) of the ancient original vessel and the replicas, they found out that, while it is indeed possible to distill alcohol using the double-chambered jar-type vessel, there is no sign that this vessel was used for distillation. Because there have also been no archaeological findings for "an alembic-like device" to date in addition to the burial goods, McGovern concluded that the correspondence between the evident knowledge of steaming and/or infusing natural products among people in Mexico with available evidence for early China points to the independent invention of a distillation apparatus there. The current lack of evidence from archaeological, archaeobotanical, and/or chemical studies prevents the development of any hypotheses, so concrete conclusions are yet to be made.[97]

All of these claims discussed above have their pros and cons. While Lumholtz and Needham suggested the possibility of East Asian stills transferred before the Spanish conquest, they did not provide concrete pieces of evidence and therefore were refuted by other scholars like Bruman and Yoshida. The most recent discussions of the origin of spirits in Mexico by Valenzuela-Zapata, McGovern, and their colleagues present some new pieces of evidence that mobilize support for these earlier claims of pre-Hispanic origins of distillation (both the pre-Columbian-transfer hypothesis and the independent-rise hypothesis) and suggest that the discussion is still open to new evidence. However, while we await new textual or archaeological evidence, we can make comparisons with the case of Korea that shed light on the process of distillation technology transfer in Korea and Mexico.

[95] Ibid., 180.   [96] McGovern et al., "Pre-Hispanic Distillation?", 1–13.   [97] Ibid., 11.

Figure 6.1 A Nahua still, south of Jalisco, Mexico. Courtesy of Ana Valenzuela-Zapata (pictures) and María de la Paz Solano-Pérez (drawings).

It is difficult to see such a possibility of the pre-Columbian transfer of the "Huichol still" discussed by several scholars, starting with Lumholtz. There is also no way to confirm that distillation developed independently in Mexico without additional evidence. However, putting aside this question of pre-Columbian influence, it is still possible that both Mongolian and Chinese types of still, both of which developed in medieval East Asia – in other words, before 1492 – found their way to the Americas after 1492 because, having spread widely across East Asia during the Mongol period – that is, the thirteenth and fourteenth centuries – they began to be transported overseas, starting with Southeast Asia. We can imagine a scenario in which East Asian stills migrated overseas, most notably to South Asia and then Mexico aboard the Manila galleons. As the research by Valenzuela-Zapata and her colleagues shows, some stills discovered in Tuxpan, Jalisco in Mexico, show surprising similarity in form with Chinese-style stills developed in Korea (see Figure 6.1; compare Figure 4.1).

As I argue elsewhere, the resemblances are too close for them to be coincidental.[98] While there is no possibility of direct influence between them, we can assume indirect connections; that is, stills that were transferred to Korea from the thirteenth century and Mexico from the sixteenth century might have evolved in the two different regions in accord

[98] Valenzuela-Zapata et al., "'Huichol' Stills," 185–186.

with a similar process. As Chapter 4 shows, in Korea, a simple still that Needham called the Mongolian type appears earliest in documentats; after that, the more complicated Chinese style – that is, the one that uses a tube to remove alcohol from the pot – became the primary still during the late Chosŏn dynasty, though the primitive still found use in some areas. We can assume that, first, the more primitive-style stills transferred from the Philippines across the Pacific because they were easier to use; however, a few Chinese-style stills were transferred together at the same time. The latter gradually grew more popular in Mexico, as happened in Korea, once people grew accustomed to the technology.

Of course, we cannot deny the independent development of stills in Mexico, for in Chapter 1 we see similar cases in China and elsewhere. Yet, as we discussed earlier, it is more important to determine when distilled alcoholic beverages – ancient distillation being limited in its application to mercury and rose water – began to be popularized. In the case of Korea, we have detected concrete evidence of popularization that narrows the time to sometime during the Mongol period, so the country offers some good insights from a comparative perspective.

In Korea, there was a long tradition of fermenting alcohol that dates back to ancient times, but there is no trace of making spirits until the idea of distillation arrived from China during the Mongol period, along with other cultural ideas and goods. Once soju as distilled liquor became popular in Korea, it began to be localized to meet the society's specific needs. For distillation in Mexico, as McGovern and his colleagues argue, the possibility of its ancient independent rise is still open, awaiting further evidence. However, we should recognize the enormous impact of new distillation techniques, such as Mongolian- and Chinese-type stills, that influenced native Mexicans, who have enjoyed a long tradition of fermenting alcohol, and who might have already possessed ancient distillation skills. Whatever the case, the Mongolian and Chinese technology migrated along with a good deal of Filipino and south Chinese culture by way of the Manila galleons, along with other cultural items such as costumes, foods, language (Acapulco dialect), and genetics (evidence of many Filipino genes in the Acapulco area). Just as Mongol soldiers and others traveling between China and Korea during the thirteenth and fourteenth centuries brought distilled liquors and their technology to the Korean peninsula and made it popular among the locals, Spanish soldiers and Asian immigrants (the majority of whom were slaves) also introduced from the Philippines the Mongolian and Chinese alcohols and stills that were already circulating throughout the Indian Ocean during the sixteenth century to Mexico, where they became popular. Indigenous people quickly assimilated Asian distillation notions, not unlike other cases we have seen; this is evident in conquerors' notebooks, which record their excitement at the ease with which Mexican indigenous groups assimilated and adapted new technologies and processes.[99] Having adopted and popularized it, locals began to experiment, adapting portable Asian techniques for the distillation of rice to agave privately, adjusting to novel environmental and cultural conditions when necessary.

This raises another question: how did the East Asian stills get to the Philippines in the first place? Various scenarios will have to be developed. As the Mongols made possible the spread of East Asian stills to such places as Korea, and even Japan through its Korean and Chinese connections, it is possible that East Asian distillation technology spread westward across such regions as Central Asia, which has already been proven by several research projects, and southward to Southeast Asia, which has yet to be conclusively shown. In the case of Korea we see rich documentation of the transfer. Since other regions are not well documented, we can use the case of Korea to draw an analogy between similar cases. While we do not have direct evidence for transfers of East Asian stills to Southeast Asia, we can assume so based on historical context. As the Mongol-ruled Yuan dynasty collapsed in 1368 and was replaced by the native Chinese Ming dynasty, many foreigners, including Muslims, moved to Southeast Asia, as did Chinese merchants seeking commerce and fleeing the anti-trade policies of the new dynasty. It is highly likely that the form of East Asian distillation technology developed in China during the Mongol period was brought back to Southeast Asia. This East Asian distillation technology could have then been transferred to Mexico via the Philippines aboard the Manila galleons, convoys conducted by the Spanish since 1500, leading the way to change in the production of fermented drink in Mexico. In this process, both Mongolian and Chinese methods migrated to Mexico together, and it is highly likely that the more complex and sophisticated Chinese-style stills became popular over time. We can also assume that, as Anderson argues, the Chinese-type stills came much later, with the enormous immigration of Cantonese to Mexico in the late nineteenth century, which also brought *saladitos* (salted Chinese apricots, *syun mui* in Cantonese, *mai* in Korean – now universal in Mexico) and popularized such items as chop suey and chow mein in Mexico.

At any rate, many sources and studies related to the world-historical development of this distillation suggest two major starting points as watersheds of great change. One is the Mongol period in the thirteenth and fourteenth centuries and the other is the expansion of Europe from

---

[99] Special thanks to Ana Valenzuela for this information.

the late fifteenth century. First, both brought significant changes in the development of distillation in the societies in East Asia. Of course, in the case of China, Korea, and Japan, the Mongol period is an important starting point. At this time, distillation spread widely in China, and spirits were first introduced in Korea and later in Japan, leading to significant changes in the culture of alcohol there. Then, as European nations arrived in East Asia, they introduced modern distillation based on Islamic alembic technology to countries there, which affected the development of modern spirits locally. The case of Mexico provides another example of larger-scale changes influenced by change in Afro-Eurasia. Here, East Asian distillation technology that had developed during the Mongol period spread to Southeast Asia; from there, Europeans introduced it to Mexico and it developed even further.

In the case of Mexico, as in Korea, the new forms of distillation technology popularized quickly once agents had introduced them from abroad, and thereafter greatly influenced the alcohol culture of the whole society. This change in drinking culture should be included among the many major changes that Europeans introduced to Mexican society. In a recent article, María de la Paz Solano-Pérez supports this view, with support from sources written in the sixteenth century. Her study shows that the pre-Columbian societies of central Mexico drank only fermented beverages, often with enhancing additives, usually at rituals in which they tried to achieve a state of ecstasy and make spiritual contact with the divine. She observes, "Very few were privileged enough to participate." Pérez adds that, in the centuries that followed the conquest of what became New Galicia, part of New Spain (the conquered territorial entity of the Spanish Empire during the Spanish colonization of the Americas, 1521–1821), liquor prohibitions came and went, and fluctuated in severity, especially during the seventeenth and eighteenth centuries. Distilled-beverage consumption affected not only policy but also social behavior. For example, the consumption of new distilled beverages offered opportunities to obtain income to support the development of New Galicia's capital, Guadalajara. Older forms of fermented drink did not disappear, but simply faded into the social background. However, in some places, they lost their sacred meaning from pre-Columbian times. Profanity prevailed, however. Pérez documents this evolution from the sacred to the profane and identifies the social forces promoting it.[100] Such

---

[100] María de la Paz Solano-Pérez, "The Early Path, from the Sacred to the Profane in Fermented Beverages in New Galicia, New Spain (Mexico), Seventeenth to Eighteenth Century," *Crossroads: Studies on the History of Exchange Relations in the East Asian World* 14 (2016): 219–256.

phenomena were the hallmark of changes in drinking culture that occurred in the wake of European conquest.

Some mescal traditions still contain aspects of Asian influence that affected the early development of tequila. These spirits of Mexico, like the soju of Korea, have entered the global economic arena. Most tequila is now manufactured using modernized machines. Yet many Mexicans still make spirits using traditional methods and distillation technology. We should not forget the development of premodern distillation and its roots in Afro-Eurasia. The Korean and Mexican cases illustrate clearly that the histories of the world's distilled liquors are interconnected, directly and indirectly, into a single global history.

Conclusion

In this chapter, we examined two cases of distillation development, Japan and Mexico, which are connected to the history of Korean soju to varying degrees. The case of Japan demonstrates a possible case of direct connection, so that Korean soju exerted direct influence once it had spread to the islands. The case of Mexico developed as a consequence of indirect connection with Korea through Chinese trade, the Philippines, and the Manila galleons. Indeed, the two cases share similar traits in their post-transfer development. Amidst the many theories about the origin, propagation, and development of Afro-Eurasian spirits in the premodern era, there lurk many unresolved mysteries about its development in places like Japan and Mexico that continue to attract the attention of scholars. Korea provides a novel point of comparison in this research. The findings on Korea presented in the book's other chapters help to shed new light on previously studied countries like Japan and Mexico.

A comparative examination of various sources available in Korea and Japan yields concrete pieces of evidence of soju's transference from Korea to Japan after it settled on the Korean peninsula and Cheju Island in the thirteenth century. Of course, distilled liquors and distillation technology probably traveled directly to Japan from China and Korea through trade in commercial goods, including liquors and books. Some Japanese scholars assume that distilled liquors and distillation methods were brought directly from Thailand to Ryukyu.[101] Bearing in mind this direct Southeast Asian connection, it is also clear that Europeans brought distilled liquors from Southeast Asia on their way to Japan, some of it produced at home through distillation techniques developed in Europe and based on Islamic alembic technology. All these transfers occurred

---

[101] Shurui sōgō kenkyūjo, *Umai sake no kagaku*, 42, 48.

after the Mongol period. Yet unlike in the case of Korea, once transferred, distilled liquors did not spread throughout Japan as fast as they did throughout Korea. The Japanese already had a well-established fermented sake industry, and this left demand for distilled liquors low. Distilled liquors flourished instead on the Ryukyu islands thanks to state sponsorship, and on Kyushu island thanks to its geographical proximity to Ryukyu, not to mention the conditions in some regions that hinder the sake brewing process, which requires cold temperatures.[102]

We must approach the relationships between Korean soju, and the various Mexican spirits, such as mescal and tequila, from a comparative perspective, but also in terms of a possible indirect connection with soju, and other Asian distillates. The only possible direct connection is through Korean stills, and the traditional stills found in Mexico, in terms of the East Asian stills which spread to Korea and Southeast Asia probably by the end of the Mongol period, and were later transported to the Americas aboard the Manila galleons. This includes the so-called Mongol type which is also unexpectedly found in Mexico as well as in Asia. The fact that more sophisticated Chinese stills grew dominant over time holds true for both places, suggesting that the spread of the distillation process in the two parts of the world developed along similar lines with similar experiences. In this book we have only compared basic traditional distillation techniques using different types of still in Mexico and Korea. A deeper study of the entire technology, including the fermentation of the raw materials, and their subsequent distillation, examining the well-documented case of Korean soju in terms of recent archaeobotanical discoveries in Mexico, might provide still more points of comparison in the future and a better understanding of what took place.

In modern Japan, sake has become more important than distilled spirits. However, Japan also experienced great booms in distilled spirits during the late twentieth century, and the most thorough research on distilled spirits has been conducted in Japan rather than Korea. Though less prominent than sake, shochu is also being promoted by Japan globally, and awamori shochu from Okinawa is now world famous as a spirit distilled using traditional methods. In Mexico, the tequila industry continues to grow long after Europeans introduced spirits to the country. Essentially, Japanese shochu and Mexican tequila are joining an emerging global spirits industry whose community is quickly expanding beyond the long-known Western varieties and welcomes Korean spirits profiled in this book as well as Chinese recipes targeting global markets but rooted in Chinese traditions. To this we can add other spirits that share soju's

---

[102] Hishinuma, *Sake*, 323.

ancient origins, like arak and arkhi in Southeast Asia, Central Asia, the Middle East, and the Mediterranean.[103] We can assume the possibility that East Asian-style spirits that developed after the Mongol period influenced the development of distillation cultures in these regions, through Mongol soldiers, merchants, and other key agents of cultural exchange.

The cases of Japan and Mexico, seen in light of the case of Korea, provide evidence that strongly suggests that distilled spirits in general spread worldwide rather suddenly in the wake of the Mongol era, enough to justify further research to fully confirm this hypothesis. In other words, the spirits popularized during the Mongol period spread to both the east and the west. This is an example of one cultural item developing through one system of exchanges and then moving further into the larger global system. In other words, the alembic tradition in West Asia, which may have affected Asian medieval distillation through interaction during antiquity and the Middle Ages, may have been stimulated by the spread of Asian-style distilled liquors after the Mongol period ended and developed into the arak tradition of Southeast Asia, Central Asia, the Middle East, and the farther Mediterranean. This alembic distillation affected Europe at the end of the Middle Ages. We should look for a more direct connection, but it is difficult to deny a certain connection. European distillation then affected distillation in Asia, including Japan and Korea, for the modern diluted shochu/soju. The distillation of Afro-Eurasia influenced the development of spirits in the new continent, mainly in Mexico, in the large flow of distillation technologies. The modern flow of spirits that connect world societies will need further research in the future to fully develop. The book's study stops here, with the expectation that this work will be done.

---

[103] An important topic that was not attempted in this book but deserves future study is the further and rapid development of arak in other parts of Afro-Eurasia, including Central Asia, Southeast Asia, the Middle East, and Europe, after the Mongol period.

Conclusion
Soju in Global Cross-cultural Exchanges

We have so far explored a rich history of soju that involves its growth – first locally, then regionally, then globally – as an alcoholic beverage of choice made possible by the development of distillation technologies and their transfer through cross-cultural exchange. In the twenty-first century, soju has expanded beyond its longtime Korean horizons and entered the global market, where it has successfully won brand recognition and gained acceptance among the world's universally popular spirits. This is taking the drink beyond not only its traditional economic sphere but its cultural sphere as well.

This, then, makes it an appropriate time to answer the often-asked question I have avoided until now: what is soju? That depends on the context. Strictly speaking, the kind of soju that has become popular in the global marketplace consists of nothing more than industrially produced soju concocted by mixing water, flavors, and ethanol spirit based on potatoes and tapioca, a recipe devised in Korea only in the last century or so thanks to the introduction of modern Western technology by way of its colonial occupier, imperial Japan. One could understandably dismiss this as fundamentally different from traditional soju. However, recent varieties are not without deep roots in tradition because the history of soju is also firmly set within a geographical context that is much wider than Korean horizons, as this book has shown. A similar process of large-scale economic integration, made possible by the creation of a Eurasian empire under the Mongols more than seven centuries ago, created the conditions for a transfer of tastes and technologies that made possible the local development of what we now call traditionally distilled soju. Once Koreans developed the know-how to make distilled spirits according to local tastes, they created soju, and for the next several centuries, from the late Koryŏ to the Chosŏn era, it grew in popularity on the Korean peninsula to ultimately become an important national drink, which in turn laid the foundation for its mass production in the twentieth century and globalization in the twenty-first.

The process is so complex and lengthy that it deserves restating: *both* the development of Korean spirits like soju and their transformation into products of global mass consumption depended fundamentally upon the development and propagation of spirits and distillation technology across Afro-Eurasia that began in ancient times; the creation of local varieties like shaojiu and arak using the technological innovations of Chinese, Arab, and other cultures; the transference of distilled spirits and their technology to the Korean peninsula thanks to a brief period of regional peace and intimate Korean–Mongol ties; a centuries-long phase of local development; and the move to mass production and global markets that began with the Japanese occupation and ends with the Korean Wave today.

Exploring a history of soju is much more complicated than simple engagement in a kind of storytelling about development, characteristics, or recipes. This exploration ultimately seeks to understand the processes that made these things possible. Doing this involves detective work that seeks to identify traces of possible transfer and influence and their trajectories across large expanses of time and space using the few remaining clues at hand. This is the task that anyone dedicated to premodern cultural history must assume. In this endeavor, lack of sources is a given, and this leaves unavoidable gaps in our understanding of the premodern development of distilled liquors. Nonetheless, we persist in reconstructing the past more credibly than at the last attempt with the few fragments that we do have. And here, additional methods such as comparative analyses can help. This helps to elucidate the links between ostensibly separate developments in distillation in all the regions of Asia – not just East, but South, Southeast, and West, as in the case of arak – that suggest connections between premodern Korea and Asia through cross-cultural contacts. By extending our comparisons even further beyond Korea, from Asia into Mexico, as this book has done – in other words, beyond soju and arak to tequila – we can also see that soju's early history connects more closely than we realized to developments not only in western Eurasia but also in the Americas, confirming the need for a global history of distillation as a whole.

This book has confirmed that even highly local cultural items like soju are linked to bigger flows of world history, and we can identify the links between the two by using interdisciplinary methods already tested in historical subfields like food or science and technology. Even those who disdain soju or eschew alcoholic beverages altogether can gain a multifaceted humanistic understanding by observing the complex processes underlying the dynamic transformations of spirits, which mankind has deemed important to such a variety of purposes from ancient times to the present. This book's conclusion will briefly return to the key points made in its chapters, reflecting on the case of this Korean cultural artifact to underscore both the importance of exchange in world history and the yet unfulfilled potential of cross-cultural comparison and connection as a method for revealing it. From this perspective, we can anticipate future directions of related research.

Rewriting the Global History of Distillation: The Importance of Cross-cultural Contacts and Comparative Approaches

Korean soju, not to mention other Korean alcoholic beverages, has been overlooked by historians who specialize in the comparative study of liquor in world history, so it has not found its way into Western-language scholarship until now. By briefly surveying the general history of Korean alcoholic beverages, this book shows that they developed not in isolation, but through exchange relationships with other societies like China since ancient times. Indeed, the rise of distilled spirits in China during the era of Mongol rule profoundly influenced new breakthroughs in the history of Korean alcoholic beverages by introducing distilled spirits like soju and arak. This demonstrates the relatively speedy transfer of distilled liquors and the technology to make them from one society to another as a consequence of Afro-Eurasian cross-cultural contacts that reached a peak of scale and intensity during the Pax Mongolica of the thirteenth and fourteenth centuries. By drawing this developmental connection between a specific cultural product, soju, and larger historical processes, the story of Korean soju takes on a new aspect that fills an important gap in the scholarship on alcohol in world history and a new perspective on the origins and spread of certain liquors.

At the same time, we had to ask why these did not develop easily in many places from ancient times before we could examine the spread of distillation and distilled liquors. The answer is simple. In order to make distilled liquors, people first had to discover the distillation principle that the alcoholic content of a base fermented wine could be increased by separating it from other contents by applying different boiling points. From there, they could devise efficient methods and technologies like the distillation apparatus called the still. This distillation idea may be simpler than most scientific theories today; nonetheless, people needed to conceptualize appropriate scientific and technological forms of knowledge to even conceive of it. Several pieces of evidence suggest that people invented basic methods for the distillation of water or mercury in several places in the ancient world, including China and Mesopotamia, as early as 3500 BCE, yet it is difficult to find similarly early evidence of innovation in making alcohol using the same basic distillation method or of distilled liquor's popularization. The first documentary and archaeological sources that prove the full-scale popularization of a spirit or the transfer of its technology are found relatively late; that is, in the thirteenth and fourteenth centuries, in the age of the Mongol Empire during which Korean soju first took form. The Mongols played a major role in spreading the distilled liquors produced by this distillation method to various regions of Eurasia on a large scale, along with other cultural items. In the course of their political expansion across Eurasia, the Mongols needed to preserve kumiss, the fermented milk of mares and cows, which they had been consuming for various reasons for a long time. To solve this problem, they adopted distillation methods from two sources: China, whose stills produced shaojiu, and West and South or Southeast Asia, whose people used another method – possibly related to the alembic – to make arak. (Interestingly the Chinese referred to arak as "Southern Barbarian shaojiu" (Nanfan shaojiu), an indication that its know-how had been transferred to China from foreign countries by way of the country's southern maritime routes.) The linguistic roots of the name *arak* reflect the spirit's historical roots, which likely connect back in time to distillation traditions in West Asia. Looking beyond, we also can detect similarities in distillation methods between arak and techniques found in the West. Before the Mongols conquered China, people consumed both Chinese shaojiu and foreign arak in some regions in the Middle Kingdom. However, it was the Mongols who promoted both distilled spirits and the technology to make them, both in China and also in neighboring countries like Koryŏ (not to mention possibly other places to the west). Therefore, we have seen that the spread of these distilled liquors to Korea is significant for more reasons than just a transfer from China to Korea, for the story of soju's transference suggests a much more complex process created by multiple Eurasian connections beyond simple Sino-Korean interactions. It would have been reasonable to perceive such a dynamic and complex process of distillation-technology transfer to various places throughout the Mongol Empire where active cross-cultural exchange flourished, yet Korea offers a concrete case study that proves such a transfer and also shows the possible modes and trajectories of distillation's spread to other regions. Before Korea's direct relationship with Yuan China formed in the thirteenth century, one finds no trace of distilled spirits recorded in Korean sources; their appearance at this point in time suggests that this active Sino-Korean exchange relationship under Mongol auspices played a decisive role in the rise of this new form of liquor and its technology in Korea. The similar form of distillation's rise in Mexico three centuries later, as told in Chapter 6, raises the possibility that similar patterns of propagation developed in other areas of world history.

The Mongols had already promoted cultural exchange among diverse societies linked together by their unprecedented mass conquest of Eurasia, and this included food, fermented alcohol, and distilled spirits.

Getting the complete picture of the spread of soju as a cultural-exchange phenomenon thus required a tight focus on the relationship between the Mongol Empire and Korea, as in Chapter 3. Shifting from the grand to the small scale makes it easier to see how the Mongol conquest of China and then Korea speedily facilitated a long-term flow of soldiers, merchants, and intellectuals that in turn facilitated a variety of exchanges such as goods exchanged through tributary and commercial channels, not just alcohol products but also, as importantly, books, which helped to disseminate expertise in distilled-liquor production and speed the popularization of this new form of alcohol in places like Korea where it had not been consumed before. By the time the Chosŏn dynasty replaced Koryŏ in the wake of the Mongol collapse in China, soju was finding its way into numerous writings that benefit us today, because they verify that its consumption had been normalized among elites and that it had become a regular feature of tribute and trade goods in Korea's politics and diplomacy on quite a large scale. After that, soju continued to spread throughout Korean society. Thanks to its preservability, soju was able to replace ch'ŏngju in summer rituals. By adopting modern technology for mass production on a large scale, soju became a cheap liquor for Koreans to consume, leading it eventually to become the most representative and popular liquor consumed by people at all levels of society in Korea, and even grew widespread abroad as a companion to Korean food. Indeed, the speed with which distilled liquor spread to a new society, how it transformed to assume local characteristics, and how it transformed local culture as it became an important part of Korean culture generally, offer historians a useful example of the great impact that such a propagation of a cultural technology can have on a society. There are many cases in history of a cultural item that developed over a long period of time (or sometimes did not develop at all) and then suddenly transferred over a relatively short period of time from one culture to another through exchange before quickly gaining influence in the recipient society. So it is with the spread of these distilled liquors. With sufficient research on the development of distilled liquors through the prism of cross-cultural exchange, seen from a global perspective, the processes of technology transfer will become clearer. This case study of Korean soju will hopefully contribute to that effort.

The many extant records of the Chosŏn dynasty allow a much clearer picture, in which we can see how a newly accepted form of alcoholic drink assimilated into Korean society and became widespread in popularity as it assumed local traits, continuing along this line of development to become, in modern times, one of Korea's cherished national drinks. This localization occurred in response to the culture of Korea, sparking changes in technology such as types of still, ingredients, recipes, and usages in accord with local exigencies. By the turn of the twentieth century, a new distillation method that utilized Western continuous stills in factories was introduced to Korea under the auspices of Japanese colonial rule, and the modern industrially produced soju that resulted has continued to develop ever since, even since independence in 1945. To put it another way, the very process by which soju developed into an identifiably Korean distilled liquor consumed worldwide thanks to globalization is the result of a long-term series of cultural exchanges. Since the early twentieth century, modern industrially produced soju continued to develop thanks to continued investments of newer technologies and organizational systems by large soju companies; contrary to expectation, traditional alcoholic beverages also turned to modern technologies to develop in order to survive in the contemporary world, which in the long run led to its revival as a national folk liquor (minsokchu). In the twenty-first century, various Korean distillers have grown confident enough to attempt to compete in the international arena with other more famous distilled spirits. In the case of distilled-liquor development, an important phenomenon is revealed: that to the human interactions among different societies that are central to world history we must add the various technologies and cultural items that people have exchanged and developed in the course of these interactions, a fact that is best explored from a comparative perspective.

More ambitiously, this book has tried to show how the developmental history of cultural goods such as soju links to a greater flow in world history. The various distilled spirits that have developed down to the present have done so in different ways in different regions, but within their variety one can still detect many components suggesting shared origins, making an important point about research methods in the field of global history. The cases of both Japan and Mexico studied in the final chapter show that exchange relationships were long capable of producing both short- and long-distance transfers of distillation technology. The case of a transfer from Korea to China, and then to Japan, appears very likely, and is easy to accept. It may be more difficult to grasp that tools and techniques of spirit manufacturing long rooted in Afro-Eurasia, which influenced soju in Korea, could have spread all the way to Mexico through a network of connections that had already reached a truly global scale in the early modern period.

If such is true for the Americas, then we must also consider the impact of Asian-style distillation on the development of Russian vodkas and other European spirits (at first even called arak). The full extent of arak's spread and development should be examined more thoroughly, and for this, the case of soju studied here could provide a useful point of comparison. Nowadays, arak enjoys popularity, and has established a place as an important alcoholic beverage in many parts of Afro-Eurasia, including Southeast Asia, Central Asia, and the Middle East, despite regional religious constraints such those imposed by Islam. The influence of distilled-liquor exchange on other kinds of beverage would also provide better insight into the world-historical significance of both.

Traditions, Transfers, Localizations, and Innovations

One of the most important features of soju's spread is the interaction between tradition and innovation in the process of a technology's transfer and its localization. The case of soju shows how a newly arrived cultural item, while maintaining some of its original form, can transform in the course of its assimilation as it adapts to its new environment and continues to change over time.

The records of the Koryŏ dynasty bear witness to two facts, that the people of the Korean peninsula had developed a variety of alcoholic drinks before the kingdom's engagement with the Mongol Empire, and that people continued to introduce many new kinds of liquor from China. As the Koryŏ and Mongol eras neared their end, distilled spirits such as arak and shaojiu began to appear on the peninsula and then spread rapidly, creating a great impact on Korean society. Shaojiu first developed in China during the Song period thanks to distillation methods there, while arak based on "foreign" (Chinese texts refer to it as "Southern Barbarian") or non-Chinese distillation methods found their way to the country by way of people traversing China's maritime routes. Flourishing in an environment in which the Mongols promoted distilleries, the two forms of distilled spirit grew popular in China and began to influence each other. The Mongols adopted pre-existing distillation methods to distill their favorite fermented drinks, including kumiss, into arkhi, which first appears in a Mongol court culinary book dating to about 1330 but soon became almost universally used as a word for distilled liquor (and remains so today, in a number of linguistic variants, throughout much of Afro-Eurasia). At the same time, Chinese distillers developed a form of shaojiu based on the grain-fermented alcoholic drinks traditional to their culture, which many would often identify as arak in the following Ming and Qing periods of Chinese history.

Distilled liquors of the Mongol Yuan dynasty were transmitted to Korea on a large scale through channels of unprecedented cross-cultural exchange between Korea and China, and they also underwent localization, a process that continued into the Chosŏn period and which incorporated many innovations in use of ingredients, still technology, and drinking culture. The inhabitants of the Korean peninsula made soju using fermented-grain wines based on rice and barley, which were popular there. In contrast, people on Cheju Island, where it is difficult to produce rice, devised a soju that uses fermented barley and foxtail millet. The Koreans also developed their characteristic still called the soju kori that includes elements of both the simple Mongolian-style and the more complicated Chinese-style stills. However, adherence to something more like the simpler Mongolian-style still, with its pot, lid, and receiving bowl, probably contributed to the rapid spread of distilled liquors during the early Chosŏn period. However simple their beginnings, distillers soon developed more complicated forms of soju kori using different materials. This Korean pattern of localization in terms of ingredients, methods, and technology can be detected in other areas where a distilled alcohol culture spread, such as Kagoshima in Japan, where people made shochu from sweet potatoes, and Mexico, where people made mescal and tequila with agave.

Notice, too, the interplay of tradition and innovation that took place in the course of cultural development in society after the introduction of distilled liquors. A variety of literature, including literary works and cookbooks by scholars and ladies in elite families, provides a rich testimony to the ongoing development of new varieties of alcoholic drink as spirits like soju moved into new cultural environments that differed from those found in places like China. The development of distilled liquors in Korea assumed different cultural characteristics. While Chinese society developed a mass industrial production system to produce distilled liquors for use at large festivals and other mass gatherings, this did not occur in Chosŏn Korea, where production remained in the court and household: the court for its own use for tributes, parties, and medicine, and households by women for the observance of family rituals, including annual ancestral worship, as well as for the entertainment of guests. As a consequence, soju diversified as each household formulated its own methods to create varieties distinctly its own, a specific culture called kayangju, or "home-brew culture." These small soju-manufacturing traditions using small, portable soju kori rapidly declined after the advent of modern large-scale distillation technology in Europe and its introduction to Korea via Japanese migrating and investing there as a part of the Japanese colonial project. Since then, and despite the various political upheavals of the twentieth century, modern industrially produced soju made by a factorized, systemized, and commercialized form of mass production became the dominant form of soju, sparking new national and then global trends toward the popularization of soju thanks to the development of new technologies and the incorporation of a range of globalized cultural items. At the same time as it was going global, however, soju brought about the resurrection of an interest in traditionally distilled soju as a part of a national folk liquor (minsokchu).

In short, the development of soju is an example of a tradition that continues to beget new traditions through innovation in new historical and cultural environments. This flies in the face of a common, though not scholarly, misconception that tradition is not only completely original but also thoroughly static. In fact, it is rare to find a completely original tradition. Tradition starts with innovation. The premodern forms of distilled liquor that we have studied here have been complemented and cultivated by a fusion of several technologies and ideas developed through cross-cultural contacts and exchanges taking place in the broader context of Afro-Eurasia as a whole, leading ultimately to modern soju in Korea. Many important elements of tradition have gained more through innovation and have often been stimulated by new cultural elements. This has been verified in the realm of food culture, and so it is not surprising that the same kind of development through technological transfer could explain the evolution of soju. The development of modern industrially produced soju, the resurgence of traditional soju, the Korean Wave (*Hallyu*) and soju's global exportation became possible in this context. Without this constant exchange and adaptation of new technologies and ideas, soju as we know it could not have come into being.

Past, Present, and Future: From National to Global and Local Foods

Why do we now examine the development and spread of these traditional techniques of the past? This case of soju helps us to understand the need to scrutinize historical change from our current vantage point in order to discover our present position and better understand ourselves. The context within which we track the course of soju's development on the Korean peninsula is the environment of technology transfer and propagation that took place during the Mongol period through a complex of cross-cultural exchanges. However, this phenomenon has manifested in a broader geographical field, so we have to go deeper and wider to fully appreciate it. Much historical research of high quality has been focused within the narrow confines of a particular period or region, so it is useful to expand our gaze to a holistic perspective of history. Soju's true historical significance could not have been obtained by focusing on a single period or place. Herodotus suggested that we think about history from the present stance.[1] From the vantage of the twenty-first century, we can think about how to reinterpret the developmental history of soju. This could be a useful method for approaching global history, a subfield of history that that has already brought us useful new interpretations and insights from the application of broader and comparative views. Let us reflect what this has taught us from our present case study.

Many people enjoy drinking. This has made alcohol an ancient and important part of food culture around the globe. Of course, there are regions that restrict or limit alcohol consumption for reasons such as religion or social custom, but in most places, alcohol remains important to everyday life. It is still common to start parties with it, to drink it with meals, and sometimes to consume it at social venues designed for its consumption, like bars. People have also used distilled alcohol in ritual and in medicine. Most recently, alcohol has served as a valuable disinfectant against epidemic disease. Korean soju companies have quickly donated ethanol to be used to produce hand sanitizer as the COVID-19 pandemic situation has worsened and become prolonged.[2] Indeed, we will feel the richness of life and history by knowing that there is history and tradition in foods and alcoholic drinks, and that they sometimes can tell us much about other aspects of history, like science and technology. We have seen in this book how liquors such as soju connect to a scientific idea such as distillation, which is linked to alchemy, a medieval form of chemistry. Due to the characteristics of cooking methods and nutrition, foodways will most likely develop greater links to science and technology in the future. This links further to the exchange of diverse cultural technologies among various societies. Although we often see traditional elements continue in many cultural items such as food, there are many cases where we look behind that tradition and observe the adoption of new outside elements and their localization. When we discover such a situation, will we not become humbler before the many things we enjoy in life? To remember that even eating and drinking are steeped in history?[3]

It is clear that Korea's soju is increasing its presence in the global market. The process of globalization is ongoing and still affects food cultures worldwide, and we have no way of knowing how much more

---

[1] Herodotus (fifth century BCE), *The Histories* (Penguin Classics Deluxe Edition), trans. Tom Holland (New York: Penguin Books, 2015), 5.

[2] Yi Chŏngŭn, "K'orona 19 sodok? 'han'gugin ŭi sul' soju ka nasŏtta" (Corona 19 Disinfection? Soju, "Korean Liquor," Came Forward to Help), *Hankook Ilbo*, March 14, 2020, www.hankookilbo.com/News/Read/202003131009759474 (accessed April 9, 2020).

[3] I bought a bottle of Andong soju from Andong Soju Museum in 2014 and asked the seller how long I could preserve it. He said that there is no expiration date, and that the longer these traditional spirits sit, the more delicious they become. Such a feature is indeed an important product of the historical process.

popular something like soju could become globally in the future. Food cannot be forced to be eaten. Any successful popularization of food, even in the age of globalization, has been possible because people in the world have chosen, accepted, changed, and developed it. It cannot be obtained automatically by a state's simple political maneuvering. Indeed, food is one item that can overcome the impulses toward traditionalism and nationalism. New generations seem inclined to accept new things, however much any new food or food culture is rooted in – indeed grows out of – traditional cultures and technologies. We can see that many ethnic foods, such as pasta and kimchi, which remain identified as national foods, have been transformed through new ideas and developed through various exchanges. Before promoting soju as a Korean national drink, it should be remembered that its existence resulted from people's exchange and their acceptance of new ideas internationally. Along with the development of soju, there have been many similarly wonderful developments in Korean food, many of which are the product of a similar historical process. Food is highly effective at interesting people in a new foreign culture, and in the course of their familiarization with that food, they will be able to partake in the rich and dynamic stories about its culture.

Perhaps a good way to overcome potential nationalist traps in our age would simply be to interest people in local foods of smaller localities, as Joo Young-ha suggests.[4] The argument of this book proposes that, in addition to this, we should remember that changing historical contexts and other related elements, such as exchange of ideas and technology, contributed to the historical development of each food culture. We can nurture a balanced understanding of how our food developed and also of how our current food culture will inevitably continue to change with many other cultural elements in our global society. In this way, food becomes one of many cultural items that are communicated, interacted with, and transformed through exchange in world history. This makes us realize the importance of cultural exchange in world history. This book encourages everyone to be interested in Korean culture and history while enjoying its foods, and also in the global history of which it has been a part.

---

[4] Chu Yŏngha (Joo Young-ha), *Ch'ap'on·chanp'on·tchamppong*, 288.

Works Cited

Primary Sources

Buell, Paul David, Eugene Newton Anderson, and Charles Perry, trans. *A Soup for the Qan: Chinese Dietary Medicine of the Mongol Era as Seen in Hu Sihui's* Yinshan Zhengyao. *Introduction, Translation, Commentary, and Chinese Text.* 2nd revised and expanded edition. Sir Henry Wellcome Asian Series, 9. Leiden: Brill, 2010.

Chang, Lady of Andong 安東張氏 (1598–1680). *Ŭmsik timibang* 飲食知味方 (Recipes for Tasty Food). Taegu: Kyŏngbuk taehakkyo ch'ulp'anbu, 2003.

Chang Tong'ik 張東翼 (Jang Dong-Ik). *Songdae Ryŏsa charyo chimnok* 宋代麗史資料集錄 (Collection of Historical Sources from the Song on Koryŏ History). Seoul: Sŏul taehakkyo ch'ulp'anbu, 2000.

Wŏndae Ryŏsa charyo chimnok 元代麗史資料集錄 (Collection of Historical Sources from the Yuan on Koryŏ History). Seoul: Sŏul taehakkyo ch'ulp'anbu, 1997.

Chen Shou 陳壽 (233–297). *Sanguozhi* 三國志 (The Record of the Three Kingdoms). Beijing: Zhonghua shuju, 1973.

Chŏn Sunŭi 全循義 (fl. mid-fifteenth century). *San'ga yorok* 山家要錄 (An Essential Record for Farming Villages). Translated by Hong Kiyong 洪起瑢 and Yun T'aesun 尹泰順. Suwon: Nongch'on chinhŭngch'ŏng (Rural Development Administration), 2004.

Chŏng Inji 鄭麟趾 (1396–1478), ed. *Koryŏsa* 高麗史 (Official History of Koryŏ) (1454). Seoul: Asea Munhwasa, 1983.

Chŏng Kwang (Chung Kwang), trans. *(Yŏkchu) wŏnbon* Nogŏltae (譯註) 原本老乞大 (An Annotated Translation of the Original *Nogŏltae*). Seoul: Pangmunsa, 2010.

Chong Yagyong. *Admonitions on Governing the People: Manual for All Administrators.* Translated by Byonghyon Choi. Berkeley: University of California Press, 2010.

Chŏng Yagyong 丁若鏞 (1762–1836). *Tasan simunjip* 茶山詩文集 (Literary Collection by Tasan). Seoul: Minjok munhwa ch'ujinhoe, 1982–1997. 10 vols.

Yŏkchu Mongmin simsŏ 譯註牧民心書 (An Annotated Translation of *Mongmin simsŏ (The Mind of Governing the People)*). Translated and annotated by Tasan yŏn'guhoe 茶山研究會. Seoul: Ch'angjak kwa pip'yŏngsa, 1978–1985. 6 vols.

Chosŏn wangjo sillok 朝鮮王朝實錄 (The Veritable Records of the Chosŏn Dynasty). http://sillok.history.go.kr.

Chōsen shuzō kyōkai 朝鮮酒造協会 (Committee of Alcoholic-Beverage Making in Korea). *Chōsen shuzōshi* 朝鮮酒造史 (History of the Production of Alcoholic Drinks in the Chosŏn Period). Keijō 京城 (Seoul): Chōsen shuzō kyōkai 朝鮮酒造協会, 1935.

Di Hao 翟灝 (?–1788). *Tongsu bian* 通俗编 (Popular Culture). Taipei: Guotai wenhua shiye youxian gongxi 國泰文化事業有限公司, 1980.

Duan Chengshi 段成式 (d. 863). *Youyang zazu* 酉陽雜俎 (The Miscellaneous Morsels from Youyang) (ninth century). Translated and annotated by Yoshio Imamura 今村与志雄. Tokyo: Heibon sha, 1980–1931. 5 vols.

Fan Ye 范曄 (398–445). *Hou Hanshu* 後漢書 (The History of the Later Han). Beijing: Zhonghua shuju, 1982.

Han'gukhak chungang yŏn'guwŏn (Academy of Korean Studies). *Han'guk minjok munhwa tae paekkwa sajŏn* (Encyclopedia of Korean Cultures). Han'gukhak chungang yŏn'guwŏn. https://encykorea.aks.ac.kr (accessed February 10, 2020).

Herodotus (5th century BCE). *The Histories*. Translated by Tom Holland. Penguin Classics Deluxe Edition. New York: Penguin Books, 2015.

Hŏ Chun 許浚 (1539–1615). *Tongŭi pogam* 東醫寶鑑 (Precious Mirror of Eastern Medicine). Seoul: Namsandang 南山堂, 1991.

Tongŭi pogam 東醫寶鑑 (Precious Mirror of Eastern Medicine). Translated by Tongŭi munhŏn yŏn'gushil (Tongŭi Literature Lab). Seoul: Pŏbin munhwasa, 2012.

Hong Mansŏn 洪萬選 (1643–1715). *Sallim Kyŏngje* 山林經濟 (Forest and Economy). Seoul: Kyŏngin munhwasa 景仁文化社, 1976.

Hu Sihui 忽思慧 (d. 1330). *Yinshan zhengyao* 飲膳正要 (Important Principles of Food and Drink). Taipei: Taiwan shangwu yinshuguan 臺灣商務印書館, 1968.

Hwang Yunsŏk 黃胤錫 (1729–1791). *Ijae nan'go* 頤齋亂藁 (Disordered Drafts by Ijae) (1786). Seoul: Han'guk chŏngsin munhwa yŏn'guwŏn, 1995.

Ibn Qutayba (828–889). *Kitāb al-Ashriba* (Book of Drinks). Edited by Yāsīn Muḥammad al-Sawwās. Damascus: al-Maṭbaʿa al-ʿilmiyya, 1420/1999.

Iryŏn 一然 (1206–1289). *Samguk Yusa: Legends and History of the Three Kingdoms of Ancient Korea*. Translated by Tae-Hung Ha and Grafton K. Mintz. Seoul: Yonsei University Press, 1972.

Jackson, Peter, trans. *The Mission of Friar William of Rubruck: His Journey to the Court of the Great Khan Möngke 1253–1255*. London: Hakluyt Society, 1990.

Jujia biyong shilei 居家必用事類 (Essential Things for Living at Home). Tokyo: Chūgoku shokukei sōsho 中國食經叢書, 1973.

Kim Bunkyō 金文京, Gen Yukiko 玄幸子, and Satō Haruhiko 佐藤晴彦, trans. *Rō kitsu dai: Chōsen chūsei no chūgokugo kaiwa dokuhon* 老乞大—朝鮮中世の中国語会話読本 (*Rō kitsu dai*: Chinese Conversation Textbook in Medieval Chosŏn). Tokyo: Heibon sha, 2002.

Kim Changsaeng 金長生 (1548–1631). *Sagye chŏnsŏ* 沙溪全書 (Complete Book of Sagye). Kwŏn 41, Ŭirye munhae 疑禮問解 (Theories about Manners), cheyre 祭禮 (Ancestral Rituals), sije 時祭 (Seasonal Rites). Han'guk kojŏn pŏnyŏgwŏn, http://db.itkc.or.kr (accessed January 5, 2019).

Kim Chongsŏ 金宗瑞 (1383–1453), ed. *Koryŏsa chŏryo* 高麗史節要 (Essentials of Koryŏ History). Seoul: Asea Munhwasa, 1972.

Kyogam Koryŏsa Chŏryo 校勘 高麗史節要 (Essentials of Koryŏ History, with Comparative Textual Analyses). Annotated by No Myŏngho. Seoul: Chimmundang, 2016.

Kim Pusik 金富軾 (1075–1151). *The Koguryŏ Annals of the Samguk Sagi*. Translated by Edward J. Shultz and Hugh H. W. Kang. Sŏngnam: The Academy of Korean Studies Press, 2011.

Samguk sagi 三國史記 (Historian's Records of the Three Kingdoms). Translated by Yi Pyŏngdo. Seoul: Ŭryu munhwasa, 1983. 2 vols.

Kim Yu (1491–1555). *Suun chappang* 需雲雜方 (Various Methods of High-Class Food Culture). Translated by Kim Ch'aesik. P'aju: Kŭrhangari, 2015.

Li Fang 李昉 (925–996). *Taiping Yulan* 太平御覽 (Readings of the Taiping Era). Shanghai: Shanghai shudian, 1985. 21 Vols.

Li Shizhen 李時珍 (1518–1593). *Bencao gangmu* 本草綱目 (Compendium of Materia Medica). Translated and annotated by the editorial team of Haengnim ch'ulp'an 杏林出版. Seoul: Haengnim ch'ulp'an 杏林出版, 1976. 3 Vols.

Compendium of Materia Medica: Bencao gangmu. Translated and annotated by Luo Xiwen. Beijing: Foreign Languages Press, 2003.

Li Tao 李燾 (1114–1183). *Xu Zizhi tongjian changbian* 續資治通鑑長編 (Long Draft of the Continuation of the *Universal Mirror for the Aid of Government*). Shanghai: Shanghai guji chubanshe, 1986.

Liu Xu 劉煦 (887–946). *Jiu Tangshu* 舊唐書 (Old History of the Tang). Beijing: Zhonghua shuju, 1975.

Mongŏ yuhae 蒙語類解 (Categorization and Translation of Mongolian Language). Seoul: Sŏul taehakkyo Kyujanggak Han'gukhak yŏn'guwŏn 서울大學校 奎章閣韓國學研究院, 2006.

Murakami Tadayoshi 村上唯吉. *Chōsenjin no ishokujū* 朝鮮人の衣食住 (Clothing, Food, and Housing in Chosŏn). Kyŏngsŏng 京城: Chōsen sōtoku fu 朝鮮総督府, 1916.

Nasrallah, Nawal, trans. *Treasure Trove of Benefits and Variety at the Table: A Fourteenth-Century Egyptian Cookbook*. Leiden and Boston: Brill, 2018.

Ouyang Xiu 歐陽修 (1007–1072). *Xin Tangshu* 新唐書 (New History of the Tang). Beijing: Zhonghua shuju, 1975.

Perry, Charles, ed. and trans. *Scents and Flavors: A Syrian Cookbook*. New York: New York University Press, 2017.

Polo, Marco (1254–1324). *The Description of the World*. Translated and annotated by Sharon Kinoshita. Indianapolis: Hackett Publishing Company, Inc., 2016.

Sharaf al-Dīn ʿAlī Yazdī (d. 1454). *Zafarnama* (Book of Victory) (1424–1428). Tihrān: Kitābkhānah, Mūzih va Markaz-i Asnād-i Majlis-i Shūrā-yi Islāmī, 1387/2008. 2 vols.

Sima Qian 司馬遷 (c. 145–86 BCE). *Records of the Grand Historian of China, Translated from the Shi chi of Ssu-ma Ch'ien*, Vol. 2: *The Age of Emperor Wu 140 to circa 100 B.C.* Translated by Burton Watson. New York: Columbia University Press, 1961.

Shiji 史記 (Records of the Historian). Beijing: Zhonghua shuju, 1959.

Sŏ Yugu 徐有榘 (1764–1845). *Imwŏn Kyŏngchechi: Chŏngjochi* 林園經濟志: 鼎俎志 (Encyclopedia of Rural Life: Records of Food and Cooking). Translated by Imwon Research Institute (Chŏng Chŏnggi (Chung Chungkee)). Seoul: Pungseok Cultural Foundation, 2020.

Sŏng Haeŭng 成海應 (1760–1839). *Yŏn'gyŏngjae chŏnjip: Oejip* 研經齋全集: 外集 (Complete Works of Yŏn'gyŏngjae: Supplementary Volume). Kwŏn 60, P'ilgiryu 筆記類, Han'guk kojŏn pŏnyŏgwŏn. http://db.itkc.or.kr (accessed February 25, 2020).

Song Lian 宋濂 (1310–1381). *Yuanshi* 元史 (The History of the Yuan). Beijing: Zhonghua shuju, 1976.

Terashima Ryōan 寺島良安. *Wakan sansai zue* 和漢三才図会 (Illustrated Sino-Japanese Encyclopedia) (1712). Tokyo: Heibonsha, 1985–1991. 12 vols.

Tuo Tuo 脫脫 (1313–1355). *Songshi* 宋史 (History of the Song). Beijing: Zhonghua shuju, 1977.

Wang Qi 王圻 (1565–1614). *Sancai tuhui* 山才圖會 (Illustrated Compilation of the Three Powers) (*c.* 1607). Shanghai: Shanghai guji, 1988.

Wang Qixing, 王啓興, ed. *Jiaobian Quan Tangshi* 校編全唐詩 (Complete Tang Poems, with Annotations). Wuhan: Hubei renmin, 2001.

Xu Jing 徐兢 (1091–1153). *A Chinese Traveler in Medieval Korea: Xu Jing's Illustrated Account of the Xuanhe Embassy to Koryŏ.* Translated, annotated, and with an introduction by Sem Vermeersch. Honolulu: University of Hawai'i Press, 2016.

Xu Song 徐松 (1781–1848). *Songhuiyao jigao* 宋會要輯稿. Draft recovered edition of the *Song huiyao.* Beijing: Zhonghua shuju, 1957.

Yamaguchi Yoshinori 山口佳紀 and Kōnoshi Takamitsu 神野志隆光, eds. *Kojiki* 古事記 (Records of Ancient Matters). Vol. 1 of *Shinpen nihon koten bungaku taikei.* Tokyo: Shōgakukan, 1997.

Yao Guangxiao 姚廣孝 (1335–1418). *Yongle dadian* 永樂大典 (Yongle Encyclopedia). Beijing: Zhonghua shuju, 1959.

Yi Chihang 李志恒 (seventeenth century). *P'yojurok* 漂舟錄 (A Record of Drifting on a Ship). http://db.itkc.or.kr (accessed February 25, 2020).

Yi Haeng 李荇 (1478–1534). *Sinjŭng Tongguk yŏji sŭngnam* 新增東國輿地勝覽 (Newly Enlarged Geographical Survey of the Eastern Country [Korea]) (1531). Seoul: Tongguk munhwasa 東國文化社, 1958.

Yi Kyubo 李奎報 (1168–1241). *Kukyŏk Tongguk Isangguk chip* 國譯東國李相國集. Translated by Minjok munhwa ch'ujinhoe (National Culture Promotion Association). Seoul: Minjok munhwa ch'ujinhoe, 1982.

Tongguk Isangguk chip 東國李相國集. Seoul: Ilchogak 一潮閣, 2000.

Tongmyŏngwang p'yŏn 東明王篇. Translated by Pak Tup'o 朴斗抱. Seoul: Ŭryu munhwasa 乙酉文化社, 1974.

Yi Kyugyŏng 李圭景 (1788–1856). *Oju yŏnmun changjŏn san'go* 五洲衍文長箋散稿 (Scattered Manuscripts of Glosses and Comments by Oju), edited by Kojŏn Kanhaenghoe. Seoul: Tongguk Munhwasa, 1959.

Yi Saek 李穡 (1328–1396). *Kugyŏk Mokŭnjip* 國譯牧隱集 (Literary Collection of Mokŭn). Translated by Im Chŏnggi. Seoul: Minjok munhwa ch'ujinhoe, 2000–. 12 vols.

Yi Sik 李植 (1584–1647). *T'aektang chip* 澤堂集 (Collection of T'aektang), Pyŏlchip 別集 (Separate Collection). Kwŏn 16, Chapchŏ 雜著 (Various Writings). Han'guk kojŏn pŏnyŏgwŏn, http://db.itkc.or.kr (accessed January 5, 2019).

Yi Sugwang 李晬光 (1563–1628). *Chibong yusŏl* 芝峰類説 (1614) (Topical Discourses by Chibong). Reproduction of original text. Seoul: Kyŏngin munhwasa 景仁文化社, 1970.

Yi Tŏkmu 李德懋 (1741–1793). *Ch'ŏngjanggwan Chŏnsŏ* 青莊館全書 (Complete Works of Ch'ŏngjanggwan). Edited by Kojŏn Kanhaenghoe. Seoul: Sol, 1997.

Yi Yukhwa, trans. *Wŏnbon* Nogŏltae shinju shinyŏk 原本老乞大 新註新譯 (A New Annotated Translation of the Original *Nogŏltae*). Seoul: Sinasa, 2015.

Yu Changwŏn 柳長源 (1724–1796). *Sangbyŏn t'onggo* 常變通攷 (General Examination of General Changes). Kwŏn 23, Cherye 祭禮 (Ancestral Rituals). Han'guk kojŏn pŏnyŏgwŏn, http://db.itkc.or.kr (accessed January 5, 2019).

Yu Chungnim 柳重臨. *Chŭngbo sallim kyŏngje* 增補山林經濟 (Expanded *Sallim kyŏngje*). Suwon: Nongch'on chinhŭngch'ŏng (Rural Development Administration), 2003–2004.

Zhang Tingyu 張廷玉 (1672–1755). *Mingshi* 明史 (The History of the Ming). Beijing: Zhonghua shuju, 1974.

Zhao Rukuo 趙汝适 (1170–1228). *Chau Ju-Kua: His Work on the Chinese and Arab Trade in the Twelfth and Thirteenth Centuries Entitled Chu-fan-chi (Description of foreign peoples)*. Translated and annotated by Friedrich F. Hirth and W. W. Rockhill. St. Petersburg: Printing Office of the Imperial Academy of Sciences, 1911.

Zhufan zhi jiaoshi 諸蕃志校释 (Description of Foreign Lands, with Annotations and Footnotes). Edited by Yang Bowen 楊博文. Beijing: Zhonghua shuju, 1996.

Zhou Qufei 周去非 (twelfth century). *Lingwai daida* 嶺外代答校注 (Notes from the Land beyond the Passes, with Annotations and Footnotes). Edited by Yang wuquan 楊武泉. Beijing: Zhonghua shuju, 1999.

Zhu Gong 朱肱 (1068–1165). *Chūgoku no shusho* 中國の酒書 (Books about Alcoholic Drinks in China) (A Japanese translation of *Jiu jing* 酒經 (Classic of Alcoholic Drinks)). Translated by Nakamura Takashi. Tokyo: Heibonsha 平凡社, 1991.

Jiu jing 酒經 (Classic of Alcoholic Drinks). Beijing: Zhonghua shuju, 2012.

Zhu Yizhong 朱翼中 (fl. eleventh century). *Beishan jiujing* 北山酒經 (Classic of Alcoholic Drinks in Northern Mountain). Zhongguo jiben gujiku 中國基本古籍庫, 3 vols. Online.

Studies

Albala, Ken, ed. *Routledge International Handbook of Food Studies*. London: Routledge, 2013.

Al-Hassan, Ahmad Y. "Alcohol and the Distillation of Wine in Arabic Sources from the 8th Century." In *Studies in Al-Kimya': Critical Issues in Latin and Arabic Alchemy and Chemistry*, 283–298. Hildesheim: Georg Olms, 2009.

Al-Hassani, Salim T. S., ed. *1001 Inventions: The Enduring Legacy of Muslim Civilization.* Washington, DC: National Geographic, 2012.

Allchin, Frank R. "India: The Ancient Home of Distillation?", *Man*, New Series 14, no. 1 (1979): 55–63.

Allsen, Thomas T. *Commodity and Exchange in the Mongol Empire: A Cultural History of Islamic Textiles.* Cambridge: Cambridge University Press, 1997.

Culture and Conquest in Mongol Eurasia. Cambridge: Cambridge University Press, 2001.

"Ögedei and Alcohol." *Mongolian Studies* 29 (2007): 3–12.

Anderson, Eugene N. *Everyone Eats.* New York: New York University Press, 2005.

Food and Environment in Early and Medieval China. Philadelphia: University of Pennsylvania Press, 2014.

The Food of China. New Haven: Yale University Press, 1988.

Asai Usuke 麻井宇介. "Supirritsu no kindai: renzokushiki jōryūki wa nani wo motarashitaka" スピリッツの近現代: 連続式蒸溜機はなにをもたらしたか (Spirits in Modern Times: How Did Continuous Stills Influence Us?). In *Shōchū higashi mawari nishi mawari* 焼酎東回り西回り (Shochu around the World), edited by Tamamura Toyo'o 玉村豊男, 131–171. Tokyo: TaKaRa Alcohol Beverage and Life Research Institute, 1999.

Atwood, Christopher P. *Encyclopedia of Mongolia and the Mongol Empire.* New York: Facts on File, 2004.

Baek, In-hwan, Byung-yo Lee, and Kwang-il Kwon. "Influence of Oxygen on Pharmacokinetics of Alcohol." *Alcoholism: Clinical & Experimental Research* 34, no. 5 (2010): 834–839.

Batjargal, Batdorj. "Probiotic Properties of Lactic Acid Bacteria Isolated from Mongolian Fermented Mare's Milk." *Crossroads: Studies on the History of Exchange Relations in the East Asian World* 14 (2016): 257–263.

Bayarsaikhan, Dashdondog. "Drinking Traits and Culture of the Imperial Mongols in the Eyes of Observers and in a Multicultural Context." *Crossroads: Studies on the History of Exchange Relations in the East Asian World* 14 (2016): 161–172.

Bearman, P. J. et al., eds. *The Encyclopaedia of Islam.* 2nd edition. Leiden: Brill, 1954–2005. 12 vols.

Bentley, Jerry H. *Old World Encounter: Cross-cultural Contacts and Exchanges in Premodern Times.* New York: Oxford University Press, 1993.

Berger, Patricia. *The Art of Wine in East Asia.* San Francisco: Asian Art Museum of San Francisco, 1985.

Biran, Michal. "The Mongol Transformation: From the Steppe to Eurasian Empire." *Medieval Encounters* 10, nos. 1–3 (2004): 339–361.

Qaidu and the Rise of the Independent Mongol State in Central Asia. Richmond: Curzon, 1997.

Blue, Anthony Dias. *The Complete Book of Spirits.* New York: HarperCollins.

Bol, Peter K. *Neo-Confucianism in History.* Cambridge, MA: Harvard University Asia Center, Distributed by Harvard University Press, 2008.

Braudel, Fernand. *Civilization and Capitalism, 15th–18th Century*, vol. 1: *The Structures of Everyday Life: The Limits of the Possible.* Translated from the French and revised by Siân Reynolds. New York: Harper & Row, 1981.

Breuker, Remco E. *Establishing a Pluralist Society in Medieval Korea, 918–1170: History, Ideology and Identity in the Koryŏ Dynasty.* Leiden: Brill, 2010.

"Koryŏ as an Independent Realm: The Emperor's Clothes?", *Korean Studies* 27 (2004): 48–84.

Brinkmann, Stefanie. "Wine in Hadith: From Intoxication to Sobriety." In *Wine Culture in Iran and Beyond.* Edited by Bert G. Fragmer, Ralph Kauz, and Florian Schwarz, 71–135. Vienna: Verlag der Österreichischen Akademie der Wissenschaften, 2014.

Brose, Michael C. "Neo-Confucian Uyghur Semuren in Koryŏ and Chosŏn Korean Society and Politics." In *Eurasian Influences on Yuan China.* Edited by Morris Rossabi, 125–158. Singapore: Institute of Southeast Asia Studies, 2013.

Subjects and Masters: Uyghurs in the Mongol Empire. Bellingham, WA: Center for East Asian Studies, Western Washington University, 2007.

Bruman, Henry J. *Alcohol in Ancient Mexico.* Salt Lake City: University of Utah Press, 2000.

"The Asiatic Origin of the Huichol Still." *Geographical Review* 34, no. 3 (July 1944): 418–427.

"Early Coconut Culture in Western Mexico." *Hispanic American Historical Review* 25, no. 2 (May 1945): 212–223.

"Some Observations on the Early History of the Coconut in the New World." *Acta Americana* 2 (1944): 200–243.

Buell, Paul D. "Korea as Part of the Mongolian World: Patterns and Differences." *International Journal of Eurasian Research* 5, no. 10 (January 2017): 137–146.

"Mongol Empire and Distillation: Technology and Popularization." An unpublished paper presented at the Science and Technology Transfer workshop held at the Hebrew University of Jerusalem, June 10–11, 2015.

Buell, Paul D., Eugene N. Anderson, Montserrat De Pablo Moya, and Moldir Oskenbay. *Crossroads of Cuisine: The Eurasian Heartland, the Silk Roads and Food.* Leiden: Brill, 2020.

Buell, Paul D., and Francesca Fiaschetti. *Historical Dictionary of the Mongol World Empire and Its Successor States.* 2nd revised and expanded edition. Lanham, MD and Oxford: The Scarecrow Press, Inc. (Historical Dictionaries of Ancient Civilizations and Historical Eras, No. 8), 2018.

Buell, Paul D., and Judy Kolbas. "The Ethos of Sate and Society in the Early Mongol Empire: Chingiz Khan to Güyük." In *The Mongols and Post-Mongol Asia: Studies in Honour of David O. Morgan,* edited by Timothy May. *Journal of the Royal Asiatic Society* 26, no. 1–2 (January, 2016): 43–64.

Buell, Paul D., and Montserrat de Pablo. "Distilling of the Volga Kalmucks and Mongols: Two Accounts from the 18th Century by Peter Pallas with Some Modern Comparisons." *Crossroads: Studies on the History of Exchange Relations in the East Asian World* 13 (2016): 115–123.

Burns, Eric. *The Spirits of America: A Social History of Alcohol.* Philadelphia: Temple University Press, 2014.

Chang Chihyŏn 張智鉉 (Chang Chi-Hyun). *Han'guk oeraeju yuipsa yŏn'gu* 韓國外來酒流入史研究 (A Study of the Influx of Foreign Alcoholic Drinks to Korea). Seoul: Suhaksa 修學社, 1989.

Chang Tongik 張東翼 (Jang Dong-Ik). *Koryŏ hugi oegyosa yŏn'gu* 高麗後期外交史研究 (A Study of Diplomatic History in the Late Koryŏ Period). Seoul: Ilchogak 一潮閣, 1994.

Cho Sŏnggi, and Yu Kimok. *Soju ŭi tosu chŏngch'esŏng hwangnip pangan yŏn'gu* (A Study on Methods of Establishing the Identity of Soju's Alcohol Content). Seoul: han'guk churyu yŏn'guwŏn [Korea Alcohol Research Center], 2010.

Cho Wŏn (Cho Won). "*Ŭmsŏnjŏngyo* wa Tae Wŏn cheguk ŭmsik munhwa ŭi Tongasia chŏnp'a" 『飲膳正要』와 大元제국 음식문화의 동아시아 전파 (*Yinshan Zhengyao* 飲膳正要 and the Influence of Yuan Culinary Culture in East Asia). *Yŏksa Hakbo* (Korean Historical Review) 233 (March 2017): 181–209.

Ch'oe Hansŏk. "Saengmyŏng ŭi mul, chŭngnyuju! Urinara ŭi chŏnt'ong soju wa kŭ ch'in'gudŭl" (The Water of Life, Spirit! Korean Traditional Soju and Their Friends). *RDA Interrobang* 168 (2016): 1–20. http: www.rda.go.kr (accessed December 9, 2018).

Ch'oe Tŏkkyŏng (Choi Duk-Kyung). "Wŏndae nongŏp ŭi paltal kwa Koryŏ e kkich'in saengwalsang ŭi yŏngyang" 元代 농업의 발달과 高麗에 끼친 생활상의 영향 (Study of the Development of Agriculture in the Yuan Dynasty and Its Influence on Goryeo Life). *Pigyo minsokhak* (Asian Comparative Folklore) 60 (August 2016): 211–254.

Ch'oe Tŏkkyŏng, Yi Chongbong, and Hong Yŏngŭi. *Ryŏ·Wŏndae ŭi nongjŏng kwa Nongsang chibyo* 麗·元代의 農政과 農桑輯要 (Agricultural Administration during the Koryŏ–Yuan Period and *Nongsang jiyao*). Seoul: Tonggang ch'ulp'ansa, 2017.

Chŏn Yonghun (Jun Yong Hoon). *Han'guk ch'ŏnmunhak sa* (History of Astronomy in Korea). P'aju: Tŭllyŏk, 2017.

Chong Dae Song 鄭大聲. *Chosen no sake* 朝鮮の酒 (Korean Alcoholic Beverage). Tokyo: Tsukigi shokan, 1987.

Chŏng Kusŏn (Chung Koo-sun). *Chosŏn wangdŭl, kŭmjuryŏng ŭl naerida* (The Kings of the Chosŏn Dynasty Imposed the Liqour Prohibition Law). Seoul: P'aendŏmbuksŭ (Fandombooks), 2014.

Chŏng Kusŏn 鄭求先 (Chung Koo-sun). "Koryŏ mal Ki hwanghu ilchok ŭi tŭkse wa mollak" 高麗末 奇皇后 一族의 得勢와 沒落 (The Rise and Fall of the Family of Empress Ki at the End of the Koryŏ Dynasty). *Dong Gook Sa Hak* 40 (2004): 167–185.

Chŏng Kwang, *Hunminjŏngŭm kwa P'asŭp'a munja* (Hunminjŏngŭm and the 'Phags-pa Script). Seoul:Yŏngnak, 2012.

Chŏng Min. *18 segi Chosŏn chisigin ŭi palgyŏn* (Discovery of the 18th-Century Korean Intellectuals). Seoul: Hyumŏnisŭt'ŭ, 2007.

Chŏng Sŭnghye et al. *Pakt'ongsa Wŏn nara Taedo rŭl kŏnilta* (Pakt'ongsa Wanders in the Yuan Dynasty's Capital Daidu). Seoul: Pangmunsa, 2011.

Chŏng T'aehŏn. "Ilche kangjŏmgi chujoŏp kwa chuse chŏrgch'aek" (The Alcohol-Producing Industry and the State Tax Policy during the Japanese Occupation). In *Han'guk ŭi sul 100 nyŏn ŭi kwaje wa chŏnmang.* Edited by Chŏng taeyŏngKu Sahoe, Chŏng T'aehŏn, Chŏng Sŏkt'ae, Kwŏn Sŏngan, Chŏng Ch'ŏl, Yi Sŏkchun, and Yi Hwasŏn, 69–87. Seoul: Hyangŭm, 2017.

Chŏng Taeyŏng, Ku Sahoe, Chŏng T'aehŏn, Chŏng Sŏkt'ae, Kwŏn Sŏngan, Chŏng Ch'ŏl, Yi Sŏkchun, and Yi Hwasŏn (eds.). *Han'guk ŭi sul 100 nyŏn ŭi kwaje wa chŏnmang* (Challenges and Prospects for 100 Years of Korean Alcoholic Beverages). Seoul: Hyangŭm, 2017.

Chu Yŏngha (Joo Young-ha). *Ch'ap'on·chanp'on·tchamppong: Tongasia ŭmsik munhwa ŭi yŏksa wa hyŏnjae* (Ch'ap'on, Chanp'on, and Tchamppong: Past and Present of East Asian Food Culture). P'aju: Sagyejŏl, 2013.

"Chŏng Dojŏn sŭsŭng Yi Saek ŭi soju sarang" (Chŏng Tojŏn's Teacher Yi Saek's Love for Soju). *Dong-A Ilbo* (East Asia Daily), June 22, 2015.

Han'gugin ŭn wae irŏk'e mŏgŭlkka: siksa pangsik ŭro pon han'guk ŭmsik munhwasa (Why Do Koreans Eat Like This? A Cultural History of Food in Korea Investigated through Its Dining Customs). Seoul: Hyumŏnisŭt'ŭ, 2018.

ed. *19 segi Chosŏn, saengwal kwa sayu ŭi pyŏnhwa rŭl yŏppoda* (Looking at the Changes in Life and Thought in Nineteenth-Century Chosŏn). P'aju: Tolbegae, 2005.

Sikt'ak wi ŭi han'guksa: menyu ro pon 20 segi han'guk ŭmsik munhwasa (A Korean History of the Dinner Table: The 20th-Century Cultural History of Food Viewed through Menus). Seoul: Hyumŏnisŭt'ŭ, 2013.

Ŭmsik chŏnjaeng munhwa chŏnjaeng (Food War and Culture War). Seoul: Sagyejŏl, 2000.

Ŭmsik inmunhak: ŭmsik ŭro pon han'guk ŭi yŏksa wa munhwa (The Cultural Anthropology of Food: Korean History and Culture Viewed from the Perspective of Food). Seoul: Hyumŏnisŭt'ŭ, 2011.

Cho Hanbyŏl. "Talk'omhan kwail soju" (Sweet Fruit Soju). *JoongAng Ilbo*, September 20, 2015. https://news.joins.com/article/18702462 (accessed December 9, 2018).

Damerow, Peter. "Sumerian Beer: The Origins of Brewing Technology in Ancient Mesopotamia." *Cuneiform Digital Library Journal* 22, no. 2 (January 2012): 1–20. www.cdli.ucla.edu/pubs/cdlj/2012/cdlj2012_002.html (accessed October 30, 2018).

Dangremond, Sam. "Here's Everything You Need to Know about Soju, the National Drink of South Korea." *Town & Country*, February 8, 2018. www.townandcountrymag.com/leisure/drinks/a16752958/soju-korean-liquor (accessed September 25, 2019).

Dardess, John. *Conquerors and Confucians: Aspects of Political Change in Late Yuan China*. New York: Columbia University Press, 1973.

Deuchler, Martina. *Confucian Transformation of Korea: A Study of Society and Ideology*. Cambridge, MA: Harvard University Press, 1992.

Dott, Brian R. *The Chile Pepper in China: A Cultural Biography*. New York: Columbia University Press, 2020.

Dreisbach, Tom. "Move over Vodka; Korean *Soju*'s Taking a Shot at America." *NPR News*, September 22, 2013. www.npr.org/sections/thesalt/2013/09/22/224522548/move-over-vodka-korean-sojus-taking-a-shot-at-america (accessed May 18, 2016).

Du Shiran 杜石然. *Zhongguo kexue jishu shi* 中國科學技術史 (History of Chinese Science and Technology). Beijing: Kexue chubanshe, 1984.

Duncan, John B. *The Origins of the Chosŏn dynasty*. Seattle University of Washington Press, 2000.

Edagawa Kōichi 枝川公一. "Jōryūshu, higashi mawari nishi mawari" 蒸溜酒、東回り西回り (Distilled Alcohol, Bound for East and West). In *Shōchū higashi mawari nishi mawari* 焼酎東回り西回り (Shochu around the World). Edited by Tamamura Toyo'o 玉村豊男, 15–71. Tokyo: TaKaRa Alcohol Beverage and Life Research Institute, 1999.

El-Asmar, Joseph. *The Milk of Lions: A History of Alcohol in the Middle East*. London: Gilgamesh Publishing, 2020.

The Encyclopædia Britannica: A Dictionary of Arts, Sciences, Literature and General Information. 11th edition, Vol. 2. New York: The Encyclopædia Britannica Company, 1910.

Feng Enxue 冯恩学. "Zhongguo shaojiu qiyuan xintan" 中国烧酒起源新探 (Preliminary Analysis of the Origin of Distilled Spirit in China). *Jilin daxue shehui kexue xuebao* 吉林大学社会科学学报 (Jilin University Journal, social sciences edition) 55, no.1 (January 2015): 163–170.

Floor, Willem. "The Culture of Wine Drinking in Pre-Mongol Iran." In *Wine Culture in Iran and Beyond*. Edited by Bert G. Fragmer, Ralph Kauz, and Florian Schwarz, 165–209. Vienna: Verlag der Österreichischen Akademie der Wissenschaften, 2014.

Forbes, R. J. *A Short History of the Art of Distillation*. 2nd edition. Leiden: E. J. Brill, 1970.

Fouquet, Pierre. *Histoire de l'alcool*. Paris: Presses universitaires de France, 1990.

Fragmer, Bert G., Ralph Kauz, and Florian Schwarz. "Introduction." In *Wine Culture in Iran and Beyond*. Edited by Bert G. Fragmer, Ralph Kauz, and Florian Schwarz, 7–10. Vienna: Verlag der Österreichischen Akademie der Wissenschaften, 2014.

Fu Jinquan 傅金泉. "Cong Lidu yizhi kan woguo baijiu shi" 从李渡遗址看我国白酒史 (Discussion of Liquor History in China through the Study of the Lidu Memorial Site). *Jiuliang keji* 酿酒科技 3 (2003): 94–95.

Gaytán, Marie Sarita. *Tequila! Distilling the Spirit of Mexico*. Stanford, CA: Stanford University Press, 2014.

Grew, Raymond. "Food and Global History." In *Food in Global History*. Edited by Raymond Grew, 1–32. Boulder, CO: Westview Press, 2000.

ed. *Food in Global History*. Boulder, CO: Westview Press, 2000.

Ha Ubong. *Chosŏn sidae haeyang kukka waŭi kyoryu sa* (A History of Contact with Other Countries through Maritime Routes in the Chosŏn Dynasty). Seoul: Kyŏngin munhwasa 景仁文化社, 2014.

Chosŏn sidae pada rŭl t'onghan kyoryu (Exchange through the Sea during the Chosŏn Dynasty). P'aju: Kyŏngin munhwasa 景仁文化社, 2016.

Hall, Joshua. "Soju Makers Aim to Turn Fire Water into Liquid Gold." *Wall Street Journal*, October 17, 2014. www.wsj.com/articles/BL-KRTB-6764 (accessed December 2, 2018).

Hames, Gina. *Alcohol in World History*. London: Routledge, 2012.

Han Pongnyŏ (Han Bok-ryo). "*Sanga yorok* ŭi punsŏk koch'al ŭl t'onghaesŏ pon p'yŏnch'an yŏndae wa chŏja" (*Sanga yorok*: Analysis of Author and

Published Dates). *Nongŏpsa yŏn'gu* (Korean Journal of Agricultural History) 2, no. 1 (2003): 13–29.

"Ŭmsiksa esŏ pon *Ŭmsik timibang*" (*Ŭmsik timibang* Examined from the Perspective of the History of Korean Foods). In *Ŭmsik timibang* 飲食知味 方 (Recipes for Tasty Food), by Chang, Lady of Andong 安東張氏 (1598–1680), 81–122. Taegu: Kyŏngbuk taehakkyo ch'ulp'anbu, 2003.

Han Yŏngo (Hahn Young-Ho), and Yi Ŭnhŭi (Lee Eun-Hee). "Ryŏmal Sŏnch'o Pon'gungnyŏk wansŏng ŭi tojŏng" 려말선초(麗末鮮初) 본국력(本國曆) 완성의 도정(道程) (Accomplishment of a Domestic Calendar System during the Late Koryŏ and the Early Chosŏn Period). *Tongbang Hakchi* (Journal of Korean Studies) 155 (2011): 31–75.

Han Yŏngo, Yi Ŭnhŭi, and Kang Minjŏng (Kang Min-Jeong), trans. *Ch'ilchŏngsan Naep'yŏn: hae wa tal, tasŏt haengsŏng ŭi ch'ŏnmunhak* (Ch'ilchŏngsan Naep'yŏn: Astronomy of the Sun, Moon, and Five Planets). Seoul: Han'guk kojŏn pŏnyŏgwŏn, 2016. Vol. 1.

Hansen, Valerie. *Open Empire: A History of China to 1800*. 2nd edition. New York: Norton, 2015.

Harrison, Joel, and Neil Ridley. *Distilled: From Absinthe & Brandy to Vodka & Whisky, the World's Finest Artisan Spirits Unearthed, Explained & Enjoyed*. London: Mitchell Beazley, 2014.

He Manzi 何滿子. *Zuixiang riyue: Zhongguo jiu wenhua* 醉鄉日月: 中國酒文化 (Drunken Sun and Moon: Chinese Wine Culture). Shanghai : Shanghai guji chubanshe 上海古籍出版社, 1991.

Henthorn, William E. *Korea: The Mongol Invasions*. Leiden: E. J. Brill, 1963.

Hishinuma, Hayato. *Sake: The History, Stories, and Craft of Japan's Artisanal Breweries*. Singapore: Gatehouse Publishing, 2015.

Hŏ Hŭngsik 許興植. *Han'guk chungse pulgyosa yŏn'gu* 韓國中世佛教史研究 (A Study of the History of Medieval Korean Buddhism). Seoul: Ilchogak 一潮閣, 1994.

Isanghyang kwa poshint'ang: yŏksahakcha ka ssŭn poshint'ang ŭl wihan pyŏnmyŏng (Utopia and Boshintang: An Excuse for Dog Meat Soup by a Historian). Seoul: Idam puksŭ, 2011.

Koryŏ pulgyosa yŏn'gu 高麗佛教史研究 (A Study of the History of Koryŏ Buddhism). Seoul: Ilchogak 一潮閣, 1986.

Hŏ Simyŏng. *Makkŏlli, nŏn nugunya? Saekkal innŭn sul, makkŏlli ŭi modŭn kŏt* (Makkŏlli, Who Are You? Alcoholic Drink with Colors, Everything about Makkŏlli). Koyang:Yedam, 2010.

Höllmann, Thomas Ottfried. *The Land of the Five Flavors: A Cultural History of Chinese Cuisine*. Translated by Karen Margolis. New York: Columbia University Press, 2013.

Hong Yŏngŭi. "Sul e ulgo uttŏn Koryŏin sam ŭi pit kwa kŭrimja (The Light and Shadow of the Life of Koryŏ People Who Laughed and Cried over Drinking)." In *Koryŏ shidae saramdŭrŭn ŏttŏk'e sarassŭlkka*, vol. 1, *Sahoe, munhwa saenghwal iyagi* (How People during the Koryŏ Period Lived, vol. 1, Stories of Social and Cultural Lives). Edited by Han'guk yŏksa yŏn'guhoe, 155–166. P'aju: Ch'ŏngnyŏnsa, 2005.

Hook, Glenn D. *Japan's International Relations: Politics, Economics, and Security*. London: Routledge, 2001.

Huang Hsing-tsung. *Science and Civilisation in China*, vol. 6: *Biology and Biological Technology*, part 5: *Fermentation and Food Science*. Cambridge: Cambridge University Press, 2000.

Ishige Naomichi 石毛直道. "Higashi yūrashia no jōryūshu: jōryūki wo moto-mete" 東ユーラシアの蒸溜酒:蒸溜器を求めて (Distilled Alcohol in East Eurasia: Seeking the Distiller). In *Shōchū higashi mawari nishi mawari* 焼酎東回り西回り (Shochu around the World). Edited by Tamamura Toyo'o 玉村豊男. Tokyo: TaKaRa Alcohol Beverage and Life Research Institute, 1999, 75–130.

Jansen, Marius B. *The Making of Modern Japan*. Cambridge, MA: The Belknap Press of Harvard University Press, 2000.

Japan Sake and Shochu Makers Association. *Nihonshu no rekishi* 日本酒の歴史 (The History of Japanese Sake). Tokyo: Japan Sake and Shochu Makers Association, 2006.

Kang Ch'anghwa (Kang Changhwa). "Cheju Pŏphwasaji ŭi kogohak chŏk yŏn'gu" (Archaeological Study of Pŏphwa Temple Ruins in Cheju). *Chejudosa yŏn'gu* (Journal of Cheju History Studies) 9 (2000) 28–33.

Kang, David C. *East Asia before the West: Five Centuries of Trade and Tribute*. New York: Columbia University Press, 2012.

Kieschnick, John. *Impact of Buddhism on Chinese Material Culture*. Princeton: Princeton University Press, 2003.

Kim Ch'anghyŏn 金昌賢. *Koryŏ ŭi yŏsŏng kwa munhwa* (Women and Culture in Koryŏ). Seoul: Shinsŏwŏn 新書院, 2007.

Kim, Djun Kil. *The History of Korea*. 2nd edition. Santa Barbara, CA: Greenwood, 2014.

Kim Hodong. *Monggol cheguk kwa Koryŏ: k'ubillai chŏnggwŏn ŭi t'ansaeng kwa Koryŏ ŭi chŏngch'ijŏk wisang* (The Mongol Empire and Koryŏ: The Birth of the Khubilai Government and the Political Status of Koryŏ). Seoul: Sŏul taehakkyo ch'ulp'an munhwawŏn, 2015 [2007].

"The Unity of the Mongol Empire and Continental Exchanges over Eurasia." *Journal of Central Eurasian Studies* 1 (2009): 15–42.

Kim Ilu (Kim Il-woo). "Monggol hwangje Sunje ŭi Cheju p'inan kungjŏn t'ŏ t'amsaek" 몽골황제 순제(順帝)의 제주 피난궁전터 탐색 (In Search of the Refuge Palace Site for the Mongol Emperor Sunje). *Monggol hak* (Mongolian Studies) 46 (2016): 27–61.

Kim, Jinwung. *A History of Korea: From "Land of the Morning Calm" to States in Conflict*. Bloomington: Indiana University Press, 2012.

Kim Kwiyŏng. "Ŭmsik ŭro ponŭn Chosŏn sidae" (The Chosŏn Dynasty Viewed through Foods). In Kim Yu (1491–1555), *Suun chappang* 需雲雜方 (Various Methods of High-Class Food Culture). Translated by Kim Ch'aesik. P'aju: Kŭrhangari, 2015.

Kim Kyŏngju (Kim Gyeong-ju). "Kogo charyo ro salp'yŏ pon Wŏn kwa T'amna" 考古資料로 살펴 본 元과 耽羅 (Study on the Yuan Dynasty and Tamla (T'amna) Based on Archaeological Materials). *T'amna munhwa* (Tamna Culture) 52 (2016): 129–160.

Kim Munsuk (Kim Moonsook). "13–14 segi Koryŏ poksik e suyong toen Monggo poksik e kwanhan yŏn'gu" (The Mongolian Costume Adopted in the Koryŏ Costume from the 13th to 14th Centuries). *Monggorhak* (Mongolian Studies) 17 (2004): 223–246.

Kim Namil. *Koryŏ mal Chosŏn ch'ogi ŭi segyegwan kwa yŏksa ŭisik: Yi Saek kwa Kwŏn Kŭn ŭl chungsimŭro* 고려말 조선초기의 세계관과 역사의식: 이색(李穡) 과 권근(權近)을 중심으로 (World Views and Historical Consciousness in the Late Koryŏ and Early Chosŏn Periods: Focusing on Yi Saek and Kwŏn Kŭn). Seoul: Kyŏngin munhwasa 景仁文化社.

Kim Posŏng (Kim Bo Sung). "19 segi Chosŏn chisigin ŭi Ilbon· Yugu e taehan insik koch'al: Oju Yi Kyugyŏng ŭi *Sigajŏmdŭng* ŭl chungsim ŭro" 19세기 조선 지식 인의 일본·유구에 대한 인식 고찰: 오주(五洲) 이규경(李圭景)의 『시가점등 (詩家點燈)』을 중심으로 (The Chosŏn Literati's Recognition of Japan and Ryukyu in the 19th Century: On the Basis of *Sigajumdeung*). *Hanmunhak nonjip* (Journal of Korean Literature in Chinese) 35 (2012): 191–235.

Kim Soyŏng and Kim Hyeju. *Sak'e, Ryu* 사케, 流 (Sake for Beginners). Seoul: Altent'e buksŭ, 2009.

Kim Yŏngje (Kim Young-jae) 金榮濟. *Koryŏ sangin kwa Tong Asia muyŏksa* (Koryŏ Merchants and the History of East Asian Trade). Seoul: P'urŭn yŏksa, 2019.

"Songdae Chungguk kwa Koryŏ saiŭi haesang kyoyŏkp'um: Tongnam Asia chiyŏk kwaŭi pigyo nŭl t'ongan kŏmt'o" 宋代 中國과 高麗 사이의 海上 交易 品: 東南아시아 地域과의 比較를 통한 檢討 (Marine Trade Commodities between China and Korea in the Song Dynasty: Review by Comparing with Southeast Asia Areas). *Yŏksa Munhwa Yŏn'gu* (Journal of History and Culture) 60 (2016): 151–189.

Kim Yunjŏng (Kim Yunjung). "14 segi Koryŏ-Wŏn kwan'gye hwakchang kwa Koryŏ ŭi Wŏn poksik munhwa suyong" (The Expansion of the Koryŏ and Yuan Relationship and Koryŏ's Acceptance of Mongolian Clothing in the 14th Century). *Yŏksa Hakbo* (Korean Historical Review) 234 (June 2017): 63–112.

Kjellgren, Björn. "Drunken Modernity: Wine in China." *Anthropology of Food* (online) 3 (December 2004): 1–13. http://journals.openedition.org/aof/249 (accessed December 3, 2018).

Ko Pyŏngik (Koh Byong-ik). *Tonga kyosŏpsa ŭi yŏn'gu* 東亞交涉史의 研究 (Studies on East Asian International Relations). Seoul: Sŏul taehakkyo ch'ulp'anbu (Seoul National University Press), 1970.

Koryŏ daehakkyo minjok munhwa yŏn'guwŏn kugŏ sajŏn p'yŏnch'ansil, ed. *Han'gugŏ taesajŏn* (Grand Korean Dictionary), vol. 1. Seoul: Koryŏ daehak-kyo minjok munhwa yŏn'guwŏn, 2009.

Kosaki Michio 小崎道雄. "Tōnan ajia no sake" 東南アジアの酒 (Alcoholic Beverages in Southeast Asia). In *Daigokai kokusai sake bunka gakujutsu kentōkai ronbunshū* 第五回國際酒文化學術研討會論文集 (Collected Volume of the Fifth Conference on International Alcohol Beverage Research). Tokyo: Nihon jōzōgakkai 日本釀造學會 (Japan Alcohol Beverage Research Association), 2005.

Kosar, Kevin R. *Whiskey: A Global History*. London: Reaktion Books, 2010.

Kranzberg, Melvin. "At the Start." *Technology and Culture* 1, no. 1 (Winter 1959): 1–10.

Kurmann, Joseph A., Jeremija L. Rasic, and Manfred Kroger. *Encyclopedia of Fermented Fresh Milk Products: An International Inventory of Fermented Milk, Cream, Buttermilk, Whey, and Related Products.* New York: Springer, 1992.

Kwŏn Sŏnghun (Kwon Seong Hun). "Cheju pangŏn sok ŭi Monggol ŏ ch'ayongŏ" (Mongolian Loanwords in Cheju Dialect). *Tongak ŏmunhak* (Dong-ak Society of Language and Literature) 70 (February 2017): 53–67.

Lee, Eun-Jeung. "Intercultural Encounter, Wolff, Chong, and Ricci." In *Dynamics in the History of Religions between Asia and Europe: Encounters, Notions, and Comparative Perspectives.* Edited by Wolkhard Krech and Marion Steinicke, 203–216. Leiden: Brill, 2012.

Lee, Ki-baik. *A New History of Korea.* Translated by Edward W. Wagner and Edward J. Shultz. Cambridge, MA: Harvard University Press, 1984.

Lee, Peter H. ed. *An Anthology of Traditional Korean Literature.* Honolulu: University of Hawai'i Press, 2017.

A History of Korean Literature. Cambridge: Cambridge University Press, 2003.

Lee, SoonGu. "The Exemplar Wife: The Life of Lady Chang of Andong in Historical Context." In *Women and Confucianism in Chosŏn Korea: New Perspectives.* Edited by Kim Youngmin and Michael J. Pettid, 29–48. Albany: State University of New York Press, 2011.

Levey, Martin. *Chemistry and Chemical Technology in Ancient Mesopotamia.* Amsterdam: Elsevier, 1959.

Lewis, James Bryant. *Frontier Contact between Chosŏn Korea and Tokugawa Japan.* London and New York: RoutledgeCurzon, 2003.

Li Liu, Jiajing Wang, and Huifang Liu. "The Brewing Function of the First Amphorae in the Neolithic Yangshao Culture, North China." *Archaeological and Anthropological Sciences* 12, no. 118 (2020): 1–15.

Li Zhengping 李争平. *Zhongguo jiu* 中国酒 (Chinese Wines). Beijing: Wuzhou chuanbo chubanshe 五洲传播出版社, 2010.

Liao Yuqun 廖育群, Zhuan Fang 傅芳, and Zheng Jinsheng 郑金生. *Zhongguo kexue jishu shi: yixue* 中国科学技术史: 医学卷 (History of Chinese Science and Technology: Medical Science). Beijing: Kexue chubanshe 科学出版社, 1998.

Lim, Jongtae. "Historiographical Dependency and a Prospect beyond It: EAHSTM's Position in Regard to Ever-Changing Trends of HPS." An unpublished paper presented at the 14th International Conference on the History of Science in East Asia (ICHSEA), the École des hautes études en sciences sociales (EHESS), Paris, France, July 6–10, 2015.

Liu Guangding 劉廣定. *Zhongguo kexue shi lunji* 中國科學史論集 (Collected Essays on the History of Chinese Science). Taibei: Guoli Taiwan daxue chuban zhongxin, 2002.

Lucia, Salvatore Pablo. *A History of Wine as Therapy.* Philadelphia: Lippincott, 1963.

Luo Feng 罗丰. "Liquor Still and Milk-Wine Distilling Technology in the Mongol–Yuan Period." In *Chinese Scholars on Inner Asia.* Edited by Xin Luo and Roger Covey, 487–518. Bloomington: Indiana University, Sinor Research Institute for Inner Asian Studies, 2012.

"Meng-Yuan shiqi de niangjiuguo yu zhengliu naijiu jishu" 蒙元时期的酿酒锅与蒸馏乳酒技术 (Wine-Making Cauldrons and Technology of Distilling

Milk Wine in the Mongolian Khanate and Yuan Dynasty Period). *Kaogu* 考古 450 (2008.5): 66–77.

McGovern, Patrick E. *Ancient Brews: Rediscovered and Re-created.* New York: W. W. Norton & Company, 2017.

Uncorking the Past: The Quest for Wine, Beer, and Other Alcoholic Beverages. Berkeley: University of California Press, 2009.

McGovern, Patrick E., Anne P. Underhill, Hui Fang, Fengshi Luan, Gretchen R. Hall, Haiguang Yu, Chen-shan Wang, Fengshu Cai, Zhijun Zhao, and Gary M. Feinman, "Chemical Identification and Cultural Implications of a Mixed Fermented Beverage from Late Prehistoric China." *Asian Perspectives* 44, no. 2 (2005): 249–275.

McGovern, Patrick E., Fabian H. Toro, Gretchen R. Hall, Theodore Davidson, Katharine Prokop Prigge, George Preti, W. Christian Petersen, and Mike Szelewski. "Pre-Hispanic Distillation? A Biomolecular Archaeological Investigation." *Open Access Journal of Archaeology and Anthropology* 1, no. 2 (2019): 1–13.

Marianski, Stanley, and Adam Marianski. *Sauerkraut, Kimchi, Pickles & Relishes.* Seminole, FL: Bookmagic, 2012.

Masson Smith, John, Jr. "Dietary Decadence and Dynastic Decline in the Mongol Empire." *Journal of Asian History* 34, no. 1 (2000): 35–52.

Matthee, Rudi. "The Ambiguities of Alcohol in Iranian History." In *Wine Culture in Iran and Beyond.* Edited by Bert G. Fragmer, Ralph Kauz, and Florian Schwarz, 137–163. Vienna: Verlag der Österreichischen Akademie der Wissenschaften, 2014.

Matus, Zahary A. *Franciscans and the Elixir of Life: Religion and Science in the Later Middle Ages.* Philadelphia: University of Pennsylvania Press, 2017.

May, Timothy. *The Mongol Conquests in World History.* London: Reaktion Books, 2012.

Michel, Wolfgang W. ミヒエル, Endō Jiro 遠藤次郎, and Nakamura Teruko 中村輝子. *Murakami ika shiryō kanzō no kusuri-bako oyobi ranbiki ni tsuite* 村上医家史料館蔵の薬箱及びランビキについて (On the Medicinal Box and the *Ranbiki* Distillation Apparatus Kept by the Murakami Archive, City of Nakatsu), Murakami Archive Series 4. Nakatsu: Nakatsu-shi kyōiku iinkai 中津市教育委員会, 2007).

Miller, Norman. "Soju: The Most Popular Booze in the World." *The Guardian*, December 2, 2013. www.theguardian.com/lifeandstyle/wordofmouth/2013/dec/02/soju-popular-booze-world-south-korea (accessed December 2, 2018).

Miya Noriko 宮紀子. *Mongoru jidai no shuppan bunka* モンゴル時代の出版文化 (The Publishing Culture of the Mongol Period). Nagoya: Nagoya University Press 名古屋大学出版会, 2006.

Morewood, Samuel. *A Philosophical and Statistical History of the Inventions and Customs of Ancient and Modern Nations in the Manufacture and Use of Inebriating Liquors.* Dublin: W. Curry Jun. and Company, and W. Carson, 1838.

Mun Pyŏnghun. *Han'guk ŭmsik, segye nŭl hyangan tojŏn* (Korean Foods' Challenge to the World). P'aju: Idambooks, 2009.

Murphie, Andrew, and John Potts. *Culture and Technology.* New York: Palgrave Macmillan, 2003.

Needham, Joseph. *Science and Civilisation in China*, vol. 5: *Chemistry and Chemical Technology*, part 5: *Spagyrical Discovery and Invention: Physiological Alchemy*. Cambridge: Cambridge University Press, 1983.

Needham, Joseph, Ho Ping-yü, and Lu Gwei-djen. *Science and Civilisation in China*, vol. 5: *Chemistry and Chemical Technology*, part 3: *Spagyrical Discovery and Invention: Historical Survey, from Cinnabar Elixirs to Synthetic Insulin*. Cambridge: Cambridge University Press, 1976.

Needham, Joseph, Ho Ping-Yü, and Lu Gwei-djen. *Science and Civilisation in China*, vol. 5: *Chemistry and Chemical Technology*, part 4: *Spagyrical Discovery and Invention: Apparatus, Theories and Gifts*. Cambridge: Cambridge University Press, 1980.

Needham, Joseph, and Lu Gwei-djen. *Science and Civilisation in China*, vol. 5: *Chemistry and Chemical Technology*, part 2: *Spagyrical Discovery and Invention: Magisteries of Gold and Immortality*. Cambridge: Cambridge University Press, 1974.

Needham, Joseph, and Wang Ling. *Science and Civilisation in China*, vol. 3: *Mathematics and the Sciences of the Heavens and Earth*. Cambridge: Cambridge University Press, 1959.

Nihon Jōzō Kyōkai 日本釀造協會 (Brewing Society of Japan). *Honkaku shōchū seizō gijutsu* 本格燒酒製造技術 (Manufacturing Technology of Distilled Shōchū). Tokyo: Nihon Jōzō Kyōkai, 2001. Korean translation: *Ilbon yangjo hyŏphoe* 日本釀造協會 (Nihon Jōzō Kyōkai). *Pon'gyŏk soju chejo kisul* 本格燒酒製造技術 (*Chŭngnyusik soju chejo kisul*) (Manufacturing Technology of Distilled Shochu (Soju)). Translated and edited by Pae Sangmyŏn 裵商冕 (Bae Sang-myun). Seoul: Pae Sangmyŏn churyu yŏn'guso 裵商冕酒類研究所, 2003.

No Myŏngho (Ro Myoungho). "*Koryŏsa* wa *Koryŏsajŏryo* ŭi chaeinsikkwa han'guksahag ŭi kwaje" (New Understanding of *Koryŏsa* and *Koryŏsajŏryo* and New Tasks of Korean History). *Yŏksa Hakbo* (Korean Historical Review) 228 (December 2015): 119–149.

Nurin, Tara. "Sake Sales Soar as Brewers around the World Defy Ancient Japanese Traditions." *Forbes*, February 28, 2017. www.forbes.com/sites/taranurin/2017/02/28/u-s-sake-sales-soar-as-brewers-around-the-world-defy-ancient-japanese-traditions/#6fbae005380e (accessed December 30, 2018).

O Yŏnggyun (Oh Young Kyun). *Kŏga p'iryong saryu chŏnjip* ŭi chŏja wa p'yŏnch'an kŭrigo p'anbon: sadaebudŭl ŭi paekkwa sajŏn 居家必用事類全集의 저자와 편찬, 그리고 판본 – 사대부들의 백과사전 (Authorship, Compilation, and Editions of the *Jujia biyong shilei quanji*: An Encyclopedia for Literati). In *Chosŏn chisigin i ilgŭn yorich'aek* (Cookbooks that Chosŏn Intellectuals Read), by Chu Yŏngha, Young Kyun Oh, Ok Yŏngjŏng, and Kim Hyesuk, 21–63. Seongnam: Academy of Korean Studies Press, 2018.

O Yŏngju (Oh Young-Ju). "Cheju chŭngnyuju 'Kosorisul' kwa Ok'inawa' Awamori' ŭi munhwa pigyo siron" 제주 증류주 '고소리술'과 오키나와 '아와모리(泡盛)'의 문화비교 시론 (A Comparative Study of the Cultures of Cheju Spirits "Kosorisul" and Okinawa "Awamori"). An unpublished paper

Works Cited 261

presented at the symposium of the Cheju-Okinawa Studies Association held at Cheju University, November 4, 2017.

Cheju ŭmsik: munhwa wa sŭt'ori t'elling (Cheju Foods: Culture and Storytelling). Cheju si: Hana ch'ulp'an, 2012.

"Tong Asia sok ŭi Cheju parhyo ŭmsik munhwa" (Jeju Traditional Food Fermentation Culture in the Environment of East Asia). *Cheju-do yŏn'gu* (Journal of Cheju Studies) 32 (2009): 157–203.

Ohnuki-Tierney, Emiko. "We Eat Each Other's Food to Nourish Our Body: The Global and the Local as Mutually Constituent Forces." In *Food in Global History*. Edited by Raymond Grew, 265–300. Boulder, CO: Westview Press, 2000.

Oskenbay, Moldir. "Fermented Dairy Products in Central Asia: Methods for Making Kazakh Qurt and Their Health Benefits." *Crossroads: Studies on the History of Exchange Relations in the East Asian World* 14 (2016): 205–218.

Pae Kyŏnghwa (Bae Kyung-Hwa). "Andong sojuŭi chŏllae kwajŏnge kwanhan munhŏnjŏk koch'al" 안동소주의 傳來過程에 관한 文獻的 考察 (Literature Review for the Transmission of the Andong Soju), MA thesis, Andong National University, 1999.

Pae Sangmyŏn 裵商冕, trans. *Chosŏn chujosa* 朝鮮酒造史: *1907–1935* (History of the Production of Alcoholic Drinks in the Chosŏn Period: 1907–1935), Korean translation of *Chōsen shuzōshi* 朝鮮酒造史. Seoul: Kyujanggak, 1997.

Pae Yŏngdong (Bae Young Dong). "16–17 segi Andong munhwagwŏn ŭmsik chorisŏ ŭi tŭngjang paegyŏng kwa yŏksajŏk ŭiŭi: *Suun chappang* kwa *Ŭmsik timibang* ŭi sarye" (The Historical Significance and Sociocultural Basis of the 16th- and 17th-Century Andong Area's Culinary Manuscripts in Korea: Examples of *Suun chappang* and *Ŭmsik timibang* ŭi sarye). *Namdo minsok yŏn'gu* 29 (2014): 135–175.

Pae Yŏnghwan (Bae Young-hwan). "Cheju pangŏn sok ŭi Monggol ch'ayongŏ e taehan yŏn'gusa chŏk kŏmt'o" (A Historical Review of Mongolian Loanwords in Cheju Dialect). *Ŏmun nonjip* (Journal of Language and Literature) 68 (December 2016): 7–36.

Paek Tuhyŏn. "Kugŏsa esŏ pon *Ŭmsik timibang*" (*Ŭmsik timibang* Examined from the History of Korean Language). In *Ŭmsik timibang* 飲食知味方 (Recipes for Tasty Food), by Chang, Lady of Andong 安東張氏 (1598–1680), 55–79. Taegu: Kyŏngbuk taehakkyo ch'ulp'anbu, 2003.

Pak Chŏngbae. *Hansik ŭi T'ansaeng* (Birth of Korean Food). Seoul: Sejong sŏjŏk, 2016.

Pak Kyŏngja (Park Kyung-Ja). "Kongnyŏ ch'ulsin Koryŏ yŏindŭl ŭi sam" 貢女出 身 高麗女人들의 삶 (Life of Korean Women as Kongnyŏ). *Yŏksa wa tamnon* 55 (April 2010): 33–64.

Pak Miyŏng. "Ap'ŭrik'a esŏdo chillo soju masinda (People in Africa Are Also Drinking Jinro Soju)." *DigitalTimes*, April 18, 2016. www.dt.co.kr/contents .html?article_no=2016041802109976798002 (accessed December 2, 2018).

Pak Yongun. *Koryŏ sidae saramdŭl ŭi ŭiboksik saenghwal* (Clothing and the Food Life of the People of the Koryŏ Period). P'aju: Kyŏngin munhwasa 景仁文化 社, 2016.

Paper, Jordan D. *The Spirits Are Drunk: Comparative Approaches to Chinese Religion*. Albany: State University of New York Press, 1995.

Park, Hyunhee. *Mapping the Chinese and Islamic Worlds: Cross-cultural Exchange in Pre-modern Asia*. New York: Cambridge University Press, 2012.

"The Rise of Soju: The Transfer of Distillation Technology from 'China' to Korea during the Mongol Period (1206–1368)." *Crossroads: Studies on the History of Exchange Relations in the East Asian World* 14, special issue (2016): 173–204.

Pérez, María de la Paz Solano. "The Early Path, from the Sacred to the Profane in Fermented Beverages in New Galicia, New Spain (Mexico), Seventeenth to Eighteenth Century." *Crossroads: Studies on the History of Exchange Relations in the East Asian World* 14 (2016): 219–256.

Perry, Charles. "The Wine Maqāma by Badī' al-Zamān al-Hamadhāni." In *Medieval Arab Cookery: Essays and Translations*. Edited and translated by Maxime Rodinson, A. J. Arberry and Charles Perry, 267–272. Totnes: Prospect Books, 2006.

Pettid, Michael J. *Korean Cuisine: An Illustrated History*. London: Reaktion Books, 2008.

Pratt, Keith. "Music as a Factor in Sung–Koryŏ Diplomatic Relations, 1069–1126." *T'oung-pao* 62, nos. 4–5 (1976): 199–218.

Qiu Yihao 邱轶皓. *Menggu diguo shiyexia de Yuanshi yu dongxi wenhua jiaoliu* 蒙古帝国视野下的元史与东西文化交流 (Studies on the History of Yuan Dynasty and Trans-Eurasian Culture Exchanges from the Perspective of Mongol World Empire). Shanghai: Shanghai guji, 2019.

Ramstedt, Gustaf John (1873–1950). *Studies in Korean Etymology*. Helsinki: Suomalais-ugrilainen Seura, 1949–1953. 2 vols.

Rawski, Evelyn S. *Early Modern China and Northeast Asia: Cross-border Perspectives*. Cambridge: Cambridge University Press, 2015.

Read, Bernard E. *Chinese Materia Medica*. Peiping: Peking Natural History Bulletin, 1931–1939. 3 vols.

"The Return of Arak." *New York Times*. December 15, 2018. www.nytimes.com /2005/01/25/travel/the-return-of-arak.html (accessed December 9, 2018).

Robinson, David M. *Empire's Twilight: Northeast Asia under the Mongols*. Cambridge, MA: Harvard University Press, 2009.

Rodinson, Maxime, A. J. Arberry and Charles Perry. *Medieval Arab Cookery: Essays and Translations*. Totnes: Prospect Books, 2006.

Rogers, Adam. *Proof: The Science of Booze*. Boston: Houghton Mifflin Harcourt, 2014.

Rogers, Michael C. "National Consciousness in Medieval Korea." In *China among Equals: The Middle Kingdom and Its Neighbors, 10th–14th Centuries*. Edited by Morris Rossabi, 151–172. Berkeley: University of California Press, 1983.

Rossabi, Morris. "Alcohol and the Mongols: Myth and Reality." In *Wine Culture in Iran and Beyond*. Edited by Bert G. Fragmer, Ralph Kauz, and Florian Schwarz, 211–223. Vienna: Verlag der Österreichischen Akademie der Wissenschaften, 2014.

ed. *China among Equals: The Middle Kingdom and Its Neighbors, 10th–14th Centuries*. Berkeley: University of California Press, 1983.

"Foreigners in China." In *China under Mongol Rule*. Edited by John Langlois, 258–295. Princeton: Princeton University Press, 1981.

Khubilai Khan: His Life and Times. Berkeley: University of California Press, 1988.

The Mongols: A Very Short Introduction. Oxford: Oxford University Press, 2012.

Sakaguchi Kin'ichiro 坂口謹一郎. *Nihon no sake* 日本の酒 (Sake: Japanese Alcoholic Beverage). Tokyo: Iwanami shoten, 2007.

Salloum, Habeeb, Muna Salloum, and Leila Salloum Elias. *Scheherazade's Feasts: Foods of the Medieval Arab World.* Philadelphia: University of Pennsylvania Press, 2013.

Sansom, Bailey G. *Japan: A Short Cultural History.* New York: D. Appleton, 1943.

Schäfer, Dagmar. "Introduction." *Crossroads: Studies on the History of Exchange Relations in the East Asian World* 14 (2016): 133–142.

Schenkkan, Joshua. "What Is 'Traditional' Soju? A Spirited Debate." *Serious Eats,* October 3, 2017, www.seriouseats.com/2017/10/what-is-traditional-soju-korea-tokki-brandon-hill.html (accessed October 2, 2018).

Schottenhammer, Angela. "Distillation and Distilleries in Mongol Yuan China." *Crossroads: Studies on the History of Exchange Relations in the East Asian World* 14 (2016): 143–160.

Schurmann, Herbert Franz. *Economic Structure of the Yuan Dynasty: Translation of Chapters 93 and 94 of the* Yuan shih. Harvard Yenching Institute Studies, 16. Cambridge, MA: Harvard University, 1956.

Schurz, William L. *The Manila Galleon.* New York: Dutton 1959.

Seijas, Tatiana. *Asian Slaves in Colonial Mexico: From Chinos to Indians.* New York: Cambridge University Press, 2014.

"Indios Chinos in Eighteenth-Century Mexico." In *To Be Indio in Colonial Spanish America.* Edited by Mónica Díaz, 123–141. Albuquerque: University of New Mexico Press, 2017.

Sen, Colleen Taylor. *Curry: A Global History.* London: Reaktion Books, 2009.

Serruys, Henry. "Remains of Mongol Customs in China during the Early Ming Period." *Monumenta Serica* 17 (1958): 475–524.

Serventi, Silvano, and Françoise Sabban. *Pasta: The Story of a Universal Food.* Translated by Antony Shugaar. New York: Columbia University Press, 2002.

Seth, Michael J. *A History of Korea: From Antiquity to the Present.* Lanham, MD: Rowman & Littlefield Publishers, 2011.

Setton, Mark. *Chong Yagyong: Korea's Challenge to Orthodox Neo-Confucianism.* New York: State University of New York Press, 1997.

Shea, Eiren. *Mongol Court Dress, Identity Formation, and Global Exchange.* New York: Routledge, 2020.

Shelach-Lavi, Gideon. *The Archaeology of Early China: From Prehistory to the Han Dynasty.* New York: Cambridge University Press, 2015.

Shin, Michael D., ed. *Korean History in Maps: From Prehistory to the Twenty-First Century.* Cambridge: Cambridge University Press, 2014.

Shultz, Edward J. *Generals and Scholars: Military Rule in Medieval Korea.* Honolulu: University of Hawai'i Press, 2000.

Shurui sōgō kenkyūjo 酒類総合研究所 (National Research Institute of Brewing). *Umai sake no kagaku* うまい酒の科学 (The Science of a Wonderful Alcohol Beverage). Tokyo: Softbank Creative, 2007.

Sin Hyŏngsik. *Han'guk kodaesa sŏsul ŭi chŏngch'ak kwajŏng yŏn'gu* (A Study of the Historiography of Korean Ancient History). Seoul: kyŏngin munhwasa, 2016.

Sin Tongwŏn (Shin Dongwon). *Chosŏn ŭiyak saenghwalsa* (A History of Medical Life in Chosŏn). P'aju: Tŭllyŏk, 2014.

Tongŭi pogam kwa Tongasia ŭihaksa (*Dongui bogam* and the History of Medicine in East Asia). P'aju: Tŭllyŏk, 2015.

Slack, Edward R., Jr. "The Chinos in New Spain: A Corrective Lens for a Distorted Image." *Journal of World History* 20, no. 1 (2009): 35–67.

Sŏl Paehwan 薛培煥 (Seol Paehwan). "Mongwŏn cheguk k'uri:t'ai (*Quriltai*) yŏngu" (A Study of the *Quriltai* in the Mongol Empire). PhD dissertation, Seoul National University, 2016.

Sørensen, Henrick. "Lamaism in Korea during the Late Koryo Dynasty." *Korea Journal* 33, no. 3 (Autumn 1993): 67–81.

Sugiyama Masaaki 杉山正明. "Kubirai Seiken to Tōhō san'ōke" クビライ政権と東方三王家 (The Khubilai Regime and the Three Kingdoms in the East). *Tōyō gakuhō* 東洋学報 54 (1982): 257–315. Reprinted in *idem, Mongoru teikoku to daigen urusu* モンゴル帝国と大元ウルス (The Mongol Empire and the Great Yuan Ulus), 62–126. Kyoto: Kyoto Daigaku Gakujutsu Shuppankai, 2004.

"Kubirai to Daito: Mongorugata 'Shūtoken' to sekai teito" クビライと大都：モンゴル型「首都圏」と世界帝都 (Khubilai and Daidu: Mongol-Style Capital City and World Imperial Capital), in idem, *Mongoru teikoku to daigen urusu* モンゴル帝国と大元ウルス (The Mongol Empire and the Great Yuan Ulus), 128–167. Kyoto: Kyoto Daigaku Gakujutsu Shuppankai, 2004.

Mongoru teikoku no kōbō モンゴル帝國の興亡 (The Rise and Fall of the Mongol Empire). Tokyo: Kōdansha, 1996. 2 Vols.

Mongoru teikoku to daigen urusu モンゴル帝国と大元ウルス (The Mongol Empire and the Great Yuan Ulus). Kyoto: Kyoto Daigaku Gakujutsu Shuppankai, 2004.

Yūbokumin kara mita sekaishi 遊牧民から見た世界史 – 民族も国境もこえて (World History Seen by the Nomads: Beyond Nations and Borders). Tokyo: Nihon Keizai shimbun sha 日本經濟新聞社, 1997.

Takahashi Kōjirō 高橋康次郎, and Tsuji Hiroshi 辻宏. "Nihon no jōryūshu, shōchū" 日本の蒸溜酒、焼酎 (The Japanese Distilled Alcohol "Shōchū"). In *Shōchū higashi mawari nishi mawari* 焼酎東回り西回り (Shochu around the World). Edited by Tamamura Toyo'o 玉村豊男, 227–246. Tokyo: TaKaRa Alcohol Beverage and Life Research Institute, 1999.

Takayama Takumi 高山卓美. "'Tōhō kenbunroku' ni okeru sake ni kansuru ichi kōsatsu" 『東方見聞録』における酒に関する一考察 (Consideration of the Alcohol Beverages in *The Travels of Marco Polo*). *Nihon jōzo kyōkaishi* 日本醸造協会誌 (Journal of the Japanese Brewery Association) 102, no. 3 (2007): 172–186.

"Tōnan ajia tairikubu no jyōryūshu" 東南アジア大陸部の蒸溜酒 (Distilled Beverages on the Southeast Asian Continent). In *Daigokai kokusai sake bunka gakujutsu kentōkai ronbunshū* 第五回國際酒文化學術研討會論文集 (Collected Volume of the Fifth Conference of the International Alcohol

Beverage Research Association). Tokyo: Nihon jōzōgakkai 日本釀造學會 (Japan Alcohol Beverage Research Association), 2005.

Tamamura Toyo'o 玉村豊男. *Shōchū higashi mawari nishi mawari* 焼酎東回り西回り (Shochu around the World). Tokyo: TaKaRa Alcohol Beverage and Life Research Institute, 1999.

Tominaga Asako 富永麻子. *Awamori wa oishi: Okinawa no aji wo sodateru* 泡盛はおいしい：沖縄の味を育てる (Awamori Is Tasty: Raising the Taste of Okinawa). Tokyo: Iwanami shoten, 2002.

Toyoshima Yuka. "1116-nen nyū-Sō Kōrai shisetsu no taiken: Gaikō, bunka kōryu no genba" (The Experience of the 1116 Koryŏ Embassy to Song: Scenes of Diplomacy and Cultural Exchange). *Chōsen gakuhō* 210 (2009): 1–56.

Valenzuela Zapata, Ana G. "East Asian Stills: Distillation Influences in Mezcal Production in Mexico." In *Tribute, Trade and Smuggling*. Edited by Angela Schottenhammer, 141–151. Wiesbaden: Harrassowitz Verlag, 2014.

Valenzuela-Zapata, Ana G., Paul D. Buell, María de la Paz Solano-Pérez, and Hyunhee Park. "'Huichol' Stills: A Century of Anthropology – Technology Transfer and Innovation." *Crossroads: Studies on the History of Exchange Relations in the East Asian World* 8 (2013): 157–191.

Valenzuela Zapata, Ana G., and Gary P. Nabhan. *Tequila! A Natural and Cultural History*. Tucson: University of Arizona Press, 2003.

Vermeersch, Sem. *The Power of the Buddhas: The Ideological and Institutional Role of Buddhism during the Koryŏ Dynasty, 918–1392*. Cambridge, MA: Harvard University Asia Center, 2008.

Waley-Cohen, Joanna. *The Sextants of Beijing: Global Currents in Chinese History*. New York: W. W. Norton & Company, 1999.

Walton, Mylie K. "The Evolution and Localization of Mezcal and Tequila in Mexico." *Revista Geografica del Institute Panamericano de Geografia Historia* 85 (1977): 113–132.

Wang Saishi 王賽时. "Zhongguo shaojiu mingshi kaobian" 中国烧酒名实考辨 (A Study on the Names of Chinese Shaojiu). *Lishi yanjiu* 历史研究 6, no. 3 (1994): 73–85.

Wang, Q. Edward. *Chopsticks: A Cultural and Culinary History*. Cambridge: Cambridge University Press, 2015.

"Encountering the World: China and Its Other(s) in Historical Narratives, 1949–1989." *Journal of World History* 14, no. 3 (September 2003): 327–358.

Wilkinson, Endymion Porter. *Chinese History: A Manual*. 2nd edition. Cambridge, MA: Harvard University Press, 2000.

Chinese History: A Manual. 4th edition. Cambridge, MA: Harvard University Press, 2015.

Williams, Ian. *Tequila: A Global History*. London: Reaktion Books, 2015.

Wu, Jiang, and Lucille Chia, ed. *Spreading Buddha's Word in East Asia: The Formation and Transformation of the Chinese Buddhist Canon*. New York: Columbia University Press, 2016.

Yang Weisheng 楊渭生. *Songli guanxishi yanjiu* 宋麗關係史研究 (Studies on the History of Relations between Song and Koryŏ). Hangzhou: Hangzhou daxue chubanshe, 1997.

Yang Yinmin 楊印民. *Diguo shang yin: Yuandai jiuye yu she hui* 帝國尚飲, 元代酒業 與社會 (The Empire's Popular Drinks: Wine Industry and Society during the Yuan Dynasty). Tianjin 天津: Tianjin guji chubanshe 天津古籍出版社, 2009.

Yi Chinhan (Lee Jin Han). *Koryŏ sidae muyŏk kwa pada* (Trade and the Sea during the Koryŏ). Seoul: Kyŏngin munhwasa, 2014.

Yi Chihyŏng. *Soju iyagi: Isūl kwa pul kwa ttam ūi sul* (A Story of Soju: Alcohol of Dew, Fire, and Sweat). P'aju: Sallim, 2015.

Yi Chŏngŭn. "K'orona 19 sodok? 'han'gugin ŭi sul' soju ka nasŏtta" (Corona 19 Disinfection? Soju, "Korean Liquor," Came Forward to Help). *Hankook Ilbo*, March 14, 2020. www.hankookilbo.com/News/Read/202003131009759474 (accessed April 9, 2020).

Yi Chongsu (Lee Jong-Soo). "13 segi T'amna wa Wŏn cheguk ŭi ŭmsik munhwa pyŏndong punsŏk" (An Analysis of the Food Culture Acculturation of Tamra in the 13th Century). *Asea yŏn'gu* (Journal of Asiatic Studies) 59 (March 2016): 143–179.

Yi Hwasŏn (Lee Hwa Seon). *Tongasia sul munhwasa* (A History of East Asian Wine Culture). Seoul: Hyangŭm, 2018.

Yi Hwasŏn, and Ku Sahoe. "Ilche kangjŏmgi chuseryŏng ŭi shilch'e wa munhwa chŏk hamūi" (The Actual and Cultural Implications of the Liquor Tax Act during the Period of the Japanese Occupation). In *Han'guk ŭi sul 100 nyŏn ŭi kwaje wa chŏnmang*. Edited by Chŏng Taeyŏng, Ku Sahoe, Chŏng T'aehŏn, Chŏng Sŏkt'ae, Kwŏn Sŏngan, Chŏng Ch'ŏl, Yi Sŏkchun, and Yi Hwasŏn, 25–66. Seoul: Hyangŭm, 2017.

Yi Ikchu (Lee Ik Joo). *Yi Saek ŭi sam kwa saenggak* (The Life and Thoughts of Lee Saek (Yi Saek)). Seoul: Ilchogak, 2013.

Yi Kaesŏk (Yi Kae-Seok). *Koryŏ Tae Wŏn kwan'gye yŏn'gu* (Studies on Koryŏ–Da Yuan relations). Seoul: Chisik sanŏpsa, 2013.

Yi Kanghan (Lee Kang Hahn). *Koryŏ ŭi chagi, Wŏn cheguk kwa mannada* (Koryŏ Porcelain and the Mongol Yuan Empire). Sŏngnam: Han'gukhak Chungang Yŏn'guwŏn ch'ulp'anbu, 2016.

Koryŏ wa Wŏn cheguk ŭi kyoyŏk ŭi yŏksa (History of the Trade Relations between Koryŏ and the Yuan Empire). Seoul: Ch'angbi, 2013.

Yi Myŏngmi (Lee Myung-mi). *13–14 segi Koryŏ·Monggol kwan'gye yŏngu: Chŏngdong haengsŏng sŭngsang puma Koryŏ kugwang, kŭ poknapchŏk wisang e taehan t'amgu* (A Study of Relations between Koryŏ and the Mongols in the 13th and 14th Centuries: An Exploration of the Complex Status of the Koryŏ King as Chŏngdong haengsŏng sŭngsang and Imperial Son-in-Law). Seoul: Hyean, 2016.

Yi Sanghun (Lee Sang-Hoon). "Urinara kangwa parhyoju ŭi chŏn'gae wa t'ūkching" (The Development and Characteristics of Korean Fortified Fermented Wine). MA thesis: Seoul Venture University, 2014.

Yi Sanghūi. *Han'guk ŭi sul munhwa* (Drinking Culture of Korea). Seoul: Sŏn, 2009. 2 vols.

Yi Sŏngu 李盛雨 (Yi Seong-wu). *Chosŏn sidae chorisŏ ŭi punsŏkchŏk yŏn'gu* 朝鮮時 代 調理書의 分析的 研究 (An Analytical Study of Cookbooks of the Chosŏn Dynasty). Seoul: Han'guk chŏngsin munhwa yŏn'guwŏn 韓國精神文化研究 院, 1982.

Han'guk sikp'um sahoesa 韓國食品社會史 (Social History of Korean Foods). Seoul: Kyomunsa 敎文社, 1984.

Yi Sŭnghan. *Honhyŏl wang, Ch'ungsŏn wang: kŭ kyŏnggyeinŭi samkwa shidae* (King Ch'ungsŏn, a Mixed-Blood King: The Life and Era of a Person at the Borderline). Seoul: P'urŭn yŏksa, 2012.

Koryŏ wangjoŭi wigi hogŭn segyehwa sidae (The Crisis of the Koryŏ Dynasty and the Era of Globalization). Seoul: P'urŭn yŏksa, 2015.

Yi Sugin (Lee Sook-in). *"Chuja karye* wa Chosŏn chunggi ŭi cherye munhwa-kyŏlsok kwa paeje ŭi chŏngch'ihak" (*Jujagarye* and Sacrificial Rites in the Middle Chosŏn: Politics of Unity and Exclusion). *Chŏngsin munhwa yŏn'gu* 29, no. 2 (2006): 35–65.

Yi Yongch'ang (Lee Yongchang). "Ilche singmin chanjae wa ch'inil munje" (Issues around the Vestiges of Japanese Imperialism and Pro-Japanese). *Kukhak yŏn'gu* 7 (December 2005): 297–328.

Yoshida, Shūji 吉田集而. "Umi wo watatta jōryūki: Mekishiko no jōryūshu" 海を渡った蒸溜器:メキシコの蒸溜酒 (Distillers Going Abroad: Distilled Alcohol in Mexico). In *Shōchū higashi mawari nishi mawari* 焼酎東回り西回り (Shochu around the World). Edited by Tamamura Toyo'o 玉村豊男, 173–226. Tokyo: TaKaRa Alcohol Beverage and Life Research Institute, 1999.

Yu Aeryŏng (Yu Ahe-Ryung). "Monggo ka Koryŏ ŭi yungnyu sigyong e mich'in yŏnghyang" (Mongol Influences on Meat Eating in Koryŏ). *Kuksagwan nonch'ong* 國史館論叢 87 (1999): 221–237.

Yun Chup'il (Yoon Ju-pil). *"Kuksunjŏn Kuksŏnsaengjŏn* ŭi uŏnjŏk tok'ae: Kajŏnŭi saeroun ihaerŭl wihayŏ" (Allegorical Reading and Understanding of *A Biography of Mr. Malt-Pure* and *A Biography of Master Malt*: For a New Comprehension of Pseudo-biography). *Han'guk Hanmunhak yŏngu* (Journal of Korean Literature in Hanmun) 47 (2011): 303–338.

Yun Ŭnsuk (Yoon Eun-sook). *Monggol cheguk ŭi Manju chibaesa: Otch'igin wangga ŭi Manju kyŏngyŏng kwa Yi Sŏnggye ŭi Chosŏn kŏn'guk* (The History of the Mongol Empire's Rule in Manchuria: Rule of the Manchurian Management of the Otchigin Royal Family and the Founding of Chosŏn). Seoul: Sonamu, 2010.

Yun Yonghyŏk (Yoon Yong-hyuk). "Cheju sambyŏlch'o wa Monggol·Tong Asia segye" (Jeju Sambyeolcho and Mongolia and East Asia). *T'amna munhwa* (Tamna Culture) 52 (2016): 105–127.

Yunoki Manabu 柚木学. *Sake zukuri no rekishi* 酒造りの歴史 (History of Brewing Japanese Sake). Tokyo: Yūzankaku, 2018.

Zizumbo-Villareal, Daniel, and P. Colunga-García Marin. "Early Coconut Distillation and the Origins of Mezcal and Tequila Spirits in West-Central Mexico." *Genetic Resources and Crop Evolution* 55, no. 4 (June 2008): 493–510.

Zizumbo-Villareal, Daniel, and P. Colunga-García Marin. "La introducción de la destilación y el origen de los mezcales en el occidente de México." In *En lo ancestral hay futuro: del tequila, los mezcales y otros agaves*. Edited by P. Colunga-García Marín, A. Larqué Saavedra, L. Eguiarte and D. Zizumbo-Villarreal, 85–112. Mexico: CICY, 2007.

Zizumbo-Villareal, Daniel, Fernando González-Zozaya, Angeles Olay-Barrientos, Laura Almendros-López, Patricia Flores-Pérez, and Patricia Colunga-GarciaMarin. "Distillation in Western Mesoamerica before European Contact." *Economic Botany* 63, no. 4 (2009): 413–426.

Zolov, Eric. *Iconic Mexico: An Encyclopedia from Acapulco to Zócalo*, vol. 2. Santa Barbara, CA: ABC-CLIO, 2015.

Index

Asian Connections

Edited by Sunil Amrith, Tim Harper and Engseng Ho